Introduction to

Geometrical Optics

Introduction to
Geometrical Optics

Milton Katz
State University of New York

World Scientific
New Jersey • London • Singapore • Hong Kong

Published by

World Scientific Publishing Co. Pte. Ltd.
5 Toh Tuck Link, Singapore 596224
USA office: Suite 202, 1060 Main Street, River Edge, NJ 07661
UK office: 57 Shelton Street, Covent Garden, London WC2H 9HE

British Library Cataloguing-in-Publication Data
A catalogue record for this book is available from the British Library.

First published in 1994 by Penumbra Publishing Co.

Illustrations: Russel Hayes and George Zikos

INTRODUCTION TO GEOMETRICAL OPTICS

ISBN 981-238-202-X
ISBN 981-238-224-0 (pbk)

Printed by FuIsland Offset Printing (S) Pte Ltd, Singapore

TABLE OF CONTENTS

PREFACE xiii

ACKNOWLEDGMENTS xiv

CHAPTER 1: LIGHT

1.1 Introduction 1
1.2 The Electromagnetic Spectrum 1
1.2.1 Wave Motion 1
1.2.2 Wavelength and Frequency Range 2
1.2.3 Measurement of the Velocity of Light 3
1.3 Light Sources 4
1.3.1 Self-Luminous 4
1.3.1.1 Incandescent Sources 4
1.3.1.2 Electric Arcs and Discharges 4
1.3.1.3 Fluorescent Lamps 4
1.3.1.4 Lasers 4
1.3.2 Light Reflecting Sources 4
1.4 Optical Media 4
1.4.1 Transparent, Colored and Translucent Materials 5
1.4.2 Reflecting and Opaque Materials 5
1.5 Point and Extended Sources 5
1.6 Rectilinear Propagation of Light 6
1.6.1 Pinhole Camera 6
1.6.2 Propagation of Light in Nonhomogeneous Media 6
1.7 The Corpuscular and Wave Theory of Light 7
1.8 Umbra, Penumbra and Eclipses 8
1.9 Real and Virtual Objects and Images 10
1.10 Stops 10
1.10.1 Field Stop 10
1.10.2 Aperture Stop 10
1.10.3 Baffles 11
1.11 Optical System 11
1.11.1 Vergence and Diopters 11
1.12 Pencils and Beams of Light 12
1.12.1 Point Source 12
1.12.2 Extended Source 12
1.13 Visual Angle 12

CHAPTER 2: REFLECTION AT PLANE MIRRORS

2.1 Reflection 15
2.2 The Law of Reflection 15
2.2.1 Mirror Rotation 16
2.2.1.1 The Galvanometer 16
2.2.1.2 Measurement of Mirror Rotation Angles 16

2.2.1.3 Hadley's Sextant 17
2.3 Image of a Point Formed by a Plane Mirror 17
2.3.1 Image of an Extended Object 18
2.3.2 Inversion and Reversion of Reflected Images 18
2.4 Constant Deviation by Two Inclined Mirrors 19
2.4.1 Applications of Constant Angles of Deviation 20
2.5. Multiple Images Produced by Two Inclined Mirrors 21
2.5.1 Construction of Images 21
2.5.1.1 Image Series J 21
2.5.1.2 Image Series K 23
2.5.2 Number of J and K Images 23
2.5.3 The Kaleidoscope 25
2.6. Ray Paths to the Eye 26
2.7. Field of View of a Plane Mirror 26
2.8 Reflection According to Newton's Corpuscular Theory 28
2.9. Reflection According to Huygens' Wave Theory 29

CHAPTER 3: REFRACTION OF LIGHT
3.1 The Law of Refraction 33
3.2 The Reversibility of Light 33
3.3 Refraction and Particle Motion 33
3.3.1 Vector Diagram - Perpendicular Component Increased by Refraction 34
3.3.2 Vector Diagram - Tangential Component Decreased by Refraction 34
3.3.3 Huygens' Construction for Refraction 35
3.4 Absolute Index of Refraction and Snell's Law 36
3.5 Construction of Refracted Ray 37
3.6 Deviation of the Refracted Ray 38
3.7 Dispersion 38
3.8 Total Internal Reflection 38
3.9 Optical Path Length 39
3.10 Fermat's Principle 40

CHAPTER 4: REFRACTION BY PLANES, PLATES AND PRISMS
4.1 Exact Ray Tracing Through a Plane Refracting Surface 43
4.2 Exact Ray Tracing Through a Parallel Plate 44
4.3 Displacement of Rays Obliquely Incident on a Tilted Plate 46
4.4 Refraction Through Prisms 47
4.5 Prism Geometry 48
4.6 Limits to Prism Transmission: Grazing Incidence and Emergence 48
4.7 Minimum Deviation by a Prism 49
4.8 Calculation of the Path of a Ray Through a Prism 50
4.9 Calculation of the Total Deviation by a Prism 50

CHAPTER 5: PARAXIAL REFRACTION AT PLANES, PLATES AND PRISMS
5.1 Paraxial Refraction at a Plane Surface 53
5.2 Paraxial Image Formation by Parallel Plates 54
5.3 Reduced Thickness 56

5.4	Thin Prisms	56
5.5	Ophthalmic Prisms, Centrads and the Prism-Diopter	57
5.7	Obliquely Combined Prisms	59
5.8	Risley Prisms	61

CHAPTER 6: REFRACTION AND REFLECTION AT SPHERICAL SURFACES

6.1	Sign Convention	65
6.2	Refraction at a Single Spherical Refracting Surface	66
6.2.1	The Paraxial Image Equation	66
6.2.2	Focal Points and Focal Lengths	67
6.2.3	Refracting Power	69
6.2.4	Construction of On-Axis Image Points	69
6.2.5	Construction of the Focal Points	70
6.2.6	Construction of Off-Axis Image Points	71
6.2.7	Image Positions as an Object Approaches the Surface	72
6.2.8	Possible Positions of the Image	73
6.2.9	Lateral Magnification	73
6.3	The Image Equation for Plane Surfaces	74
6.4	Newtonian Equations	74
6.5	The Focal Planes and Focal Images of a Spherical Refracting Surface	75
6.6	The Smith-Helmholtz Formula or Lagrange Invariant	76
6.7	Reflection by Spherical Mirrors	76
6.7.1	The Paraxial Image Equation	76
6.7.2	The Focal Points of a Spherical Mirror	78
6.7.3	Construction of Axial Image Points for Spherical Mirrors	79
6.7.4	Construction of Focal Points of a Spherical Mirror	79
6.7.5	Construction of Off-Axis Image Points Formed by a Mirror	79
6.7.6	Image Positions as an Object Approaches a Spherical Mirror	80
6.7.7	Possible Positions of the Image Formed by a Spherical Mirror	81
6.7.8	Lateral Magnification by Spherical Mirrors	81
6.7.9	Newtonian Equations for a Spherical Mirror	81
6.7.10	The Field of View of a Spherical Mirror	82
6.8	Angular Magnification	83
6.9	Vergence, Power and Curvature	84
6.9.1	Vergence and Power	84
6.9.2	Curvature and Vergence	85
6.9.3	Curvature and Sagitta	86
6.9.4	The Correction of Lens Measure Readings	87

CHAPTER 7: THIN LENSES

7.1	Convex and Concave Lenses	93
7.2	Lens Nomenclature	94
7.3	The Optical Center of a Lens	94
7.4	The Thin Lens	95
7.5	The Thin Lens Equation	95
7.6	Focal Lengths	97
7.7	Construction of the Off-Axis Image Point	97

7.8 Lateral Magnification 98
7.9 Vergence Equations and Power of a Thin Lens 99
7.9.1 Thin Lens Power in Non-Uniform Medium 99
7.10 Object and Image Formation by a Thin Lens 100
7.10.1 Positive Lenses 100
7.10.2 Negative Lenses 100
7.10.3 Applications of Lenses with Various Conjugates 102
7.10.4 The Minimum Separation Between a Real Object and Its Real Image 102
7.11 Newtonian Equations 103
7.12 The Focal Planes and Focal Images of a Thin Lens 103
7.13 Prismatic Power of a Thin Lens 104
7.14 Stops and Pupils of Lens Systems 105
7.14.1 Procedure for Finding the Stops, Pupils and Field of View of a Thin Lens 106

CHAPTER 8: ROTATIONALLY SYMMETRICAL SYSTEMS
8.1 Introduction 111
8.2 Paraxial Construction of the Image 111
8.3 Paraxial Calculation of Images Through an Optical System 112
8.4 The Vergence Form of the Refraction and Transfer Equations 115
8.4.1 Iterative Use of the Vergence Equations 115

CHAPTER 9: ASTIGMATIC LENSES
9.1 Astigmatic Images 119
9.2 Curvature and Power in Principal and Normal Sections 121
9.3 Power in Oblique Meridians of a Cylinder 123
9.3.1 The Oblique Power Equation 123
9.4 Thin Astigmatic Lenses 124
9.4.1 Forms of Astigmatic Lenses 125
9.5 Transposition of Flat Prescriptions 127
9.6 The Circle of Least Confusion and the Spherical Equivalent Lens 128
9.7 The Circle of Least Confusion for Any Object Distance 129
9.8 The Lengths of the Focal Lines and the Diameter of the CLC 130
9.9 Obliquely Combined Cylinders 130

CHAPTER 10: THICK LENS SYSTEMS: PART I
10.1 Introduction 139
10.2 The Cardinal Points 139
10.2.1 The Principal Planes and Points 139
10.2.2 The Nodal Points 140
10.2.3 Construction of the Nodal Points 141
10.2.4 Other "Cardinal" Points 143
10.3 The Newtonian Equations for a Thick System 143
10.4 The Refraction Equations for a Thick System 144
10.5 The Lateral Magnification of a Thick Lens 144
10.6 Summary of Vergence Equations for a Thick Lens System 145
10.7 Lateral, Axial and Angular Magnification 145
10.7.1 Lateral Magnification: Y 145

10.7.2 Axial Magnification: X 145
10.7.3 Angular Magnification: Ma 146

CHAPTER 11: THICK LENS SYSTEMS: PART II
11.1 The Gullstrand Equations 149
11.1.1 Equivalent Power 149
11.1.2 The Position of the Second Focal Point **F'** 151
11.1.3 The Position of the Second Principal Point **H'** 151
11.1.4 The Positions of the First Focal and Principal Points F and H 151
11.2 Distances Referred to Principal Planes 152
11.2.1 Summary of the Gullstrand Equations 152
11.2.1.1The Positions of the Nodal Points N and N' 153
11.2.2 The Effect of Lens Thickness on the Positions of the Principal Planes 154
11.2.3 The Effect of Lens Shape on the Position of the Principal Planes 155
11.2.4 Telephoto and Wide-Angle Lenses 155
11.3 Vertex Power 157
11.4 Vertex Power and Ophthalmic Lenses 157
11.5 Effective Power 158
11.5.1 The Relationship of Effective Power to Vertex Power 159
11.6 Combination of Two Thick Lens 159
11.7 The Gullstrand Schematic Eye 161
11.7.1 Procedure for Calculating Cardinal Points of the Schematic Eye 161
11.8 Purkinje Images 164
11.8.1 Calculation of Purkinje Image I 164
11.8.2 Calculation of Purkinje Image II 165
11.8.3 Calculation of Purkinje Image III 166

CHAPTER 12: STOPS, PUPILS AND PORTS
12.1 Types of Stops 173
12.2 Pupils 173
12.3 The Field Stop and the Entrance and Exit Ports 174
12.4 Field of View and Vignetting 175
12.5 Procedure for Finding the Stops, Pupils, Ports and Field of View 176

CHAPTER 13: NUMERICAL APERTURE, f-NUMBER, AND RESOLUTION
13.1 Numerical Aperture 183
13.2 f-Number 183
13.3 Angular Resolution 185
13.4 Linear Resolution 187
13.5 Resolution Charts 187
13.6 Snellen Charts 188
13.7 Modulation Transfer Function 189
13.8 The Numerical Aperture of a Fiber Optic 191

CHAPTER 14: MAGNIFIERS AND MICROSCOPES
14.1 Introduction 195
14.2 Visual Angle and Angular Magnification 195

14.3 The Magnifier or Simple Microscope 195
14.3.1 Nominal Magnification 196
14.3.2 Maximum Magnification 197
14.3.3 A General Equation for Magnification 198
14.3.4 The Accommodation Required to View the Image 199
14.3.5 The Field of View of Magnifiers 199
14.3.6 Types of Magnifiers 200
14.4 The Compound Microscope 200
14.4.1 Magnification by a Microscope 201
14.4.2 Maximum Magnification by a Microscope 202
14.4.3 The Compound Microscope as a Simple Magnifier 202
14.4.4 The Stops of a Microscope 202
14.5 Numerical Aperture and Resolution of Microscopes 203
14.6.1 Maximum Usable Magnification of a Microscope 204
14.6.2 Resolution, Wavelength and the Electron Microscope 205
14.6.3 Depth of Focus of Microscope 205

CHAPTER 15: TELESCOPES
15.1 Introduction 209
15.2 Refracting Telescopes 210
15.2.1 The Astronomical or Keplerian Telescope 210
15.2.1.1 Angular Magnification 210
15.2.1.2 The Field of View of the Astronomical Telescope 212
15.2.1.3 Field Lenses 213
15.2.2 The Terrestrial Telescope 214
15.2.2.1 Relay Lenses 214
15.2.3 The Galilean Telescope 214
15.2.3.1 Field and Opera Glasses 215
15.2.4 The Prism Binocular 216
15.2.4.1 Stereoscopic Range Through Binoculars 217
15.3 Reflecting Telescopes 218
15.3.1 Newtonian Telescope 218
15.3.2 Gregorian Telescope 219
15.3.3 Cassagrainian Telescope 219
15.4 Catadioptric Telescopes 219
15.5 The Maximum Useful Magnification of a Telescope 219

CHAPTER 16: CAMERAS AND PROJECTORS
16.1 Pinhole Camera and Camera Obscura 223
16.2 Types of Cameras 223
16.3 Camera Lenses 225
16.3.1 Field of View of Cameras 225
16.3.2 Depth of Field and Focus 225
16.3.3 Hyperfocal Distance 227
16.4 The Paraxial Design of a Zoom Lens 229
16.5 Paraxial Design of a Telephoto Lens 230
16.6 Optical Projection Systems 231

CHAPTER 17: OPHTHALMIC INSTRUMENTS

17.1 Introduction 235
17.2 Ophthalmometer 235
17.2.1 Optical Principle 235
17.2.1.1 Fixed Mire-Variable Image Size Ophthalmometers 236
17.2.1.2 Variable Mire-Fixed Image Ophthalmometers 236
17.2.2 Radius of Curvature 236
17.2.3 Refractive Power of the Cornea 237
17.3 Doubling Principle 238
17.3.1 Fixed Mire Size Ophthalmometer 239
17.3.2 Adjustable Mire Size Ophthalmometer 241
17.4 Badal Optometer 242
17.4.1 Badal Target Position and Ametropia 243
17.4.2 Alternate Arrangement of Badal Optometer 244
17.4.3 Telecentric Systems 244
17.5 The Lensometer 245
17.5.1 Lensometer Target and Reticle 247
17.6 Slit Lamp or Biomicroscope 247

CHAPTER 18: DISPERSION AND CHROMATIC ABERRATION

18.1 Introduction 251
18.2 Wavelength and Index of Refraction 251
18.2.1 Index of Refraction 252
18.2.2 Dispersion 253
18.3 Optical Glass 254
18.3.1 Crown and Flint Glasses 254
18.4 Chromatic Aberration of Thin Prisms 255
18.5 Achromatic Prisms 256
18.6 The Longitudinal Chromatic Aberration of Lenses 257
18.6.1 Longitudinal Chromatic Aberration in Diopters 258
18.6.2 Linear Longitudinal Chromatic Aberration 258
18.7 Achromatic Lenses in Contact 260
18.8 Transverse Chromatic Aberration 261
18.8.1 Linear Transverse Chromatic Aberration 261
18.8.2 Transverse Chromatic Aberration in Prism Diopters 263
18.9 Achromatic Combination of Two Air-Spaced Lenses 263
18.9.1 The Huygens Eyepiece 264
18.9.2 The Ramsden Eyepiece 266
18.9.3 The Kellner and Other Eyepieces 267

CHAPTER 19: TRIGONOMETRIC RAY TRACING

19.1 Introduction 271
19.2 Ray Tracing Sign Conventions and Nomenclature 271
19.3 Ray Tracing Equations 272
19.3.1 Transfer Equations 275
19.4 Ray Tracing Through a Series of Surfaces 276

19.5 Longitudinal and Transverse Spherical Aberration 277
19.6 Ray Trace Calculations for a Pencil of Rays 278
19.7 A Format for Tracing Rays Through a Lens 278
19.8 Image Evaluation 278
19.8.1 Spherical Aberration Curves 278
19.8.2 Ray Intercept Curves 280
19.8.3 Spot Diagrams 280
19.8.4 Radial Energy Diagrams 281
19.8.5 Knife Edge Distributions 282
19.9 Graphical Ray Trace 282

CHAPTER 20: MONOCHROMATIC ABERRATIONS
20.1 Introduction 285
20.2 Third Order Theory 285
20.3 Third Order Aberration Polynomial 286
20.3.1 The Spherical Aberration Terms of the Polynomial 286
20.3.2 Spherical Aberration of a Thin Lens 287
20.3.3 Spherical Aberration and Lens Bending 288
20.3.4 Correction of Spherical Aberration 290
20.4 The Coma Terms of the Aberration Polynomial 291
20.4.1 Coma of a Thin Lens 293
20.4.2 The Sine Condition 293
20.5 Astigmatism 295
20.5.1 Astigmatism by a Spherical Refracting Surface 295
20.5.2 Astigmatism by a Thin Lens 297
20.6 Petzval Curvature 297
20.6.1 Petzval Curvature of a Refracting Surface 297
20.6.2 Petzval Curvature of a Series of Surfaces 298
20.6.3 Petzval Curvature of a Thin Lens in Air 299
20.6.4 The Petzval Condition for Flattening the Field 299
20.7 Astigmatism and Petzval Curvature Polynomial Terms 299
20.8 Distortion 300
20.8.1 The Third Order Distortion Polynomial Term 300
20.8.2 The Tangent Condition 301

BIBLIOGRAPHY 305

INDEX 309

PREFACE

This work was motivated by the need to provide an intensive course in paraxial optics for students of optometry and vision science. Its organization reflects my experiences in teaching Geometrical Optics in a changing optometry curriculum. Where once optics was central to the training of optometrists, it is now one of many areas of knowledge to be mastered. As a result, course hours in optics have contracted, although the subject material has expanded with the development of sophisticated optical instrumentation, and increasing diversity in ophthalmic materials and lenses.

The challenge, then, was to determine the most effective and efficient way to teach the essentials of geometrical optics, within these constraints, to students who, most commonly, are not physics or mathematics majors. The goal was for the students to retain a lifelong basic understanding of image formation by lenses and mirrors.

To work toward this goal, I organized this work so that the single spherical refracting surface is the basic optical element. Spherical mirrors are treated as special cases of refraction, with the same applicable equations. Thin lens equations follow as combinations of spherical refracting surfaces. The cardinal points of the thick lens make it equivalent to a thin lens. Ultimately, one set of vergence equations is applicable to all these elements.

As recently as 40 years ago we thought that everything important about optics had been discovered. We relegated optics to classical physics. Physicists primarily were interested in atomic and nuclear physics. Today optics is in the forefront of scientific and technological study. This metamorphosis is the result of the invention of the laser, and the development of holography, optical fibers for telecommunications, thin film technology and optical memory for computers.

We divided classical optics into geometrical and physical branches, but today the subdivisions are many times more numerous, as can be gleaned from the titles of books: Optical Holography, Optical Physics, Optical System Design, Optical Engineering, Applied Optics, Modern Optics, and the list goes on.

This book covers only a tiny fraction of geometrical optics, mainly paraxial optics. The laws of geometrical optics go back to ancient times for reflection of light, and the 17th century for the refraction of light. In fact, just one law, Snell's Law, **[n sin a = n' sin a']** encapsulates most of the geometrical optics in this course. Furthermore, we mainly will consider a simplification of this law for the paraxial case: **[n a = n'a]**.

Although the ideas are not very complicated, the reader's first challenge will be to reacquire some basic knowledge and skills in using geometry, trigonometry and algebra. Then it is essential to use only the sign convention and nomenclature consistently to solve problems. Failure to solve problems is frequently caused by confusion whether a distance is positive or negative. The second most important habit to develop is to draw clear and fully labeled diagrams.

To test your understanding of the material you must do the homework problems. They should be done regularly. Cramming at exam time has been the undoing of many students. Furthermore, regularly doing homework will enable you to raise timely questions in class.

Milton Katz

Acknowledgments

I am indebted to J. P. C. Southall's long out of print book, "Mirrors, Prisms and Lenses" for introducing me to Geometrical Optics. His emphasis, in all drawings, on clearly showing the direction of measurement of a distance according to the sign convention, is one that I have adopted, along with his nomenclature. Other influential optics books were The Fundamentals of Optics by Jenkins and White, and Modern Optical Engineering by W. J. Smith. A list of references is appended.

I am grateful for the generous help of Mr. Richard Feinbloom, President of Designs For Vision, Inc., who sponsored the preparation of the illustrations and, consequently, made this book possible. Mr. Russel Hayes and my graduate student, George Zikos, produced the illustrations.

I also thank the many students who identified errata, pointed out topics that needed clarification and expressed appreciation for this work.

CHAPTER 1: LIGHT

1.1 INTRODUCTION

Electromagnetic radiation is composed of oscillating electric and magnetic fields emitted by vibrating charged particles. It transports energy and travels through empty space with a constant velocity **c**, where **c** equals the product of wavelength and frequency. The *electromagnetic spectrum* comprises cosmic rays to long wavelength electrical oscillations, as shown in Figure 1.1. *Light* is that very narrow band of the electromagnetic spectrum to which the eye is specifically sensitive.

The study of light falls into two classes of phenomena: **physical optics** and **geometrical optics.** Our interest is in the latter but we will touch briefly on the former.

Physical Optics deals with theories of the nature of light and its interaction with matter. The nature of light has been of fundamental interest to physicists. Two answers to the question, "What is light?" have vied for acceptance. Isaac Newton's **corpuscular theory of light** was initially dominant but was supplanted by the **wave theory** originated by Christian Huygens and developed by Thomas Young and Augustus Fresnel. In this century, Planck showed that radiation is emitted or absorbed in discrete packets or quanta. Einstein extended this idea to explain, where the wave theory was unable to do so, the photoelectric effect. He said that electromagnetic radiation existed as quanta. A *photon* is a quantum of light. Contemporary optics treats photons as having the properties of a particle and a wave.

Geometrical Optics is concerned with how light is propagated, reflected, and refracted, and the formation of images. Light is assumed to consist of *rays*. Rays are merely the paths taken by light. Experimental facts have resulted in some basic postulates of geometrical optics:

1) light is propagated in straight lines in a homogeneous medium,
2) the angle of reflection equals the angle of incidence.
3) the ratio of the sines of the angle of incidence to the angle of refraction is a constant that depends only on the media, and
4) two independent beams of light that intersect each other will in no way affect one another.

1.2 THE ELECTROMAGNETIC SPECTRUM

1.2.1 WAVE MOTION

A pebble dropped into a pond produces ripples that spread out in ever widening circles. These ripples, or waves are disturbances of particles of the water medium. The particles move up and down. The properties of a train of waves are velocity, frequency, wavelength, amplitude and phase. See Figure 1.2. If we observe a fixed point in the wave train, we will find that waves pass it at regular intervals of time. The number of waves that pass per unit time is the *frequency* **f**. The peak-to-peak distance between waves is the *wavelength* λ. More generally, it is the distance between two particles m_1 and m_2 that occupy corresponding positions in two successive waves. Two such particles have the same displacement and are moving in the same direction. They, therefore, are in the same *phase*. The maximum height of the wave is its *amplitude* **a**. The time it takes one wave to pass the observation point is the *period* T. It is the reciprocal of the frequency. The *velocity* of any wave is given by: $\mathbf{v = \lambda f}$.

Figure 1.1 (a) The total electromagnetic specturm

WAVELENGTH λ	TYPE OF RADIATION	FREQUENCY v (hz)
10^{-9} μm = 10^{-6} nm	COSMIC RAYS	3 X10^{23}
10^{-8} μm = 10^{-5} nm	GAMMA RAYS	
10^{-7} μm = 10^{-4} nm		
10^{-6} μm = 10^{-3} nm		3 X10^{20}
10^{-5} μm = 10^{-2} nm	X-RAYS	
10^{-4} μm = 10^{-1} nm		
10^{-3} μm = 1 nm		3 X10^{17}
10^{-2} μm = 10 nm	ULTRA-VIOLET	
10^{-1} μm = 100 nm	VISIBLE	
10^{0} μm = 1 μm	INFRARED	3 X10^{14}
10^{1} μm = 10 μm		
10^{2} μm = 100 μm	HEAT	
10^{3} μm = l mm		3 X10^{11}
10^{4} μm = 10 mm	RADAR	
10^{5} μm = 100 mm		
10^{6} μm = l meter		3 X10^{8}
10^{7} μm = l0 meters		
10^{8} μm = l00 meters	FM RADIO	3 X10^{5}
10^{9} μm = l km	AM RADIO	
10^{10} μm = l0 km		
10^{11} μm = l00km		
10^{12} μm = l,000 km		3 X10^{2}
10^{13} μm = l0,000 km	ELECTRICAL	

(b) The visible spectrum

WAVELENGTH λ (nm)	TYPE OF RADIATION	FREQUENCYv (hz)
300		
	ULTRAVIOLET	
350		8.5 X10^{14}
	VIOLET	
400		
	INDIGO	
450		
	BLUE	
500		
	GREEN	
550		
	YELLOW	
600		5 X10^{14}
	ORANGE	
650		
	RED	
700		
750		
	INFRARED	
800		4 X10^{14}

1.2.2 WAVELENGTH AND FREQUENCY RANGE

Electromagnetic radiation ranges from electrical waves with a wavelength of 10^{10} cm, to cosmic rays of wavelength 10^{-13} cm. Energy is transported by all electromagnetic radiation at a constant velocity in vacuum of **c** = 2.998 x 10^{10} cm/sec. Since $\mathbf{c = \lambda f}$, the frequencies range from 3 Hz to 3x10^{23} Hz.

The visible spectrum is barely one octave wide. It ranges from 400 nm violet light to 760 nm red light. Beyond red lies the infrared region and below violet is the ultraviolet region of the spectrum.

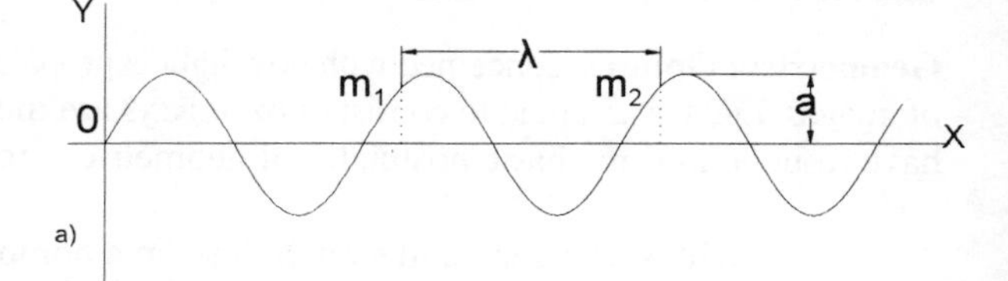

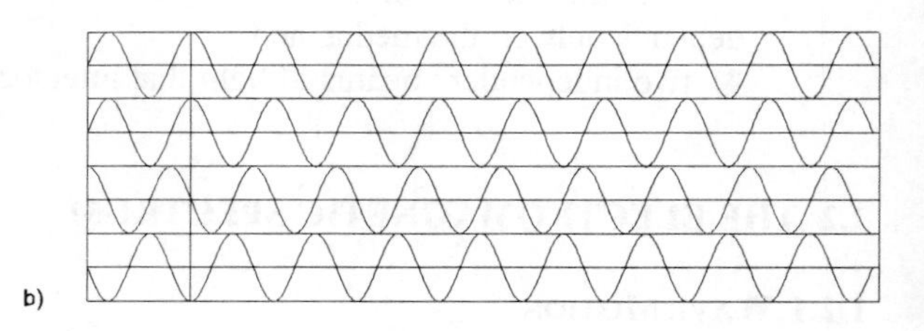

Figure 1.2 Wavelength (λ), amplitude (a) and phase. Points m_1 and m_2 are in phase since they occupy positions of identical amplitude on the wave.

We specify optical wavelengths most commonly in: Angstroms (Å), nanometers (nm), micrometers (μm), millimeters (mm), centimeters (cm), and meters (m). Where:

1 Å = 10^{-10} m, 1 nm = 10^{-9} m, 1 μm = 10^{-6} m, 1 mm = 10^{-3} m, 1 cm = 10^{-2} m.

1.2.3 MEASUREMENT OF THE VELOCITY OF LIGHT

The first evidence that light travels at a finite speed was provided by the Danish astronomer Römer about 1675. He observed that a moon of Jupiter revolved about Jupiter in an average period of 42.5 hours. In other words every 42.5 hours the moon would disappear (be eclipsed) behind Jupiter. However, the eclipse occurred earlier than predicted as the earth approached Jupiter, and later than predicted as the earth moved away from Jupiter as shown in Figure 1.3. The difference in time was about 1,000 seconds. Römer inferred that the 1,000 seconds were needed for light to travel across Earth's orbit. Subsequently, the diameter of Earth's orbit was found to be 3×10^8 km. Huygens in 1678 calculated the velocity of light (c) by dividing this distance by 1000 seconds, thus, c = 300,000 km/sec.

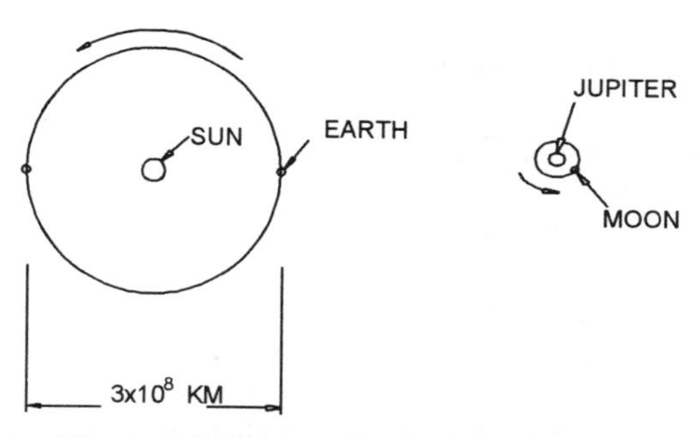

Figure 1.3 Römer's method for inferring the velocity of light.

A. A. Michelson, the first American physicist to win the Nobel Prize, did so by measuring the speed of light. Figure 1.4 illustrates his method. A brief flash of light is reflected from side 1 of an octagonal mirror (on Mt. Wilson) to a fixed mirror 35.4 km away (on Mt. San Antonio). The light, on returning to the octagonal mirror, is then reflected by side 7 into the slit and the eye only when the faces of the octagonal mirror are in the positions shown in Figure 1.4a. If the octagonal mirror is rotated slightly as the brief flash of light travels to the distant mirror, the returning light will strike face 7 at a different angle and the reflected light will miss the slit. See Figure 1.4b. The light will enter the slit and be visible only if the mirror rotates exactly 1/8th of a revolution in the time it takes the light to travel from Mt. Wilson to Mt. San Antonio and back. Faces 2 and 8 will then occupy the same positions as faces 1 and 7 did when the light was flashed. Knowing the angular velocity of the octagonal mirror in revolutions per second (rps), and the distance that the light traveled, Michelson could calculate its velocity. He found that the required angular velocity was about 528 rps. The round trip took one-eighth of a revolution or about 0.000236 sec. He found the velocity of light to be 299,853 ±10 km per sec.

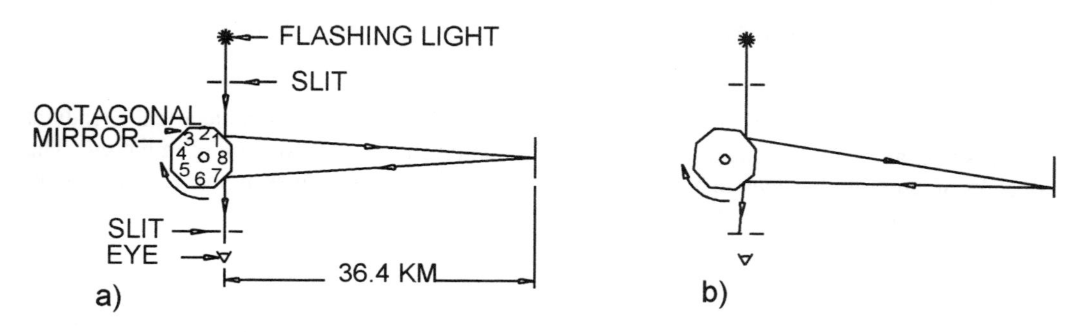

Figure 1.4 Michelson's method for measuring the velocity of light.

1.3 LIGHT SOURCES

1.3.1 SELF-LUMINOUS

1.3.1.1 Incandescent sources

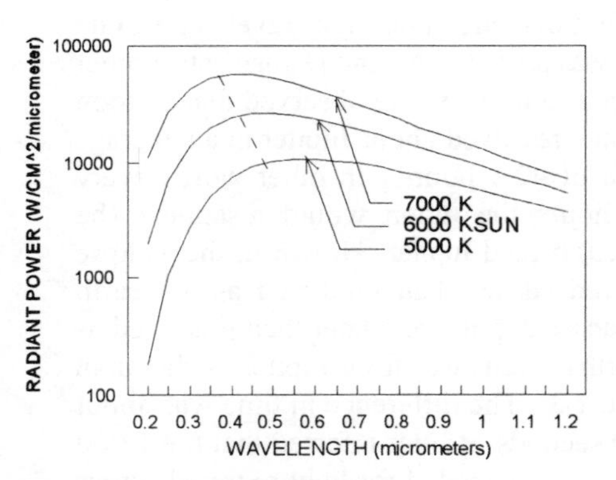

Figure 1.5 Blackbody curves.

Thermally induced motions of the molecules of solids produce incandescent light. Among incandescent sources are tungsten lamps, candles, and the sun. Incandescent sources emit all visible wavelengths in a continuous spectrum. Special incandescent sources or blackbodies are complete radiators. Figure 1-5 illustrates the spectral radiance of three blackbodies, including the sun at 6000K, at the indicated color temperatures. How much red, yellow, green and blue wavelengths are in the emitted light depends on the temperature of the solids. As the temperature of a solid is increased to incandescence it begins to have a reddish glow, then yellow and at very high temperature it predominantly emits blue light. In other words the wavelength of maximum energy becomes shorter with increasing temperature. The dashed line shows the shift of the peaks of the three radiators toward shorter wavelengths. Of course we may heat objects, such as an iron, below the temperature that produces incandescence. They then principally emit infrared energy.

1.3.1.2 Electric Arcs and Discharges. Passing an electric current through gases produces a disturbance of the atoms, which then emit light. The emitted light is not a smooth continuous spectrum, but concentrated at wavelengths that are characteristic of the gases. Sodium vapor produces nearly monochromatic light at 589 and 589.6 nm.

1.3.1.3 Fluorescent lamps. The light produced by fluorescent lamps is emitted by the fluorescent powders or phosphors that coat the inside of glass tubes filled with a mercury vapor. When an electric current is passed through the tube, the mercury vapor emits ultraviolet energy. The coating absorbs the UV and reradiates at longer (visible) wavelengths. The color of the light produced by these lamps depends on the mixture of phosphors that coat the tube. Because the phosphors emit at various narrow wavelengths, the spectrum is not smooth.

1.3.1.4 Lasers. Lasers consist of solids, liquids or gases. They emit light because either electrical or radiant energy stimulates their atoms. Laser stands for light amplification by stimulated emission of radiation. The unique properties of lasers are that their light is coherent, i.e., they emit all the waves in phase, and the light has a sharply a defined wavelength and a low beam divergence angle.

1.3.2 LIGHT REFLECTING SOURCES

Substances that reflect light may serve as sources. The moon illuminates the earth by reflecting sunlight. This page reflects light, as does the black print. White surfaces reflect all wavelengths almost equally. Colored surfaces reflect wavelengths selectively. Blue paper efficiently reflects blue light and absorbs longer visible wavelengths.

1.4 OPTICAL MEDIA

1.4.1 TRANSPARENT, COLORED AND TRANSLUCENT MATERIALS

Transparent substances, such as polished glass, transmit about 96% of the light normally incident at each surface. A clear optical glass plate absorbs very little light, consequently, such a plate transmits 0.96^2 or 92% of the light. Transparent optical glass is **homogeneous** and **isotropic**, i.e., has the same optical properties in all directions. Glass and water are isotropic substances. Certain crystals such as calcite and quartz are transparent but anisotropic. They have different optical properties in different directions. Specifically, they polarize light, or they produce double refraction, consequently the source would appear double when viewing it through the crystal.

The addition of various oxides in the manufacture of glass will color it. For example, a red filter glass transmits red or long wavelength light and absorbs blue and green. Other chemicals are used to produce neutral density filters. They absorb light of all colors almost uniformly.

If a glass surface is roughened by sand blasting or acid etching the transmitted light is diffused and the glass is *translucent.*

1.4.2 REFLECTING AND OPAQUE MATERIALS

The polished glass surface that transmits 96% will reflect 4% of the normally incident light. A transparent glass plate will transmit 0.96^2=92% of the light that perpendicularly strikes it. The reflected light is regularly or *specularly* reflected. A white wall may reflect 70% of the light. Since the wall texture is rough, it diffusely reflects the light.

Red paint absorbs most of the short wavelengths and reflects red. The use of green light to illuminate plants is not a very effective way to promote photosynthesis since the plant reflects green. It must absorb light for photosynthesis to occur.

A substance is *opaque* if it only absorbs and reflects light. Ordinary glass is opaque to short ultraviolet light; black glass is opaque to visible light.

In short, light is transmitted, reflected and absorbed. *Transmittance* (t) is the ratio of transmitted to incident light; *reflectance* (r) is the ratio of reflected to incident light; and *absorptance* is the ratio of absorbed to incident light.

EXAMPLE 1:
A two cm thick glass plate reflects 4% of the light at each surface and absorbs 2% per cm thickness. Find the light transmittance of the plate.

The transmittance by each surface is 1-r = 1-0.04 = 0.96. Given *two* surfaces, $t_s = 0.96^2 = 0.9216$. The internal transmittance by one cm of glass is 1-a = 1-0.02 = 0.98. Given two cm of glass, $t_i = 0.98^2 = 0.9604$. Total transmittance $t = (t_s)(t_i) = 0.8851$.

1.5 POINT AND EXTENDED SOURCES

Luminous sources that subtend negligibly small angles are called *point sources.* A distant star and a pinhole are examples of point sources. When the source of light subtends a finite angle, it is an extended source. The big dipper is an extended source as is the sun, the moon, a fluorescent lamp and most objects that we see.

1.6 RECTILINEAR PROPAGATION OF LIGHT

Euclid (300 BC) observed that light travels in straight lines in a homogeneous medium. This is a fundamental postulate of geometrical optics. Rays are drawn as straight lines to represent rectilinear propagation. They do not physically exist, but are merely a useful way to depict light.

1.6.1 PINHOLE CAMERA

The pinhole camera is simply a light-tight box with a pinhole aperture in the center of its front face, as shown in Figure 1.6. It shows that light travels in straight lines. A ray drawn from point M at the lower end of the object through the aperture A will strike the back face of the camera. Rays drawn through the pinhole from all points on the object between M and Q will strike corresponding points at the back end of the box and produce an inverted image. The size of the image is given by

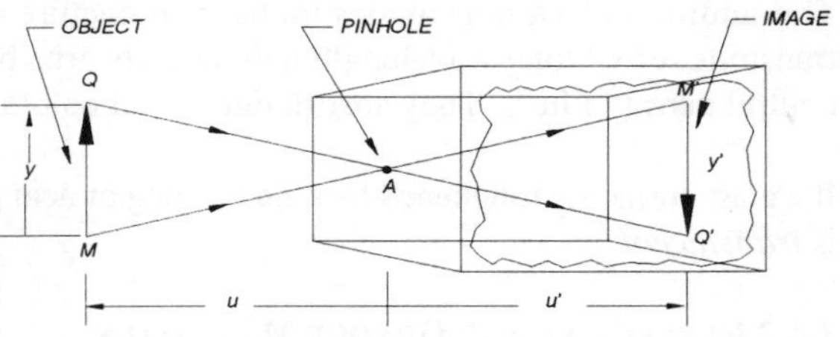

Figure 1.6 The pinhole camera.

$$\frac{y'}{y} = \frac{u'}{u} \qquad 1.1$$

where:

y =MQ = height of the object
u = AM = distance of object from the aperture,

y' = M'Q '= height of the image
u'= AM' = distance of image from the aperture

A pinhole has a finite diameter, and admits a small cone of rays from each point on the object, as illustrated in Figure 1.7. The corresponding "images" are small blurs, and the image of the object will be fuzzy. Attempts to isolate a ray by decreasing the diameter of the pinhole ultimately fail because, although the geometrical blur gets smaller, the blur due to diffraction gets larger. See Chapter 13. Diffraction is best explained with the wave theory, but may be thought of as the bending of light rays as they pass through an aperture. The optimum diameter of a pinhole is that which produces a geometrical blur equal to the diffraction blur.

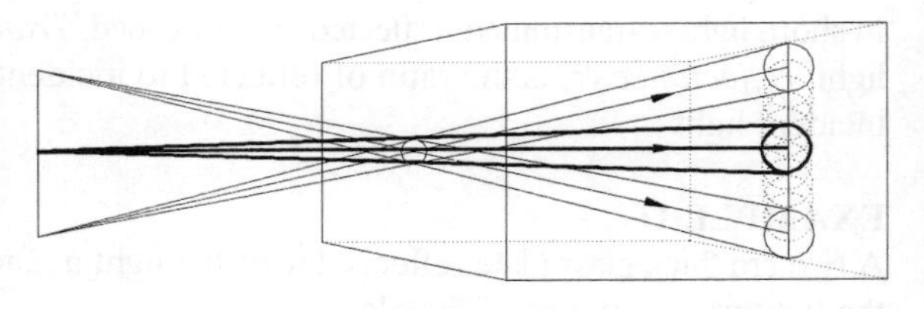

Figure 1.7 Geometric blur produced by a pinhole camera with a large aperture.

1.6.2 PROPAGATION OF LIGHT IN NONHOMOGENEOUS MEDIA

Earth's atmosphere is not homogeneous. Air varies in density with altitude and temperature. Consequently, its refractive index varies. (When light leaves a vacuum and enters a new medium its velocity is reduced. The *refractive index* of the new medium is equal to the ratio of the velocity of light in a vacuum to its velocity in the new medium). The variations in the index of air cause light from the sun and stars, especially when they are at the horizon, to curve on entering the atmosphere. Thus, a star appears higher in the sky than it actually is and is still visible after it is below the horizon. Mirages and looming occur when hot and cool layers of air form over the terrain. See Figure 1.8. (Does the curved path contradict the idea of rectilinear propagation?). In all these phenomena we see images projected along straight lines based on the direction that the rays actually enter the eye.

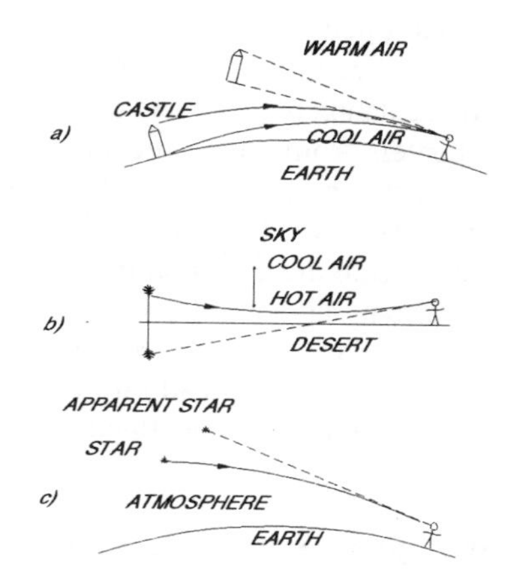

Figure 1.8 a) Looming produces "castles in the air", b) Mirages cause the blue sky to appear as lakes in the desert, c) Earth's atmosphere appears to elevate the position of stars.

1.7 THE CORPUSCULAR AND WAVE THEORY OF LIGHT

Newton considered light to be streams of *corpuscles* spreading out from a point source. The corpuscular theory dominated scientific thinking for decades. According to this theory, light sources emit particles that travel in a straight line because Newton's first law of motion requires this of bodies traveling in a uniform medium, which are not acted on by any forces. That light travels in straight lines would account for the sharp shadows cast by opaque objects. Waves, however, would spread out behind the object and so would not produce sharp shadows. Reflection and refraction were very simply explained by attractive and repulsive forces acting on these particles. The corpuscular theory was less successful in explaining interference, diffraction and polarization.

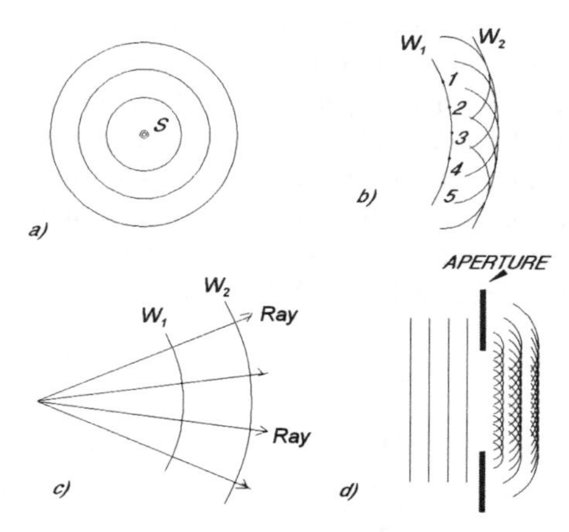

Figure 1.9 Propagation of light according to Huygens' Principle.

These deficiencies led to the wave theory of light.
According to this theory, light is a transverse vibration that travels outward from the source like ripples on the surface of water. The theory could account for reflection and refraction, and easily explained interference, diffraction, and polarization.

For the *wave theory* to gain acceptance, it had to show that light travels in straight lines. Huygens postulated that a point source would produce a wavelike disturbance that would spread out as an ever growing sphere. See Figure 1.9a. At any instant all points on the surface of this sphere would have the same state of excitation or phase and constitute a wavefront. All points on the wavefront then become new sources or centers from which secondary waves or wavelets spread out. These wavelets overlap and interfere with each other. As shown in Figure 1.9b, the envelope of these wavelets from points 1-5 on w_1 is the new wavefront w_2. It corresponds to a surface tangent to each wavelet. The normals to each point of tangency indicate the direction in which the light is traveling. Since the normals to these expanding spherical wavefronts all radiate from a common center, the normals can be thought of as rays. Thus, light travels in straight lines. Figure 1.9c. Close examination revealed that shadows were not perfectly sharp. Light waves, in fact, did spread into the shadow by *diffracting* or bending away from the edge of the aperture. Figure 1.9d. The wave theory dominated optical thinking until the 20th century.

1.8 UMBRA, PENUMBRA and ECLIPSES

The sharp shadows produced by point sources are called **umbras**. Figure 1.10 shows light from point source **S** being obstructed by an opaque globe **B** and producing a sharply defined shadow that increases in diameter with distance from **B**. The region between u and u' receives no light from any points on the source. When the source **S** is extended, as in Figure 1.11, the sharp boundary between an umbra and the illuminated regions adjoining it will be replaced by a region of partial shadow called a **penumbra**. Between u and p the shadow gradually fades, as increasing numbers of points in the source contribute light to the screen. If **S** is smaller than **B**, the umbra continues to infinity. In Figure 1.12, **S** is *larger* than **B**. The umbra is a cone that constricts with distance **d** between the opaque object and the screen, becoming zero at $\mathbf{d_o}$. Concomitantly, the penumbra is an expanding annulus. When the screen is moved beyond $\mathbf{d_o}$, the umbra disappears; only penumbra is formed.

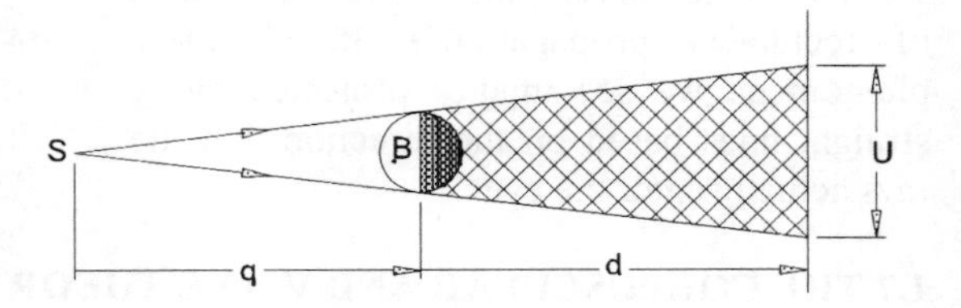

Figure 1.10 Umbra of a sphere formed with a point source.

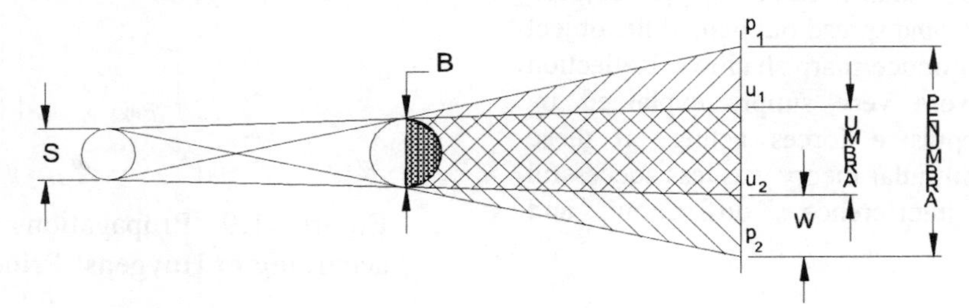

Figure 1.11 Shadows produced by an extended source. p_1p_2 denote the outer diameter (OD) of the penumbra; u_1u_2 denote the outer diameter of the umbra.

A solar eclipse occurs when the moon moves between the sun and the earth. The eclipse is full if we are in the umbra cast by the moon, partial when we are in the penumbra. A lunar eclipse occurs when the moon passes through earth's shadow.

The diameter of the umbra **U** = u_1u_2 is given by

$$U = B + (B - S)\frac{d}{q} \qquad 1.2$$

where, S = diameter of the source
B = diameter of an opaque object
q = distance between the source and opaque object
d = distance between the opaque object and the screen

The distance $\mathbf{d_o}$ at which the umbra terminates is found from similar triangles in Figure 1.12, or by setting U = 0 in Eq. 1.2.

$$d_o = \frac{qB}{(S - B)} \qquad 1.3$$

A negative value for **U** means $d > d_o$. The width of a penumbra ring **W**, where $W = p_1u_1 = u_2p_2$, is given by

$$W = \frac{dS}{q} \qquad 1.4$$

Finally, the outside diameter of the penumbra is given by $\mathbf{D = p_1u_1 + u_1u_2 + u_2p_2 = 2W + U}$. 1.5

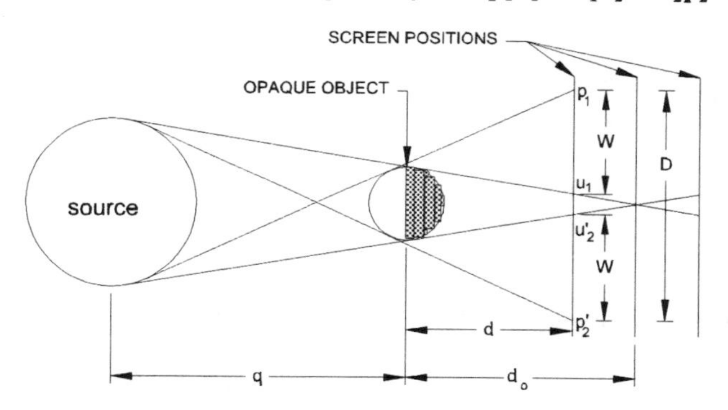

Figure 1.12 Length of shadow (d_0) when the diameter of the source is greater than the diameter of the opaque body.

EXAMPLE 2: A circular source 24 cm in diameter is 2.4 meters from a screen. A coin 2 cm in diameter is held 12 cm from the screen. Find the diameter of the umbra, the width of the penumbra ring and the outside diameter of the penumbra.

Given: S = 24 cm; B = 2 cm; q+d = 240 cm; d = 12 cm. Therefore, q = 228 cm. Substitute in Eq. 1.2 .

$U = \frac{228(2) + (2-24)(12)}{228} = 0.842\ cm$. According to Eq. 1.4, $W = \frac{12(24)}{228} = 1.263\ cm$.

The diameter of the penumbra from Eq.1.5 is: D = 2(1.263) + 0.842 = 3.368 cm.

Find the distance to the end of the umbra. From Eq. 1.3: $d_o = \dfrac{228\ (2)}{24-2} = 20.727 \quad cm.$

Draw the diagram and use the geometry to solve this example.

1.9 REAL AND VIRTUAL OBJECTS AND IMAGES

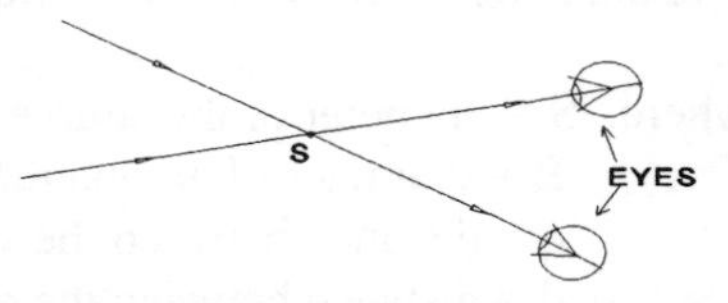

Figure 1.13 Two intersecting rays are necessary for locating point S. Two eyes indicate position.

One light ray reaching the eye will tell us the direction of the point source from which it originated. However, to locate a point source two rays are necessary. The intersection of the two rays will locate the source or its image if the rays travel through a homogeneous and isotropic medium. See Figure 1.13.

Objects and images are *real* if the rays coming from the object or going to the image actually intersect. If the rays intersect when extended backwards or forward the image and object, so formed, are *virtual*. Figures 1.14 and 1.15.

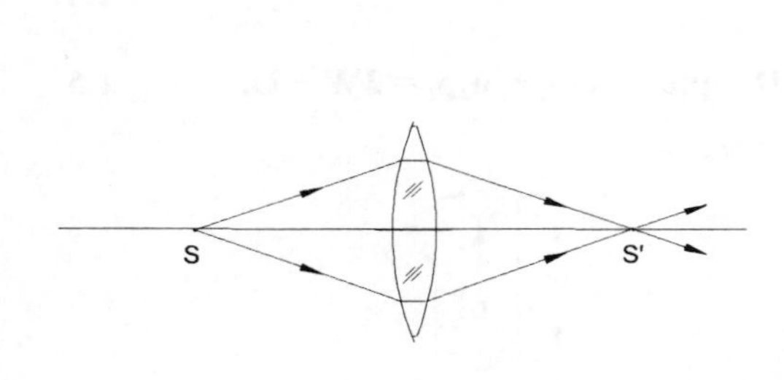

Figure 1.14 A real image is formed when two rays actually intersect.

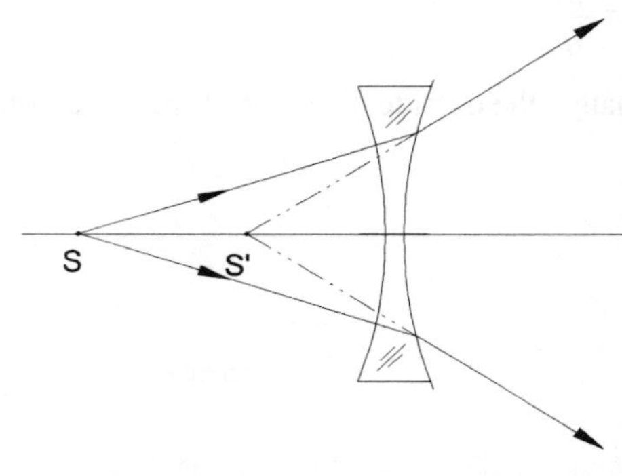

Figure 1.15 A virtual image is produced when two rays from source S must be projected back in order to intersect at S'.

1.10 STOPS

1.10.1 FIELD STOP. The field of view presented to an eye looking through a window depends on the size of the window and the position of the eye. The window acts as a field stop. It limits the extent of the landscape beyond the window that is visible. See Figure 1.16.

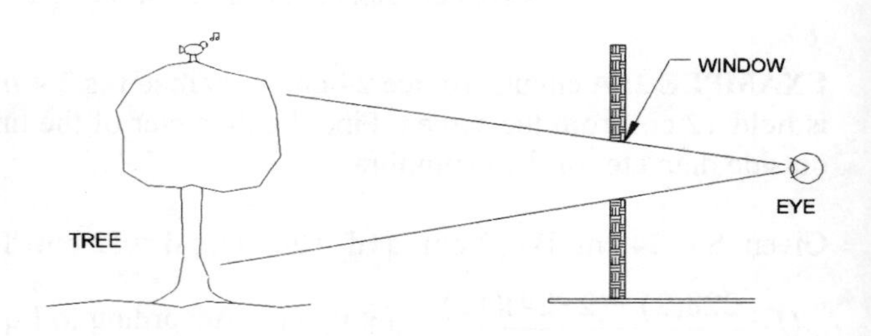

Figure 1.16 The field stop (window) prevents the bird from being seen.

1.10.2 APERTURE STOP. Each point on the landscape emits light in all directions. The pupil of the eye, however, can only admit a small solid angle of light from each point. The iris diaphragm of the eye acts as an aperture stop.

1.10.3 BAFFLES. Baffles are diaphragms within an optical system that block stray light from reaching the image. In Figure 1.17 a), the detector receives stray light from rays reflected by the inside walls of the camera. b). Baffles block the stray light.

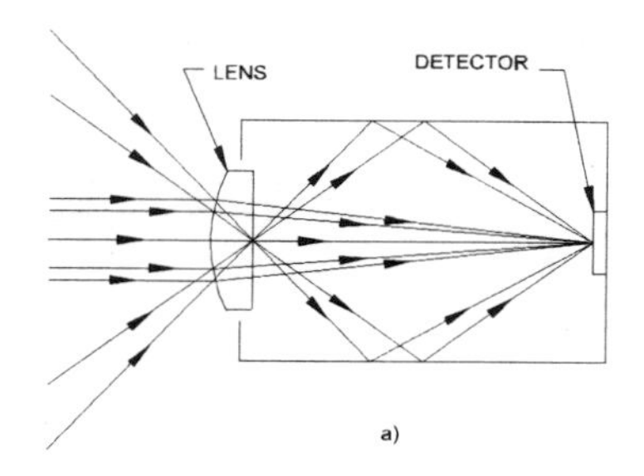

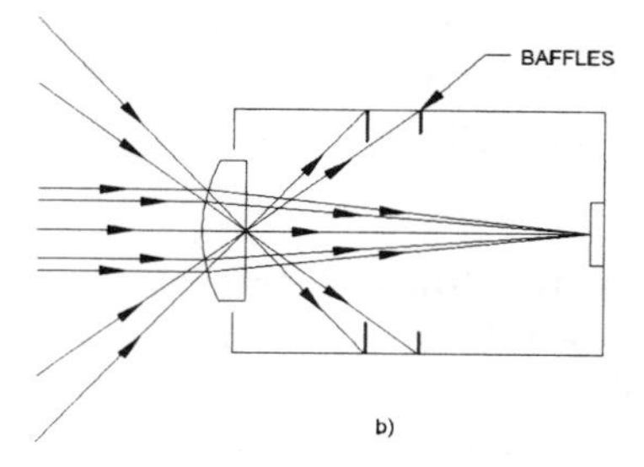

Figure 1.17 Baffles block stray light from reaching the detector.

1.11 OPTICAL SYSTEM

Generally, an optical system consists of a series of spherical reflecting and refracting surfaces that are rotationally symmetrical about an optical axis. Only a few rays from a point object enter an optical system, the rest of the rays are *vignetted*. Before entering the system, they are object rays. Upon leaving the system, they become image rays. In *paraxial optics*, all rays from an object point meet at an image point. An *astigmatic* system will result in image rays that meet to form a line focus.

1.11.1 VERGENCE AND DIOPTERS. If all object rays from a point meet in an image point, the image is perfect or ideal, and the image point is *conjugate* to the object point. Rays from a real object *diverge or have negative vergence*. To form a real image the optical system must produce *positive vergence or converge* the image rays. See Figure 1.18. Vergence is specified in *diopters* as the reciprocal of the distance of the object or image in meters. For example, light from a point source 0.5 meters away has a vergence of -2.00 diopters.

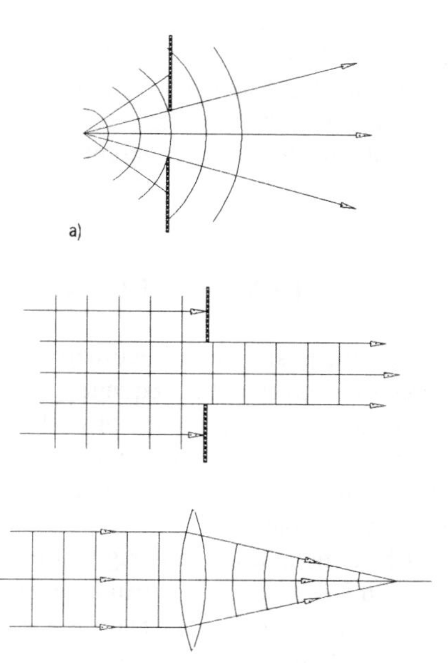

Figure 1.18 a) divergent, b) parallel or collimated, and c) convergent pencils of rays.

1.12 PENCILS AND BEAMS OF LIGHT

1.12.1 POINT SOURCE. A point source at a finite distance emits *divergent* light in all directions. When very distant, the light from a point source is *collimated*, i.e., the rays are parallel. A sufficiently strong positive lens will focus the rays to an image point, in which case the light is *convergent*. A section through a bundle of rays that contain the *chief* ray is called a *pencil* of rays. The chief or central ray of each pencil goes through the center of the aperture stop of a system.

1.12.2 EXTENDED SOURCE. An extended source produces overlapping pencils of rays called beams. One postulate of geometrical optics is that intersecting independent bundles and beams of light do not affect one another when they intersect. Figure 1.19.

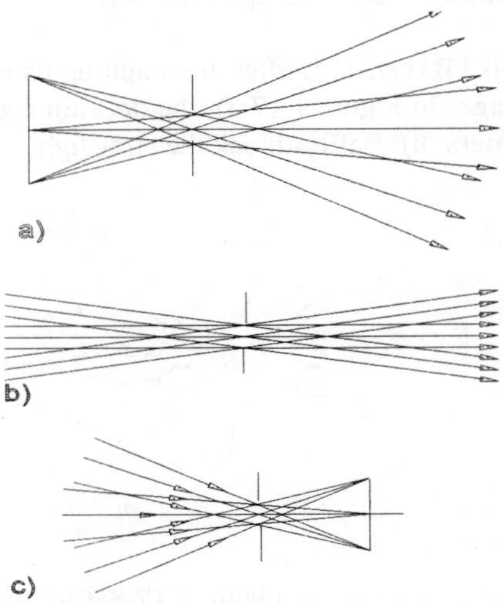

a)

b)

c)

Figure 1.19 Beams of light produced by: a) Near real extended object; b) Distant extended object; c) virtual extended object.

1.13 VISUAL ANGLE

Objects subtend *visual angles* at the eye. The tangent of the visual angle is the *apparent size* of the object. Given a tree with a height **h**, at a distance **d** from the eye, the apparent size = **tan h/d**. Figure 1.20. Another tree, half as tall and at half the distance will have the same apparent size.

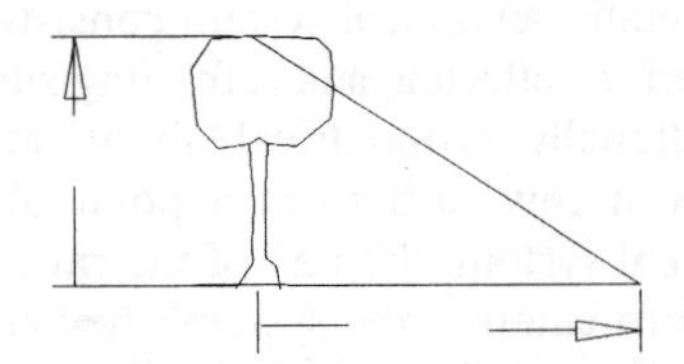

Figure 1.20 Apparent size or visual angle = h/d.

REVIEW QUESTIONS

1. What are the postulates of Geometrical Optics?
2. Define: ray, wavelength, frequency, velocity of light.
3. What is the wavelength range of visible light?
4. What is the difference between a fluorescent and an incandescent source?
5. Explain why light may take a curved path through the atmosphere. Does this violate the postulate of rectilinear propagation?
6. Describe a method for showing that light travels in straight lines.
7. What is the difference between an aperture stop and a field stop?
8. A star is a point source. The sun is a star, is it a point source?
9. Where must an object be to produce collimated light?
10. How are rays accommodated by the wave theory of light?
11. What is meant by conjugacy of object and image?
12. An astigmatic image of a point is a line. What does a stigmatic image look like?
13. Distinguish between a ray, a pencil and a beam.

PROBLEMS

1. Convert a) infrared light of 1.5 μm wavelength to nm, b) 10,000 Å to mm; c) 508 nm to inches.
Ans.: a) 1,500 nm; b) 0.001 mm; c) $2x10^{-5}$ in.

2. Yellow light has a wavelength of 590 nm. If the speed of light is $3x10^8$ m/sec what is the frequency of the light?
Ans.: $5.08x10^{14}$ Hz.

3. The sun is 93 million miles away. How long does it take sunlight to reach the earth? There are 1.609 km/mile.
Ans.: 8.31 minutes.

4. Assume in Michelson's experiment that the octagonal mirror had to make 530 rps before light was seen through the slit. What is the velocity of the light?
Ans.: 300,192 km/sec.

5. a) Light is successively transmitted through two glass plates, each 2 cm thick. The glass absorbs 2% of the light per cm of thickness, and reflects 4% per surface. What percent of the light will be transmitted? b) Repeat this problem for 10% reflection per surface and 10% absorption per cm.
Ans.: a) 78.34%; b) 43.05%

6. A pinhole camera is used to photograph a 10 ft. tall statue, located 20 ft. away. The image is 6 inches tall. How long is the camera?
Ans.: 1 ft.

7. A circular source 24 cm in diameter is 48 cm from a 12-cm diam. opaque disk. The disk is 48 cm from a screen. Find a) the diameter of the umbra, b) the width of the penumbra ring, c) the outside diameter of the penumbra, and d) the length of the shadow. Draw a scale diagram.
Ans.: a) 0 cm; b) 24 cm; c) 48 cm; d) 48 cm.

8. Repeat Problem 7 for a 16-cm diameter disk. Ans.: a) 8 cm; b) 24 cm; c) 56 cm; d) 96 cm.

9. Repeat Problem 7 for a 24-cm diameter disk. Ans.: a) 24 cm; b) 24 cm; c) 72 cm; d) ∞.

10. The diameters of the sun, earth and moon are 864,000, 7,927, and 2,160 miles, respectively. The mean distance of the sun from the earth is 93 million miles; the mean distance of the moon is 238,857 miles. a) How long is the earth's shadow? b) what is the diameter of earth's shadow where the moon passes through it? c) what is the corresponding outside diameter of the penumbra?
Ans.: a) 861,154 mi; b) 5728 mi; c) 10,166.4 mi.

11. The apparent size of a 500 ft high tower is 5°. How far away is it? Ans.: 5715 ft.

12. Find the vergence of the light in diopters (D), a) if a real object is three meters away, b) if the light is collimated, c) if a real image is 25 cm away, d) if virtual object is 100 mm away, e) if a virtual image is 10 inches away.
Ans.: a)-3.0 D., b) zero, c) +4.0 D., d) +10 D., e) -3.94 D.

NOTES:

CHAPTER 2

REFLECTION AT PLANE MIRRORS

2.1 REFLECTION

When light strikes a smooth polished surface, some light is *specularly reflected*. Most of the light will be *transmitted* if the medium behind the surface is transparent. Colored glass, for example red glass, will transmit red light and *absorb* the other visible wavelengths. Black glass is *opaque* and will *absorb* all visible wavelengths. Reflection is greatly enhanced by coating the surface with a film of metal, such as silver, gold or aluminum, thus, making a mirror. Highly polished, specularly reflecting mirrors are nearly invisible. Scratches, roughness or dust on their surfaces will *scatter* or *diffusely reflect* light and render the surfaces visible.

2.2 THE LAW OF REFLECTION

The *law of reflection* is an empirical law. It says that *the reflected ray lies in the plane of incidence, and the incident and reflected rays make equal but opposite angles with the normal.* Let us examine this statement with reference to the concave and convex reflecting surfaces shown in Figure 2.1. **NA** is a **normal** (perpendicular) to that surface. **RA** is a ray incident on the surface at point **A**. The plane containing the normal and the incident ray is called the **plane of incidence**. The angle of incidence, that is, the angle α that the incident ray makes with the normal, is *the acute angle through which the normal has to be turned to bring it into coincidence with the incident ray.* Counterclockwise rotation of the normal connotes a positive angle; clockwise rotation means a negative angle. Thus angle **NAR** = α, and in Figure 2.1 it is positive. The sequence of letters **NAR** shows the sense of rotation. α is the angle obtained by rotating the line **NA** about the point **A** until it coincides with the line **AR**.

The reflected ray **AR'** makes an angle with the normal called the angle of reflection α'. Angle **NAR'** = α'. As shown in Figure 2.1 it is negative because it is a clockwise rotation of the normal. According to the law of reflection, in absolute angles, $\alpha = \alpha'$. However, our definition requires $\alpha = -\alpha'$

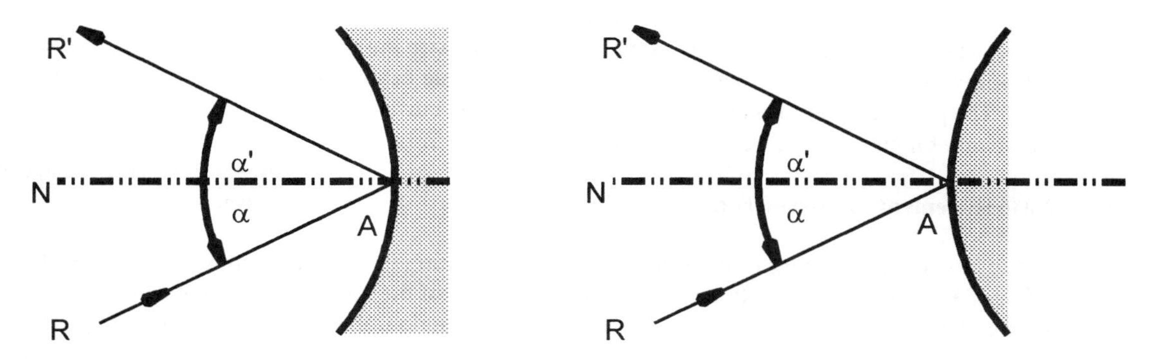

Figure 2.1 Angles of incidence (α) and reflection (α') about the normal (N).

The law of reflection may be verified by mounting a small flat mirror perpendicularly at the center of a circular protractor, marked in degrees, as illustrated in Figure 2.2a. Let an incident ray strike the mirror at angle $\alpha = +15°$. The reflected ray will cross the 30° mark on the protractor, as required by the law of reflection.

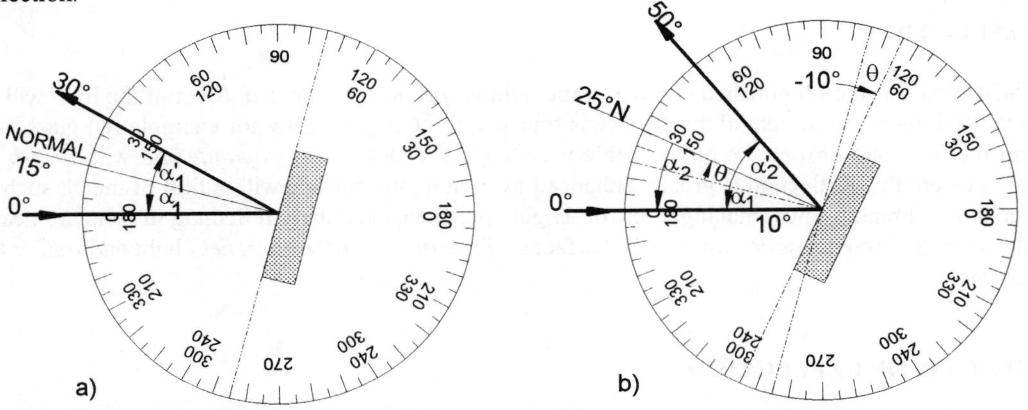

Figure 2.2 (a) Demonstration of the Law of Reflection. (b) The reflected ray is deviated through twice the angle of rotation of the mirror.

2.2.1 MIRROR ROTATION

If, in the previous experiment, the mirror is rotated through angle $\theta = -10°$, as shown in Figure 2.2b, the angle of incidence becomes $\alpha - \theta = +25°$, and the reflected ray crosses the 50° mark. *This shows that when a mirror is turned through some angle θ, the reflected ray is turned through 2θ.*

2.2.1.1 The Galvanometer

This relationship is used in the galvanometer to measure extremely small electric currents. In the D'Arsonval galvanometer a coil of fine wire is suspended in the magnetic field of a horseshoe magnet. The wire twists in proportion to the current in it. By attaching a small mirror to the wire the twist can be measured. A projected cross-hair is reflected across a scale calibrated in amperes. The position of the cross-hair on the scale shows the current. The reflected beam of light acts as a weightless pointer. See Figure 2.3.

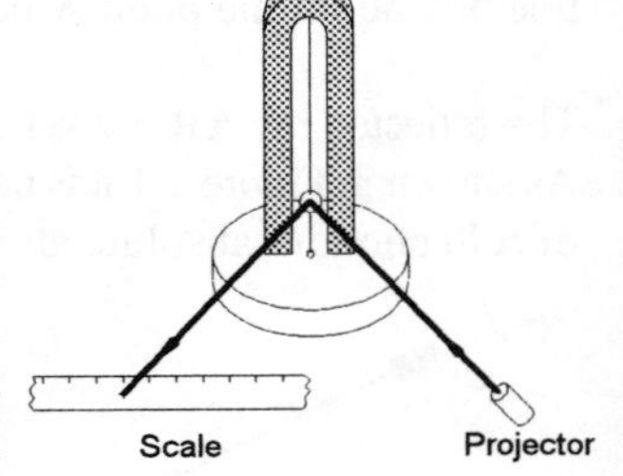

Figure 2.3 The D'Arsonval Galvanometer.

2.2.1.2 Measurement of mirror rotation angles

A method to measure the angular rotation of a mirror is by sighting through a low power telescope at the image of a scale seen in the mirror. In the zero rotation position of the mirror the telescope line of sight is toward point **O.** Figure 2.4. After the mirror is rotated through angle θ the line of sight is toward point **P.**

To determine angle θ, note that **tan 2θ = x/d**, where, x = OP, and d is the distance between the scale and the mirror.

2.2.1.3 Hadley's Sextant

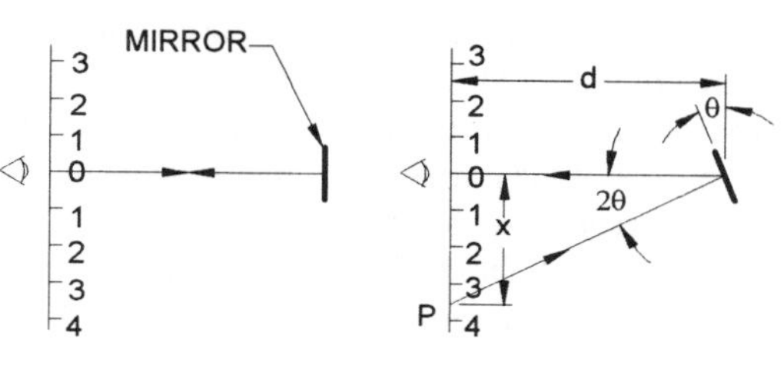

Figure 2.4. The scale mark at P, seen in the mirror, indicates tilt angle.

Another very important application of mirror rotation is in Hadley's sextant (Figure 2.5a). The sextant is an instrument that enables a navigator to measure the elevation of a celestial body above the horizon. By measuring the elevation of the sun, for example, finding local noon is possible (when the sun is at its highest point), or one's latitude. The sextant produces coincidence of two angularly separated objects such as the sun and the horizon. It consists of a fixed horizon half-mirror and a rotatable index mirror. In Figure 2.5b, the index mirror is parallel to the horizon mirror and only the horizon is seen. The vernier reads 0°. In Figure 2.5c the index mirror has been rotated to view a star. When the vernier index reads about 40°, the star is seen on the horizon. The mirror was rotated through half this angle to produce coincidence. The scale is a sixty-degree sector, therefore the name sextant, indexed at twice the angle of the rotation to read off the elevation of the star directly.

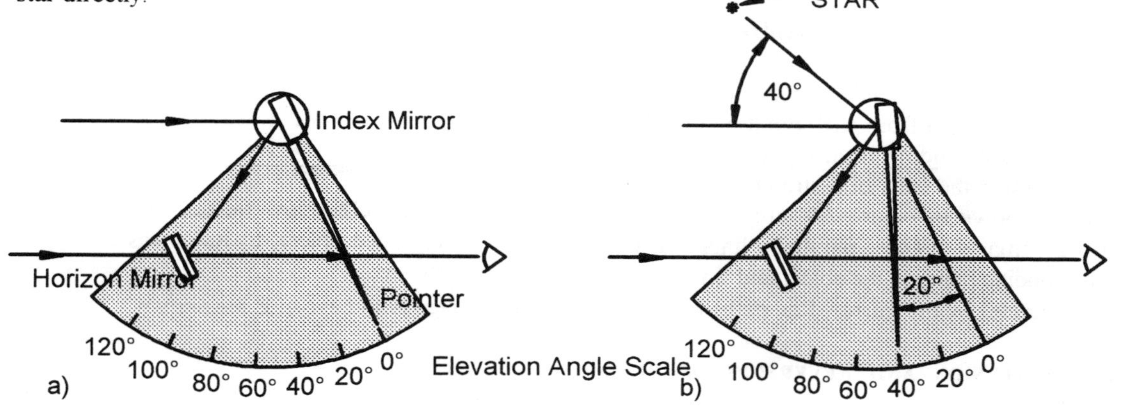

Figure 2.5 The sextant. (a) The Index and Horizon mirrors are parallel. (b) The star at 40° elevation appears on the horizon when the index mirror is rotated 20°.

2.3 IMAGE OF A POINT FORMED BY A PLANE MIRROR

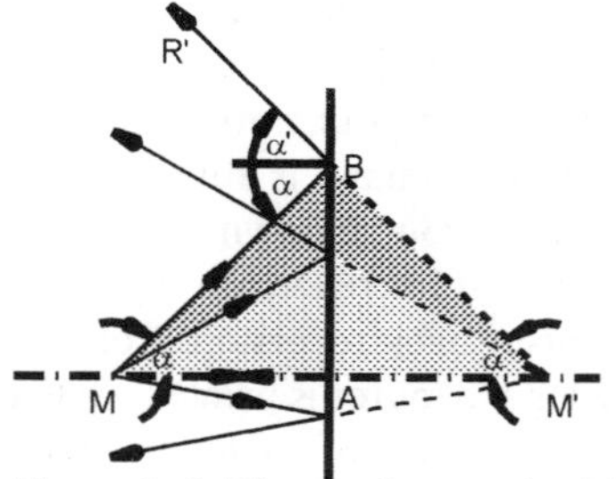

Figure 2.6 All rays from point M reflect as if from point M'.

In Figure 2.6, an object **M** is on a normal to the mirror at **A**. Point **M** emits rays in all directions. One such ray, **MA**, coincides with the normal and reflects back upon itself. Ray **MB** strikes the mirror at point **B,** making an angle of incidence **α** with the normal at point **B,** and reflecting along the direction **BR'**, at an angle α'. If the rays **MA** and **BR** are projected behind the mirror, they meet at point **M'**. The triangles **MAB** and **M'AB** are congruent, consequently, **M'** lies at a distance behind the mirror equal to the distance that the real object **M** is in front of the mirror. All other rays from **M** to the mirror will be reflected such that their projections behind the mirror will result in congruent triangles with the same bases. Therefore, all these ray projections meet at **M'**. Because the rays do *not* actually go through the point **M'** it is a *virtual image* of **M**. Also, see Figure 2.7a.

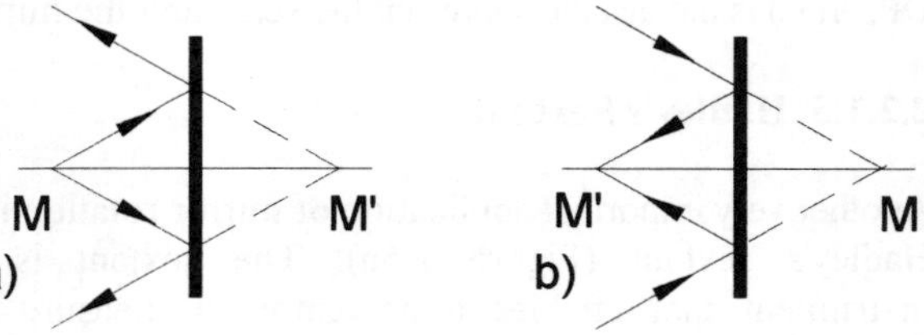

Figure 2.7. Reflection by a plane mirror. a) Real object/virtual image. b)virtual object/real image.

If rays converge toward the mirror so that their projections meet at **M,** behind the mirror, the point **M** is a *virtual object.* See Figure 2.7b. The point **M** belongs to *object space* although it appears to be behind the mirror. A *real image* at **M'** is formed by the reflected rays. **M'** is in *image space.*

2.3.1 IMAGE OF AN EXTENDED OBJECT

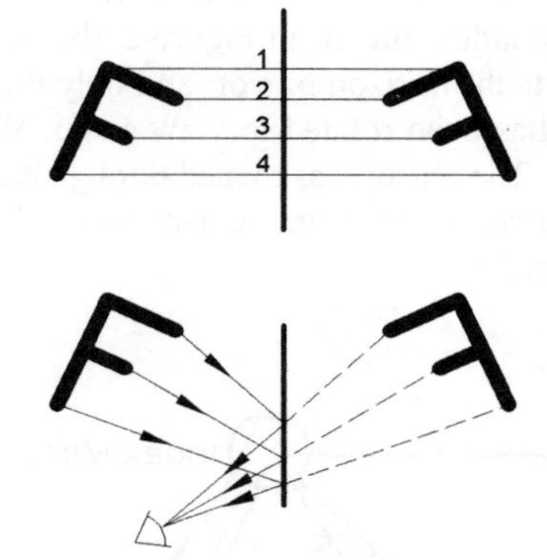

Figure 2.8 a) Image of an extended object, b) Ray paths to the eye.

The image of an extended object can be formed by dropping a normal from each point on the object to the mirror, and extending it an equal distance behind the mirror. Only the ends of a straight object, such as the letter **F**, shown in Figure 2.8, are needed to find the mirror image. The parallel lines **1-4** joining corresponding points in the object and image are bisected at right angles by the mirror. Ray paths from the ends of the image to the eye at **E** are shown.

2.3.2 INVERSION AND REVERSION OF REFLECTED IMAGES

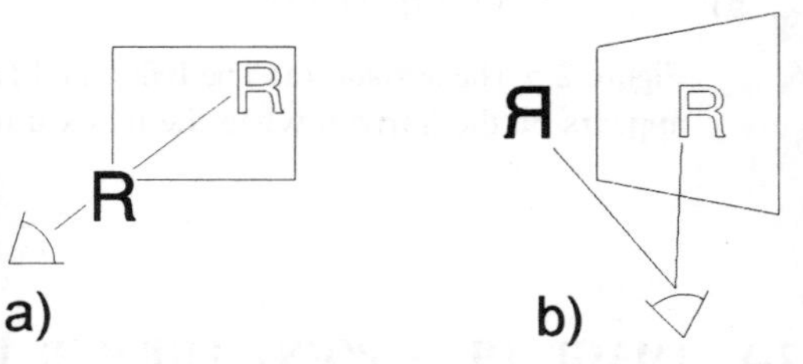

Figure 2.9 a) Front-to-back reversion; b) Side to-side reversion.

When you look at your own image in a mirror it seems that your right and left sides have been switched. Extend your right hand to shake, and your image seems to extend its left hand. If you part your hair on the left, it is seen on the right of your mirror image. Concluding incorrectly that your image is reverted from side-to-side is easy. Suppose you lie down parallel to the mirror. There is no head-to-foot reversion. In fact, the mirror produces a front-to-back reversion. Consider the letter **R** in front of the mirror in Figure 2.9a. The side of the letter facing the mirror is painted white and the reverse side is black. If you stand, facing the mirror, and look at the black side of the letter it will be a normal letter **R**. Its image, observed in the mirror will be a normal white letter **R.**

However, if you move to a position between the letter and the mirror, from where you directly view the white side of the letter and its image in the mirror, there is a side-to-side reversion. See Figure 2.9b. Defining the position of the observer is essential, to describe the action of the mirror.

Images produced by optical systems are termed *right or left handed.* The letter **R** and an inverted and reverted image of the letter **R** are right-handed. A single reversion or inversion produces a left-handed image. See Figure 2.10.

R ᴚ Я ʁ

Figure 2.10 The first two letters are right-handed, the last two are left-handed.

2.4 CONSTANT DEVIATION BY TWO INCLINED MIRRORS

Two inclined plane mirrors that meet at an edge to form a dihedral angle γ, will successively reflect a ray of light, whatever its initial angle of incidence, so that the emerging ray is deviated through a constant angle δ equal to 2γ. Figure 2.11 represents a section perpendicular to the edge of the mirrors, or a principal section. The angle γ is formed by rotating the initially struck mirror into the second mirror. The incident ray travels from **R** to **A**, reflects to **B**, and reflects again to **R'**. At point **S** the emergent ray intersects the incident ray. The angle of deviation δ is obtained by rotating the incident ray segment **SA** about the point **S** to bring it into coincidence with the emergent ray segment **SR'**.

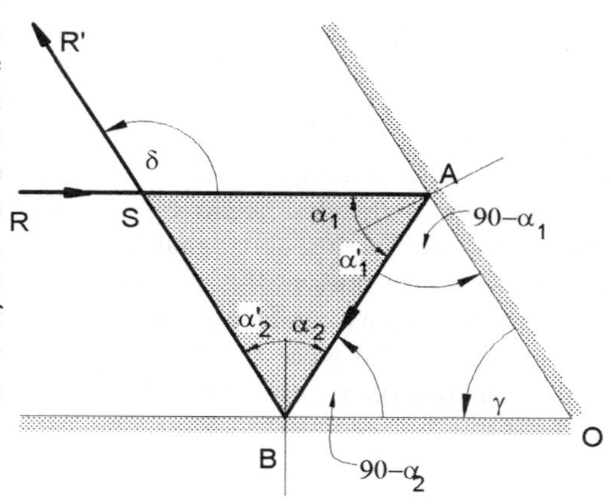

Figure 2.11 Constant deviation of a ray by two inclined mirrors.

Figure 2.11 shows that $\delta = 2\gamma$. Angle δ, the external angle of the shaded triangle, equals the sum of the two opposite internal angles of the triangle. Therefore,

$$\delta = 2\alpha_1 + 2\alpha_2 = 2(\alpha_1 + \alpha_2) \tag{2.1}$$

From triangle **AOB**:

$$180° = \gamma + (90-\alpha_1) + (90-\alpha_2)$$

Thus,

$$\gamma = \alpha_1 + \alpha_2 \tag{2.2}$$

Substitute Eq. 2.2 into 2.1:

$$\boxed{\delta = 2\gamma} \tag{2.3}$$

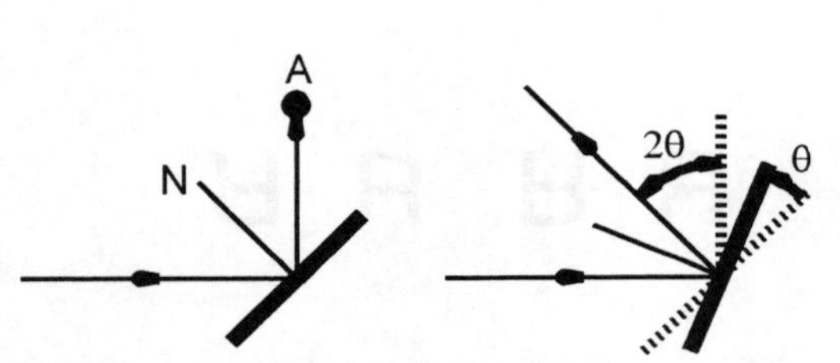

Figure 2.12 Deviation by a plane mirror equals twice the tilt angle.

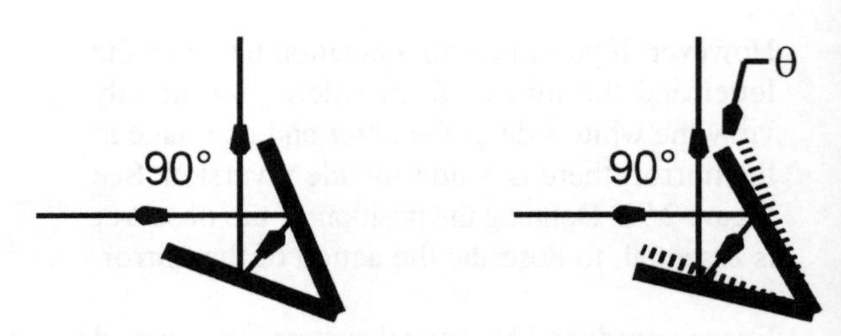

Figure 2.13 Deviation by two inclined mirrors is not affected by tilt about an axis perpendicular to a principal section.

2.4.1 APPLICATIONS OF CONSTANT ANGLES OF DEVIATION

Although a single plane mirror may be used to deviate light through some particular angle, if the mirror mount is subject to mechanical and thermal stresses the mirror may become misaligned. For example, if we wanted to deviate rays through a 90° angle, we could place a plane mirror at 45° to the rays, as in Figure 2.12a. However, any tilt in the mirror angle would introduce an error in deviation of double the tilt angle. See Figure 2.12.b. By using two mirrors with a dihedral angle of 45° the deviation will always be 90° whatever the orientation of the two mirrors. See Figure 2.13. Instead of two mirrors we can incorporate the 45° dihedral angle in a solid piece of glass with five sides called a **pentaprism**. See Figure 2.14.

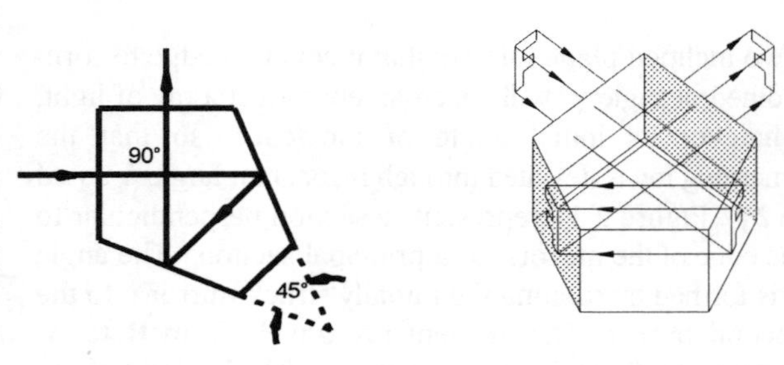

Figure 2.14 The pentaprism incorporates two inclined reflecting surfaces in a solid element.

Two mirrors set 90° apart will deviate light through 180°, as shown in Figure 2.15a. Mirror faces may be permanently set at 90° by making a **Porro prism**, (the prism section is an isosceles triangle comprising 90°, 45° and 45° angles). See Figure 2.15b. Note: the object is left-handed and so is the reflected image. Two Porro prisms are used in each half of a binocular to produce erect right-handed images. By folding the path through 360°, the prisms shorten the length between the objective lenses and the eyes. See Figure 2.16

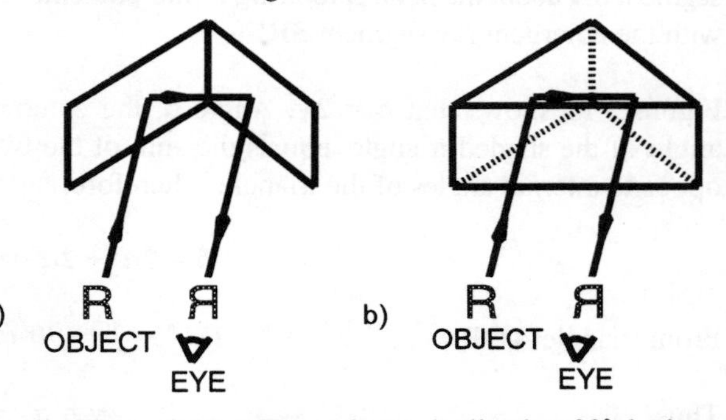

Figure 2.15 a) Two mirrors inclined at 90° deviate light through 180°. A left-handed R remains left-handed. b) The Porro prism is a solid 180° deviator.

2.5 MULTIPLE IMAGES PRODUCED BY TWO INCLINED MIRRORS

2.5.1 CONSTRUCTION OF IMAGES

When an object is placed between two inclined plane mirrors, the rays are reflected back and forth and produce a succession of mirror images. Two series of such images are produced, depending on which mirror *first* reflects the rays. We will determine where, and how many images are formed.

Mirrors **OA** and **OB** are inclined to each other with a dihedral angle **c**. An object *point* **M** placed between the mirrors lies at a distance **OM** = **r**, from the edge. It is at an angle **a** from mirror **OA**, and **b** from mirror **OB**. See Figure 2.17a.

Thus, $\mathbf{c} = \mathbf{a} + \mathbf{b}$. 2.4

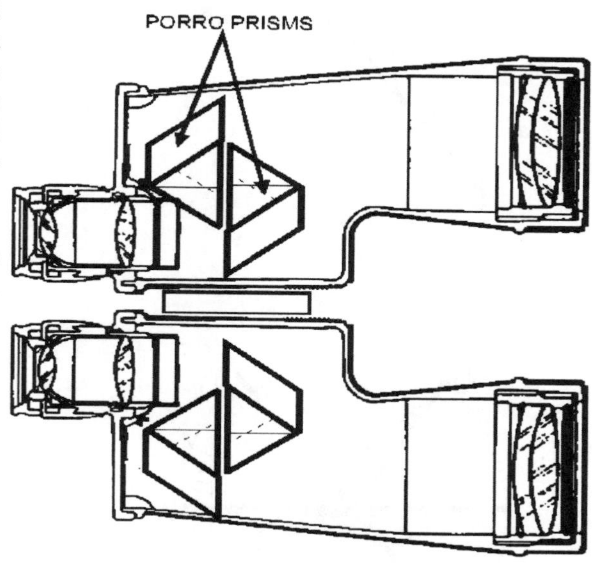

Figure 2.16 The prism binocular incorporates two Porro prisms in each barrel to re-invert and re-erect the image.

2.5.1.1 Image Series J

If rays from **M** *initially* strike the horizontal mirror **OA**, the **J** series of images will be formed. The first image is $\mathbf{J}_1$. It is found by dropping a normal to the mirror **OA** and extending it an equal distance behind **OA** as **M** is in front of it. The shaded triangles **OAM** and $\mathbf{OAJ}_1$ are congruent, therefore, $\mathbf{J}_1$ is a distance **r** from the edge. See Figure 2.17b.

$\mathbf{J}_1$ lies in front of the oblique mirror **OB** and becomes an object for it. By dropping a normal from $\mathbf{J}_1$ to that mirror and extending it an equal distance behind it, we can locate image $\mathbf{J}_2$. The shaded triangles show that this image is also a distance **r** from the edge. *All images will lie on a circle of radius **r**.* See Figure 2.17c. Image $\mathbf{J}_2$ serves as an object for the horizontal mirror.

Image $\mathbf{J}_3$ is found by again dropping a normal to mirror **OA**. See Figure 2.17d. This procedure is repeated with the oblique mirror to obtain image $\mathbf{J}_4$. See Figure 2.17e. Images will continue to be formed until one is formed in the sector bounded by the dashed line extensions of the mirrors, as shown in Figure 2.17f. Image $\mathbf{J}_4$ is *behind* both mirrors, consequently, it is the *final* image in the **J** series. It should be noted that the odd numbered images were produced by the horizontal mirror **OA**, and the even numbered images were produced by the oblique mirror **OB**.

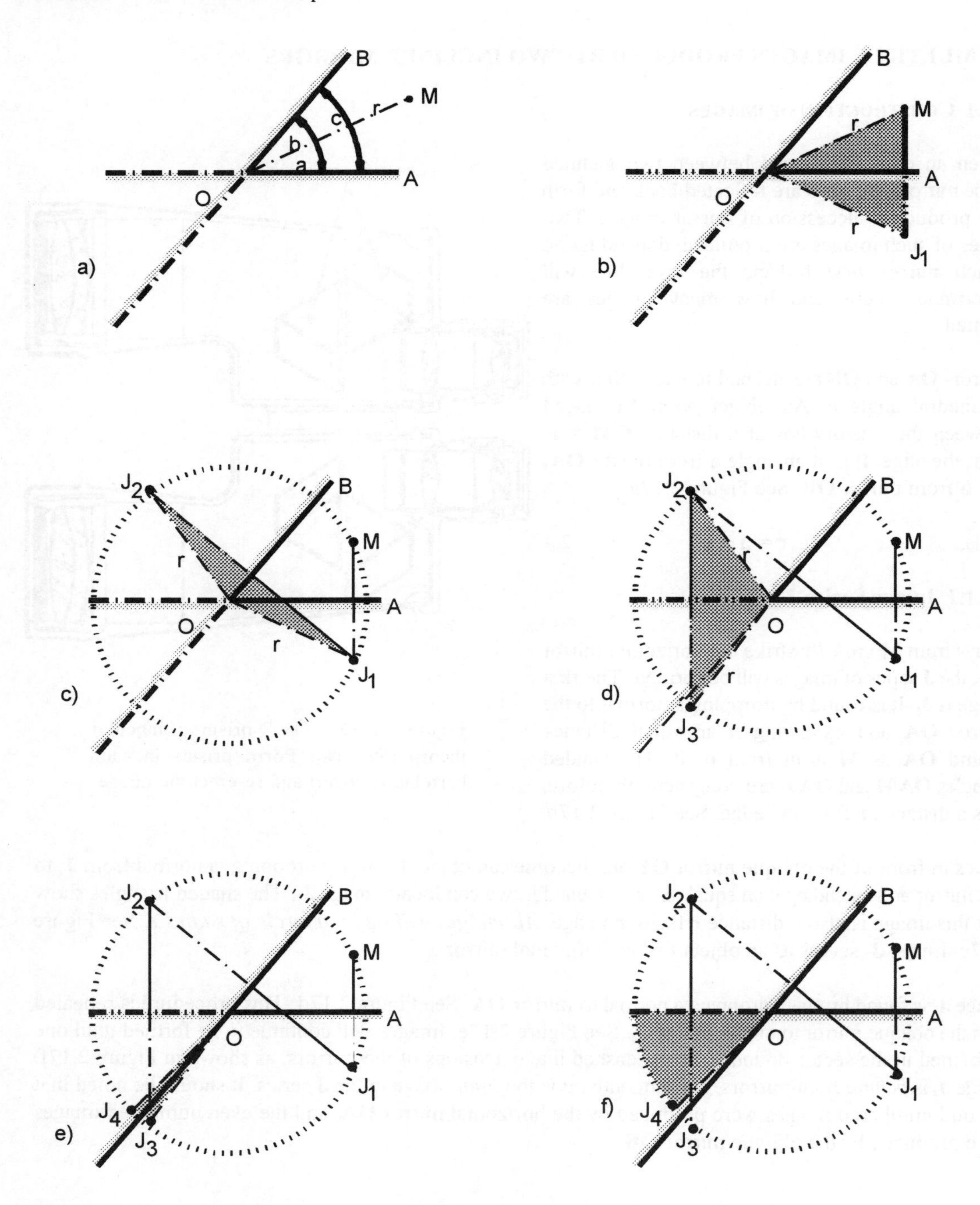

Figure 2.17 Multiple images produced by two inclined mirrors. The J series is due to an initial reflection from mirror OA.

2.5.1.2 Image Series K

A **K** series of images of *point* **M** is produced by *first* reflecting from the oblique mirror **OB**. See Figure 2.18. The odd-numbered **K** images are due to reflection from the oblique mirror; the even-numbered **K** images are reflected from the horizontal mirror. The final image falls in the sector behind both mirrors.

2.5.2 NUMBER OF J AND K IMAGES

Southall shows that the following conditions apply to images formed by successive reflections:

1. The total number of **J** images is given by the *integer next higher than*

$$\frac{180° - a}{c} \tag{2.5}$$

2. The total number of **K** images is given by the *integer next higher than*

$$\frac{180° - b}{c} \tag{2.6}$$

Eliminating angle b from this equation is useful. According to Eq. 2.4, **b = c-a**. Substitution into Eq. 2.6 results in the number of **K** images being given by

$$\frac{180° + a}{c} - 1 \tag{2.6a}$$

3. When (180°-a)/c or (180°-b)/c equals an *integer* the final image falls *on* a mirror. Do *not* increase to the next higher integer.

4. When **180°**/c equals an *integer*, the number of **J** images equals the number of **K** images. The final **J** and **K** images will *coincide* and so we will *see one image fewer than are formed.*

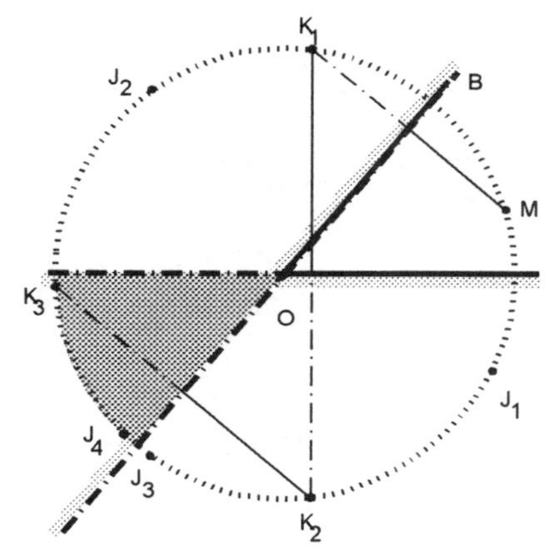

Figure 2.18 Construction of K series images.

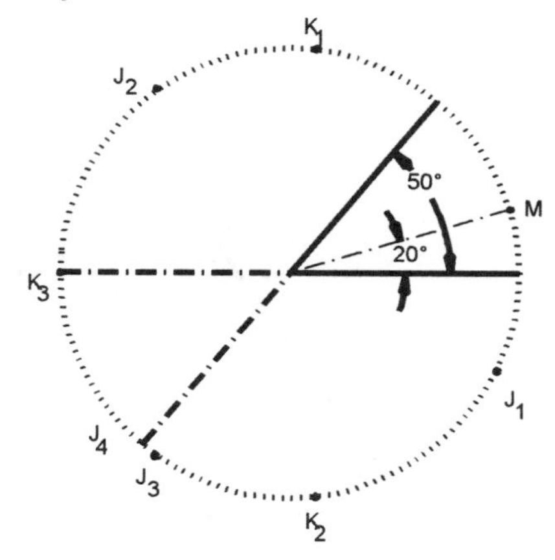

Figure 2.19 a) J and K series images of Example 1.

EXAMPLE 1. Given: c = 50°, a = 20°. Construct and label J and K images. Compute the number of images formed.

Solution: See Figure 2.19a.

From Eq. 2.4 b = 50°-20°= 30°

From Eq. 2.5 No. J images: (180°-20°)/50 = 3.2 --> 4

From Eq. 2.6 No. K images: (180°-30°)/50 = 3.0 --> 3

Note: the last K image falls on mirror OA. Seven images are formed and seen.
Since, 180°/50 = 3.6 is not an integer the final A and B images do not coincide.

EXAMPLE 2. Given: c = 60°, a = 20°. Construct and label J and K images. Compute the number of images formed.

Solution: See Figure 2.19b

From Eq. 2.4 b = 60°-20°= 40°

From Eq. 2.5 No. J images: (180°-20°)/60 = 2.67 --> 3

From Eq. 2.6 No. K images: (180°-40°)/60 = 2.33 --> 3

However, 180°/60 = 3.0 is an integer. The last A and B images coincide. Six images are formed but only five are seen.

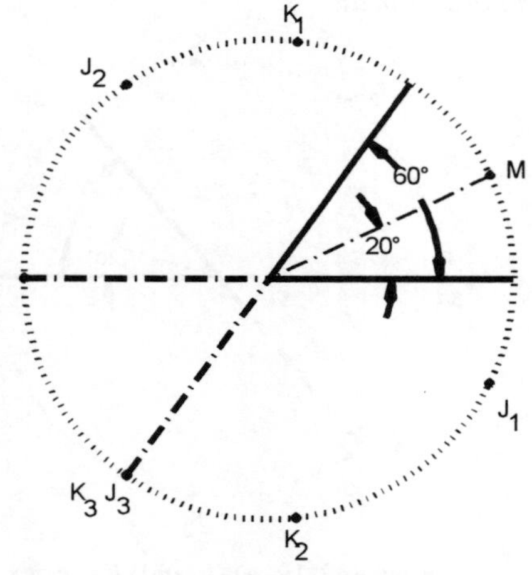

Figure 2.19 b) Example 2 image positions.

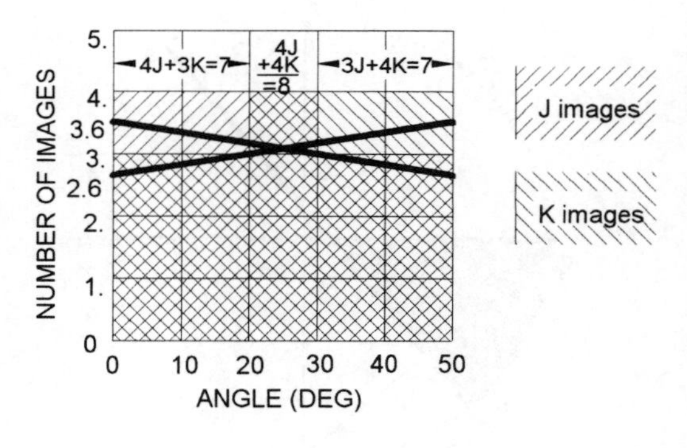

Figure 2.20 Graphical solution to Example 3.

EXAMPLE 3

Two mirrors are inclined with a dihedral angle of 50°. A point source is placed between the mirrors. At what angles from a mirror must the source be placed to produce either 7 or 8 images.

Problems of this type are easily solved by setting **a** equal to 0 and **c,** and solving Eqs. 2.5 and 2.6a to find the minimum and maximum numbers of **J** and **K** images when the object is at either mirror, and the angular position at which the numbers suddenly change.

No. **J** images:

a=0° $\frac{180-0}{50} = 3.6 \rightarrow 4$

a=50° $\frac{180-50}{50} = 2.6 \rightarrow 3$

Find the value of **a** that gives exactly 3 images. Answer: 30°. Consequently, *four* **J** images will be formed when the source is between 0 and *30 °*. *Three* **J** images will be formed for positions between 30-50°.

No. of **K** images:

a=0° $\frac{180+0}{50} -1 = 2.6$

a=50° $\frac{180+50}{50} -1 = 3.6$

Exactly *three* **K** images are formed when **a** equals 20°, consequently, three **K** images are formed when the source is between *0-20 °*. *Four* are formed when the source is between 20-50°.

As a result, seven images are formed when the source is between 0-20° (four **J** and three **K** images). Eight images are formed when the source is between 20-30° (four **J** and four **K** images). When the source is between 30-50°, three **J** and four **K** images are formed.

A graphical solution is shown in Figure 2.20. The straight line between 3.6 and 2.6 corresponds to the **J** solutions. This line crosses the value 3 at 30°. The **K** solutions correspond to the line from 2.6 to 3.6, which crosses the value 3 at 20°. The portions of these lines that are above 3 correspond to 4 images. The portions below 3, correspond to 3 images.

2.5.3 The Kaleidoscope

The kaleidoscope is based on the principle of successive reflections between two mirrors. Long narrow mirrors are mounted in a tube, generally with a dihedral angle of 60°, but some devices use 45°. A variety of colorful beads and oddly shaped bits of glass or plastic tumble in a cell at the far end of the tube, and are diffusely illuminated when the tube is rotated and pointed at a bright source. Beautiful symmetric patterns are seen through a peephole, situated between the mirrors at the other end of the tube.

2.6. RAY PATHS TO THE EYE

Figure 2.21 shows an eye viewing image J_3, by looking into mirror **OA**. No rays are behind the mirror, therefore, no rays can be coming directly from J_3. Rays must originate at the source **M**. To find its path, tracing the ray backwards is necessary, image-by-image. The ray merely appears to come from J_3. Its real part travels from point **T** to the eye, or path **TE**. The dashed prolongation of the ray to J_3 is virtual. J_3 is an image of J_2, and **RT** is the real ray path corresponding to that image. J_2, in turn, is an image of J_1, and **SR** is the real ray path from J_1. Finally, it is the ray **MS** from the source that completes the ray path to the eye.

The eye's position between the mirrors determines the ray path. The mirror into which the eye looks determines which images may be seen.

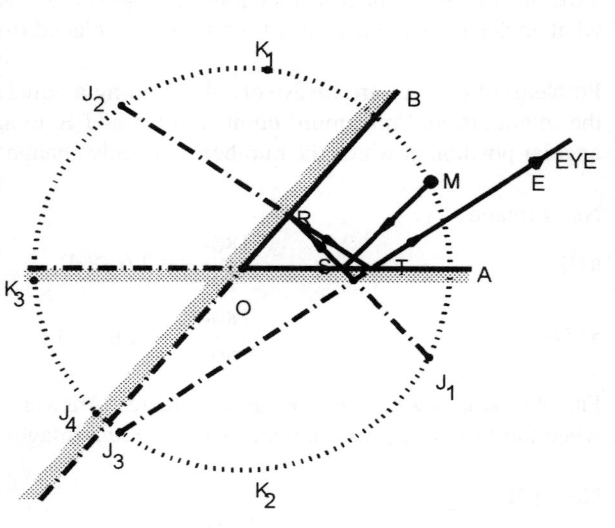

Figure 2.21 Construction of ray paths to the eye from image J_3.

2.7. FIELD OF VIEW OF A PLANE MIRROR

The field of view of an optical system depends on two tangible openings in the system called the *aperture stop* and the *field stop*. The aperture stop controls how much light is transmitted through the system from each point on an object. The field stop determines the extent of the field of view. In a camera, the iris diaphragm is the aperture stop, and the film frame is the field stop.

To find out the limits of the field of view of a plane mirror, as shown in Figure 2.22a, we will consider the optical system to consist of a plane mirror with an aperture diameter equal to **UV**, and an eye with an aperture diameter equal to **S'T'**, centered at **E'.** The transverse plane through **M** is an object plane. We will *assume* that the eye is the aperture stop and the mirror is the field stop. The *entrance pupil* of the system is the image of the aperture stop formed by all optical elements that precede it. Since light from the object strikes the mirror before entering the eye the entrance pupil is the image of the eye formed by the mirror. Rays heading toward the mirror, and in particular toward **SET** must reflect toward **S'E'T'**. The field of view is determined by the rays that go through the edge of the field stop and toward the center of the entrance pupil. After reflection, these rays actually go through the center of the aperture stop, and in this case through the center of the exit pupil. Angle **UEV** is the real angular field of view; angle **UE'V** is the apparent angular FOV. The extent of the object within the angle UEV is the real linear field of view. The apparent linear field of view is the extent of the image within the angle UE'V.

For an object to be within the field of view of an optical system it must send rays toward the entrance pupil via the field stop. Point **P** in the object plane lies within the FOV. Its image **P'** can be constructed by drawing rays from **P** to the entrance pupil **SET**, which reflect from the field stop to corresponding points in the exit pupil. The prolongations of the reflected rays intersect at the image point **P'**, which lies in *the image plane*. The *exit pupil* is the image of the aperture stop formed by all optical elements that follow it. Since the eye is the last element, the exit pupil coincides with the eye.

In Figure 2.22b the object point **P** lies outside the field of view of the mirror. This is seen by noting that the rays that strike the mirror or field stop do not head toward the entrance pupil and, thus, do not reflect into the eye. Secondly, the pencil of rays shown entering the entrance pupil evidently misses the field stop.

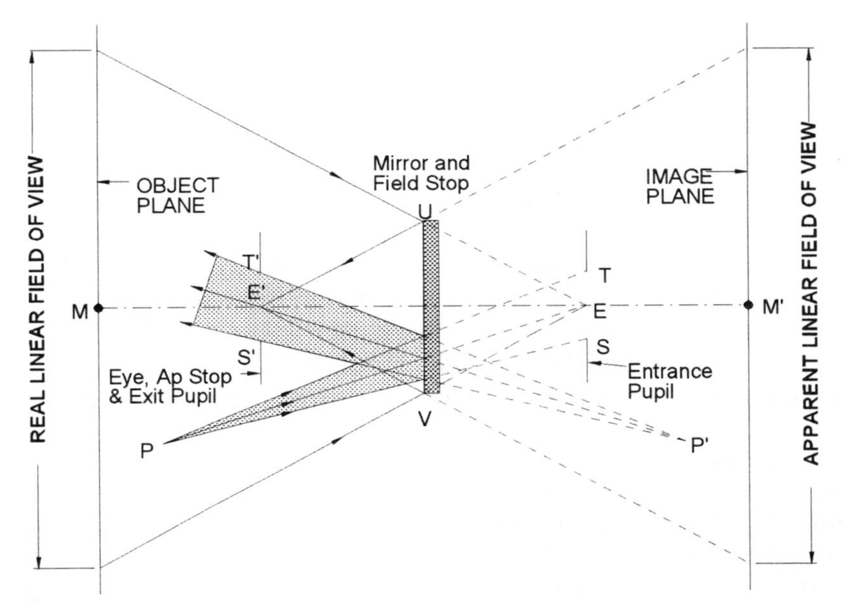

Figure 2.22 a) The field of view (FOV) of a plane mirror. Object P lies within the FOV. Rays headed toward the entrance pupil reflect into the eye.

To decide whether the mirror or the pupil of the eye is the aperture stop we must first find the entrance pupil. This is done by looking into the mirror from the object point **M** and viewing the image of the eye and the mirror. Whichever of the two subtends the smaller angle will be the entrance pupil. If it is the image of the pupil of the eye, then the pupil of the eye is the aperture stop, and by default the mirror is the field stop. If the mirror subtends the smaller angle, it will be the aperture stop as well.

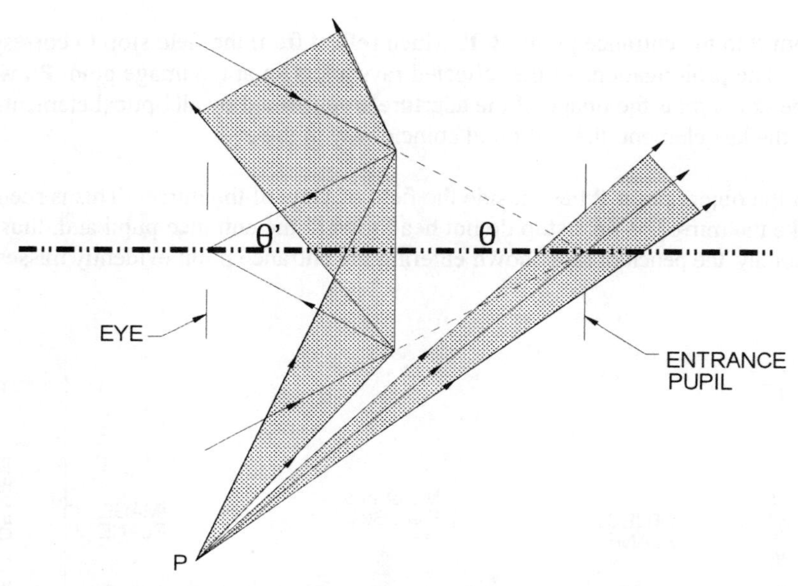

Figure 2.22 b) Object P lies outside the FOV. Rays headed toward the entrance pupil miss the mirror, therefore cannot reflect into the eye. Rays headed toward the mirror miss the entrance pupil, therefore, they miss the eye also.

2.8 REFLECTION ACCORDING TO NEWTON'S CORPUSCULAR THEORY

Newton's corpuscular theory explained reflection by his laws that describe the motions of bodies. A particle or corpuscle of light is a body in motion. Reflection is analogous to bouncing a billiard ball off the edge of the table. The ball, ideally, undergoes an elastic collision. It loses no energy, therefore, its velocity before (**v**) and after (**v'**) the collision must be the same, where **v** and **v'** are vectors. However, the perpendicular component of the velocity **v'** is reversed.

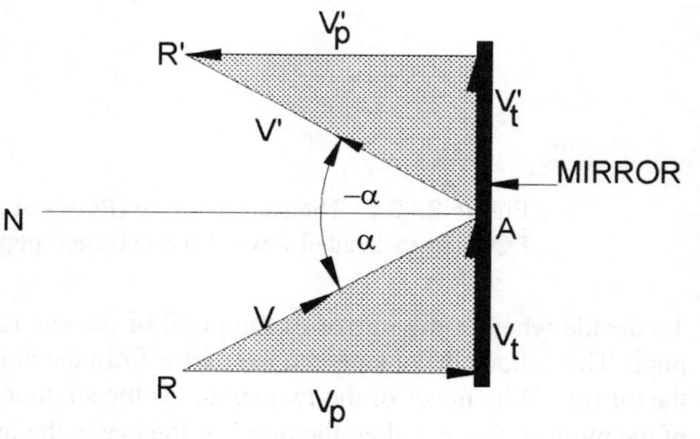

Figure 2.23 The law of reflection is satisfied by Newton's corpuscular theory.

Figure 2.23 is a vector diagram illustrating the tangential $\mathbf{v}_t$ and perpendicular $\mathbf{v}_p$ velocity components of the incident light particle, traveling along direction **RA**. The vector components of the velocity of the particle, after reflecting along direction **AR'**, are $\mathbf{v'}_t = \mathbf{v}_t$, and $\mathbf{v'}_p = -\mathbf{v}_p$. The shaded triangles are congruent, consequently, the angles of incidence and reflection are equal (but of opposite signs).

2.9. REFLECTION ACCORDING TO HUYGENS' WAVE THEORY

In order for Huygens' wave theory to supplant the corpuscular theory, among other things, it had to account for the law of reflection. *Huygens postulated that each point on a wave was a source of secondary wavelets, the wavefront is tangent to all the wavelets, and rays are normal to the wavefront.* Collimated light, i.e., light from a distant point, proceeds as a train of plane wavefronts with a velocity **v** toward a plane mirror. See Figure 2.24a, Assume that Rays 1, 2 and 3 are perpendicular to the wavefront. At a given instant, the edge of the wavefront $\mathbf{A_1A_3}$ strikes the mirror at point $\mathbf{A_1}$. In the time it takes a wavelet from $\mathbf{A_3}$ on the opposite edge of the wavefront (or Ray 3) to reach the mirror at point $\mathbf{B_3}$, a wavelet at point $\mathbf{A_1}$ (or Ray 1) would have reached point $\mathbf{B_1'}$ if there was no mirror in the way. Due to the mirror, Ray 1 is reflected and travels a distance equal to $\mathbf{A_1B_1'}$, that is, equal to the radius of the wavelet centered on point $\mathbf{A_1}$. Note that this wavelet extends on both sides of the mirror. The question is where, on the portion of the wavelet that is in front of the mirror, will the reflected wavefront be?

Meanwhile Ray 2 will strike point $\mathbf{C_2}$ on the mirror. A wavelet from point $\mathbf{C_2}$ will reflect back into the medium in front of the mirror. In the time it takes Ray 3 to travel from $\mathbf{C_3}$ to $\mathbf{B_3}$, Ray 2 will travel a distance equal to the radius of the wavelet centered on point $\mathbf{C_2}$. Again, where on this wavelet is the reflected wavefront?

According to Huygens, the reflected wavefront corresponds to the tangent to the wavelets, namely, the line $\mathbf{B_1D_2B_3}$. All other secondary wavelets in the incident pencil of light will be tangent to this line, and the reflected rays will be perpendicular to the wavefront if they intersect the wavelets at the points of tangency. Therefore, the direction of reflected Ray 1 will be along $\mathbf{A_1B_1}$, Ray 2 will reflect along $\mathbf{C_2D_2}$, and Ray 3 will be parallel to these rays.

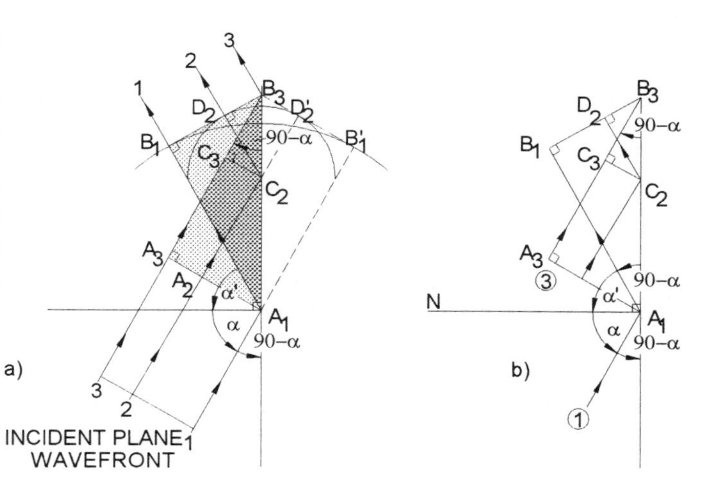

Figure 2.24 Huygens' wavefront theory satisfies the law of reflection. a) Construction of path of plane wavefront. b) Congruent triangles from a), with a common side that corresponds to the reflected wavefront, prove the construction.

First let us show that the plane reflected wavefront is tangent to all the wavelets. Two pairs of congruent right triangles taken from the previous figure are shown in Figure 2.24b. Each pair has a common hypotenuse, and the two common hypotenuses are coincident. Therefore, sides $\mathbf{B_1B_3}$ and $\mathbf{D_2B_3}$ of the triangles must coincide and, thus, form the reflected wavefront.

Next, showing that this construction satisfies the law of reflection is necessary. Ray 1 makes an angle of incidence $\boldsymbol{\alpha}$ with the normal **NA** to the surface at **A**. The angle between Ray 1 and the mirror is **90-α**. Ray 3 is parallel to Ray 1, therefore, angle $\mathbf{C_3B_3C_2} = \mathbf{90\text{-}\alpha}$. Due to congruency, angle $\mathbf{B_1A_1B_3}$ must equal **90-α**, also. Therefore, reflected Ray 1 makes an angle with the normal equal to α. Finally, the incident and reflected rays and the normal clearly all lie in one plane.

REVIEW QUESTIONS

1. State the law of reflection.
2. Define the angles of incidence and reflection.
3. How can you make a weightless pointer?
4. What is the advantage of a pentaprism over a plane mirror for deviating light?
5. Define the entrance pupil, aperture stop and exit pupil. Can the pupils coincide with the aperture stop?
6. Distinguish between the functions of the aperture and field stops.
7. What did Huygens postulate regarding the wave nature of light?
8. Some clinic rooms have semi-transparent mirrors to allow instructors to observe student clinicians and patients. Why is the student unable to see the instructor?
9. Your hair stylist proudly shows you the back of your head. Draw the path of a ray to your eye.

PROBLEMS

1. In Fig. 2.12a assume that the reflected ray strikes the point "A". Show that the incident ray takes the shortest path to "A".

2. One of a pair of identical twins parts his hair on his right side. His mirror image looks exactly like his brother. On which side does his brother part his hair?

3. In Problem 2, the mirror is replaced by two plane mirrors set at 90° to each other. On which side of his head will he see his part?

4. A plane mirror is inclined 10° to the horizontal. Collimated rays, traveling vertically downward, strike the mirror.Find a) the angle of incidence, and b) the angle between the reflected ray and the horizontal?

Ans.: a) 10°, b) 70°

5. The inclination of the mirror in the above question is changed to 20°. a) What is the angle between the reflected ray and the horizontal? b) through what additional angle has the ray been deviated?

Ans.: a) 50°, b) 20°.

6. A patient sits 15 feet from a mirror on a wall. On the opposite wall, 5 feet behind the patient, is an eye chart. How far from the patient does the image of the eye chart appear to be? Ans. 35 feet.

7. If, in the above problem, the mirror is 2 feet in diameter, what is the total apparent field of view?

Ans.: 7.63°

8. Two mirrors are inclined to each other. A ray traveling parallel to mirror #1 is incident on mirror #2. After reflecting once at each mirror, the ray travels parallel to mirror #2. What is the dihedral angle of the mirrors?

9. Two plane mirrors are inclined at an angle of 10°, and M is a point in one mirror. What must be the

angle of incidence of a ray from M on the second mirror so that, after three reflections, it leaves parallel to the first mirror?
Ans.: 60°

10. Can the eye see image J_4 by looking into the oblique mirror in Figure 2.21? Trace the ray path from J_4 to the eye.

11. The ordinary kaleidoscope displays a hexagonal pattern. What is the angle between the mirrors?

12. The dihedral angle between two mirrors is 40°. A point object is placed 10° from one mirror. a) How many images are formed? b) how many are seen?
Ans.: a) 9. b) 9.

13. Change the dihedral angle to 45° in the above problem.
Ans.: a)8. b) 7.

14. Change the angle to 15°, and repeat.
Ans.: a) 24. b) 23.

15. The dihedral angle between two mirrors is 54°. What range of angles may "a" take to produce a) six images? b) seven images? c) Construct the images when **a** = 20°. d) Construct images when **a** = 40°.
Ans.: a) 18 to 36° b) 0 to 18° and 36 to 54°

16. The dihedral angle between two mirrors is 100°. What range of angles may "a" take to produce a) four images? b) three images? Construct the images when **a** = 90°. Place an object at angle **a = 15°**. Construct the ray path from the object to the eye that allows viewing $\mathbf{J}_2$. (Note: you must choose the eye position).
Ans.: a) 20 to 80° b) 0 to 20° and 80 to 100°

17. Two mirrors, separated by 100 mm, face and are parallel to each other. An object is 20 mm from one mirror. Treat images that are first formed at this mirror as the **J** series. Calculate the distance of the $\mathbf{J}_3$ and $\mathbf{K}_4$ images from the object. Construct the path of the ray from the $\mathbf{K}_4$ image to the eye.

18. Draw two plane mirrors, so arranged that a ray will be constantly deviated through 120°, after successive reflections.

19. An eye, with a 4-mm pupil, views an object plane through a plane rear-view mirror that is 20 mm in diam. The eye is 200 mm from the mirror, and the object plane is 1 meter from the mirror. a) Identify the aperture stop and find the entrance and exit pupils; b) Find the diameters of the entrance and exit pupils; c) Find the real and apparent angular and linear fields of view.
Ans.: a) the eye is the aperture stop; the entrance pupil is 200 mm behind the mirror; the exit pupil coincides with the eye; b) both pupils are 4 mm in diameter; c) the total real and apparent FOVs are equal, 5.72°, and 120 mm.

20. A plane mirror, 4 mm in diameter, reflects light from an object plane 100 mm away, to a diaphragm. The diaphragm is 20 mm in diameter and 20 mm from the mirror. a) Identify the aperture stop and find the entrance and exit pupils; b) Find the diameters of the pupils; c) Find the real and apparent angular and linear fields of view.
Ans.: a) the mirror is the aperture stop, the entrance and exit pupils coincide with the mirror; b) both pupils are 4 mm in diameter; c) the total real and apparent FOVs are equal, 53.13°, and 100 mm.

NOTES:

CHAPTER 3

REFRACTION OF LIGHT

3.1 THE LAW OF REFRACTION

Light rays are bent when they cross the boundary separating optical medium **a** from medium **b**. In Figure 3.1, refraction from a rarer to a denser medium is shown. The incident ray makes an angle **α** with the normal **NB**; the angle of the refracted ray is **α'**; where, **α** = angle **NBA,** and **α'** = angle **N'BC.** The refracted ray is bent toward the normal.

Willebrord Snell was the first to discover the relationship between **α** and **α'**. Snell's law says that the *ratio of the sines of the angles of incidence and refraction is equal to a constant which depends on the nature of the two media and the wavelength of the light, and the incident ray, refracted ray and normal all lie in the plane of incidence.*

$$\frac{\sin\alpha}{\sin\alpha'} = n_{ab} = \frac{v_a}{v_b} \qquad 3.1$$

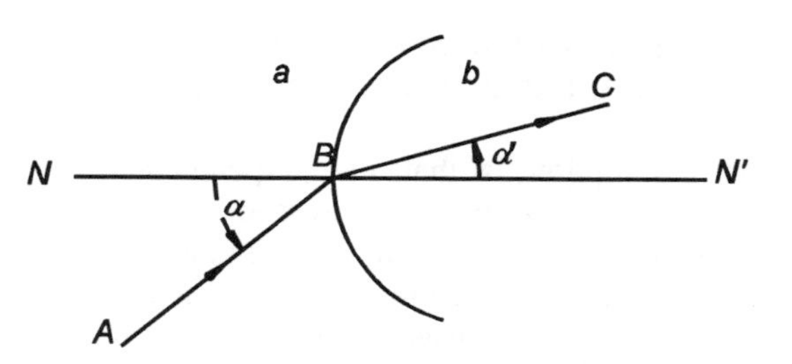

Figure 3.1 Refraction of a ray from rare medium **a** into dense medium **b**, by a spherical surface.

The constant n_{ab} is equal to the *relative index of refraction* and that, in turn, equals the ratio of the velocity of light in medium **a** to the velocity in medium **b**. For example, in going from air to water $\mathbf{n}_{aw} = 4/3$; the relative index of air-to-glass is $\mathbf{n}_{ag} = 3/2$; and water-to-glass has a relative index $\mathbf{n}_{wg} = 9/8$.

3.2 THE REVERSIBILITY OF LIGHT

If a ray is refracted through several media finally striking a mirror with normal incidence, the reflected ray will retrace the original path. The relative indices from **a** to **b** and from **b** to **a** are reciprocals of each other.

$$\frac{\sin\alpha}{\sin\alpha'} = n_{ab} \qquad \frac{\sin\alpha'}{\sin\alpha} = n_{ba} \qquad 3.2$$

Thus, $\mathbf{n}_{aw} = 4/3$, but $\mathbf{n}_{wa} = 3/4$, and $\mathbf{n}_{ga} = 2/3$, etc.

When $\mathbf{n}_{ab}$ is greater than unity the second medium is more highly refracting and light is bent toward the normal.

3.3 REFRACTION AND PARTICLE MOTION

Any theory of light must explain refraction if it is to gain acceptance. Figure 3.2 illustrates a mechanical analogy to the corpuscular theory that shows how bending toward the normal can occur. Imagine a ball rolling on a smooth horizontal surface from A to B with a velocity v. The ball enters the ramp obliquely.

Gravity accelerates the ball in a direction that is perpendicular to the ramp, but does not affect the ball's velocity in the tangential direction (i.e., along the ramp). When the ball reaches the lower horizontal surface its velocity v' is greater than v, and its path is bent toward the normal. Applying this analogy to light particles means that somehow the denser medium is applying an attractive force that increases the perpendicular component of its velocity. The following vector discussion is based on Goldwasser.

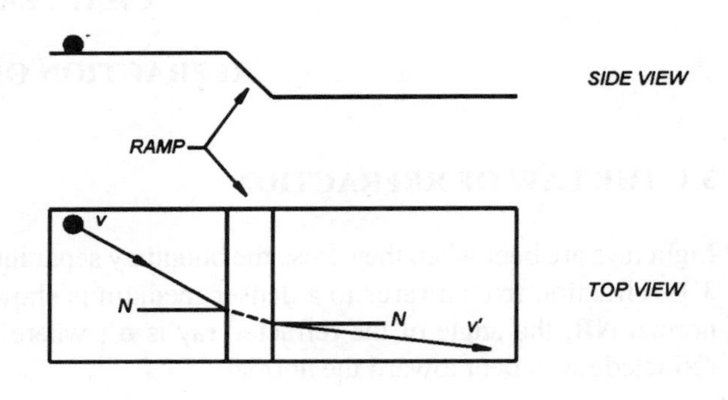

Figure 3.2 Mechanical analog of refraction.

3.3.1 Vector diagram - perpendicular component increased by refraction

The assumption is that the perpendicular velocity component is increased and the tangential component is not affected in crossing from one medium to another. If the refraction is from a rarer to a denser medium, Snell's law requires that the rays be bent toward the normal. In Figure 3.3 the incident ray velocity is **v**, and the perpendicular and tangential components are $\mathbf{v}_p$ and $\mathbf{v}_T$. After refraction the components are $\mathbf{v'}_p$ and $\mathbf{v'}_T$, where

$$\mathbf{v'}_P > \mathbf{v}_P, \text{ and } \mathbf{v'}_T = \mathbf{v}_T$$

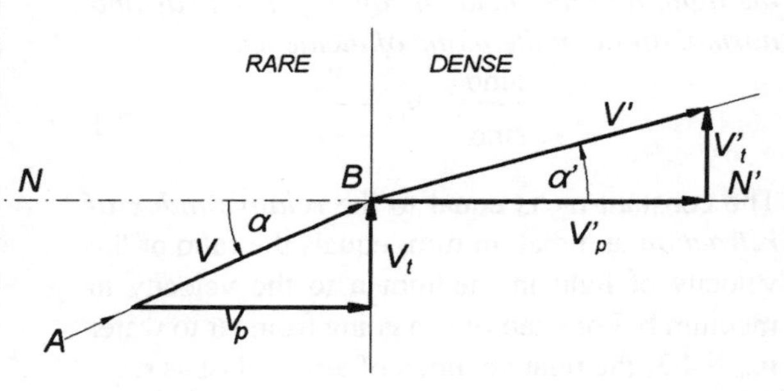

Figure 3.3 Refraction toward the normal is obtained if the perpendicular component of the velocity is increased. However the refracted ray has a greater velocity in the denser medium.

The ray is bent toward the normal but travels with a greater velocity in the second medium than it had before refraction, i.e., **v'> v**. Does this satisfy Snell's law? According to Figure 3.3,

$$\sin\alpha = \frac{v_T}{v} \qquad \sin\alpha' = \frac{v'_T}{v'} = \frac{v_T}{v'}$$

therefore,

$$\frac{\sin\alpha}{\sin\alpha'} = \frac{v'}{v} = \frac{v_b}{v_a} = constant$$

Although this diagram satisfies Snell's law, the constant is the inverse of Eq. 3.1; it requires light to travel with a *greater* velocity in the *denser* medium. It was not known in Newton's and Huygens' day whether this was correct.

3.3.2 Vector diagram - tangential component decreased by refraction

Alternatively, to satisfy Snell's law we could require that the tangential component be shortened and the perpendicular component be unchanged, in going from rare to dense media (e.g., air and glass). As seen in Figure 3.4 this would bend light toward the normal. Since, $v_P = v'_P$, but $v'_T < v_T$

It follows that v'< v, or the light will travel more *slowly* after refracting into the glass. If the perpendicular component of the velocity is unchanged on crossing an interface, by having the ray strike a second face perpendicularly the tangential component to the velocity will be zero. According to our hypothesis the perpendicular component will be unchanged on refraction. Thus, the light will emerge into the air with the same velocity it had in the glass. It will travel more slowly in air than it did initially. By setting a series of prisms at appropriate angles we may repeat this process. See Figure 3.5. Each prism will cause the light to emerge with a reduced velocity. This is not found to occur experimentally. Furthermore, if the velocity v' after traversing a prism is equal to v/1.5 (where 1.5 is the index of the glass) 25 prisms will reduce the velocity of light to about 7½ miles per hour. An absurd result.

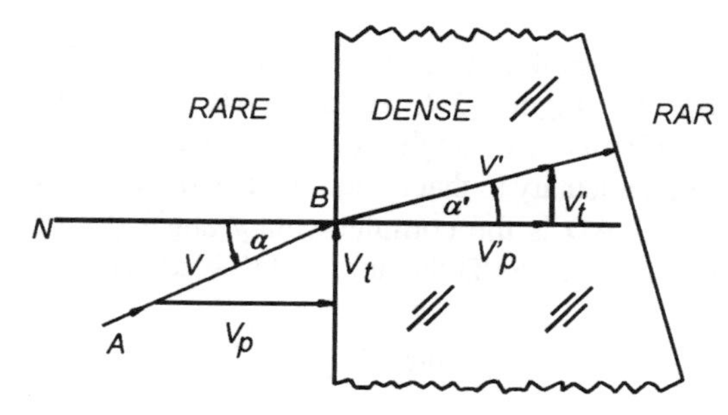

Figure 3.4 Refraction toward the normal is obtained by decreasing the tangential component of the velocity. The ray velocity is decreased in the denser medium.

3.3.3 HUYGENS' CONSTRUCTION FOR REFRACTION

In Figure 3.6, the plane wavefront **AB**, in a rare medium, has reached the refracting surface at the point **A** at time zero. Point **A** becomes a new origin of wavelets entering the dense medium. Huygens has the wavelets travel more slowly in the denser medium. Consequently, in the time it takes the wavefront to go from **B** to **D** with a velocity v_a, in the rare medium, the wavelet will travel with a velocity v_b in the denser medium and expand to some radius **AC.** The time taken to go from **B** to **D**, and from **A** to **C** is the same, therefore,

$$t = \frac{BD}{v_a} = \frac{AC}{v_b}$$

or the radius of the wavelet equals

$$AC = \frac{v_b}{v_a} BD$$

A similar construction for an intermediate point **Q** will result in a wavelet with a radius, at the instant the wavefront reaches **D,** of $QR = \frac{v_b}{v_a} KD$.

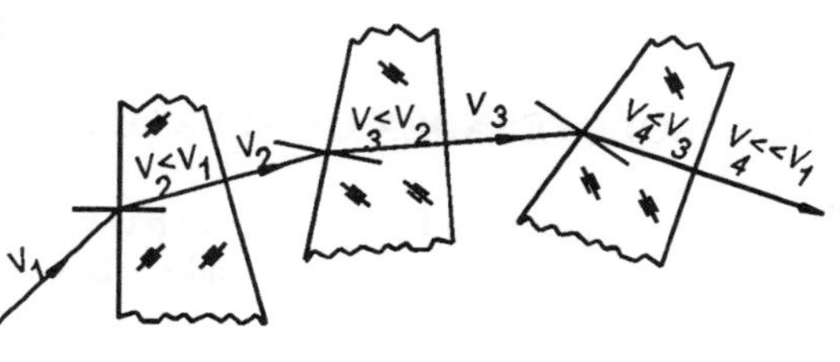

Figure 3.5 Decreasing the tangential component of the velocity causes light to emerge from each prism with reduced velocity.

According to Huygens the new wavefront **CD** is tangent to all the wavelets and the rays travel in a direction that is normal to the wavefronts.

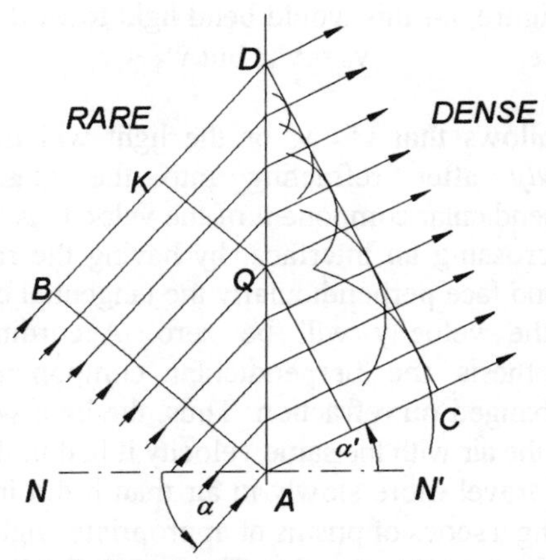

Figure 3.6 Huygens' construction for refraction of a plane wave at a plane surface.

To see whether this construction satisfies Snell's law, consider Figure 3.7. Angle NAR = α, BAN is the complementary angle (90-α), and angle DAB is complementary to that or equals α. Similarly, angle N'AC = α', CAD is the complementary angle (90-α'), and in right triangle ACD the angle ADC = α'. Therefore,

$$\sin\alpha = \frac{BD}{AD} = \frac{v_a t}{AD}$$

and

$$\sin\alpha' = \frac{AC}{AD} = \frac{v_b t}{AD}$$

Therefore,

$$\frac{\sin\alpha}{\sin\alpha'} = \frac{v_a}{v_b} = constant \qquad 3.3$$

This constant is identical to the *relative index of refraction* n_{ab}. Huygens construction satisfies Snell's law and correctly has light traveling more slowly in the denser medium. Compare this constant with that of Eq. 3.1.

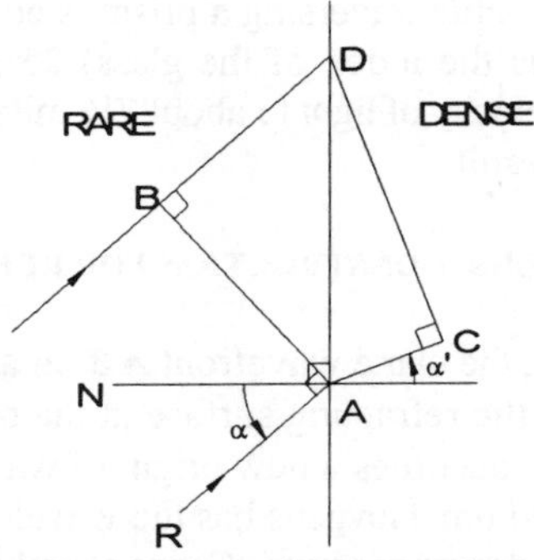

Figure 3.7 Huygens' construction - detail.

3.4 ABSOLUTE INDEX OF REFRACTION AND SNELL'S LAW

Suppose, as shown in Figure 3.8, we have three optical media **a**, **b**, and c. The corresponding velocities of light in those media are v_a, v_b, and v_c. The relative indices are:

$$n_{ab} = v_a/v_b \qquad 3.4$$

$$n_{ca} = v_c/v_a \quad \text{or} \quad v_a = v_c/n_{ca} \qquad 3.5$$

and

$$n_{cb} = v_c/v_b \quad \text{or} \quad v_b = v_c/n_{cb} \qquad 3.6$$

Substituting Eqs. 3.5 and 3.6 into 3.4 we obtain $n_{ab} = \dfrac{n_{cb}}{n_{ca}}$.

That is, the relative index from medium **a** to **b** equals the relative index from **c** to **b** divided by the relative index from **c** to **a**. In other words the *relative* index between any two media can be found if we know the *relative* index of each medium with respect to some standard medium **c**. Vacuum is an ideal standard medium. Light of all wavelengths travels with a constant and maximum velocity in a vacuum. Therefore, we will consider *relative* indices with respect to vacuum (medium **c**) as *absolute indices of refraction.*

Relative index n_{ca} becomes absolute index n_a, and n_{cb} becomes n_b. Then the relative index, on going from medium **a** to **b**, is equal to the ratio of the absolute index of medium **b** to the absolute index of medium **a**.

$$n_{ab} = \frac{n_b}{n_a} = \frac{\sin\alpha}{\sin\alpha'}$$

For example, let medium **a** be water, $n_a = 4/3$, and let medium **b** be glass, $n_b = 3/2$. From Eq. 3.7 we find that the relative index n_{ab} *from water to glass* equals 9/8.

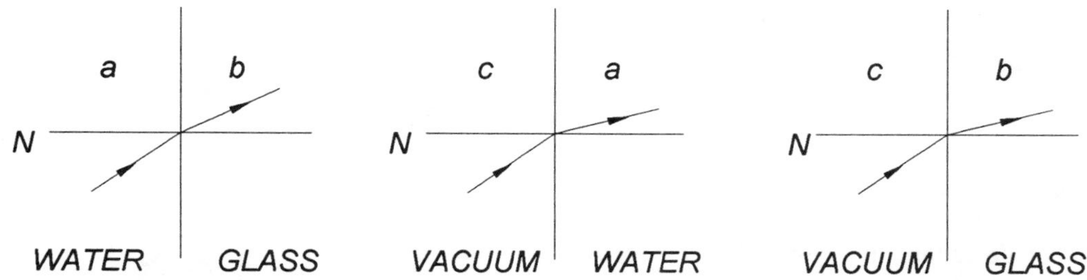

Figure 3.8 The refractive index for water/glass is relative; for vacuum/water and vacuum/glass it is absolute.

The nomenclature that we will use for the absolute index of refraction when refracting rays through a surface separating two optical media is **n** for object space (the medium to which the object belongs) and **n'** for image space. Thus, the usual form of Snell's Law is

$$n\sin\alpha = n'\sin\alpha' \tag{3.7}$$

3.5 CONSTRUCTION OF REFRACTED RAY

Ray **JA** is incident on a surface separating two media of indices **n** and **n'** (e.g., air and glass) in Figure 3.9. Draw two arcs, centered on the point of incidence with radii proportional to **n** and **n'**. Specifically, let

$$r_1 = n \qquad\qquad r_2 = \frac{n'}{n} r_1$$

Mark the point **J** where the incident ray intersects the arc corresponding to the object space medium. Translate the normal to the surface to make it intersect **J**, and mark the point **J'** on the arc corresponding to image space. The direction of the *refracted* ray is along the projection of line **J'A**. The construction works because it satisfies Snell's Law. According to Figure 3.9

$$\sin\alpha = \frac{h}{r_1} \qquad\qquad \sin\alpha' = \frac{h}{r_2} = \frac{nh}{n' r_1}$$

thus,

$$\frac{\sin\alpha}{\sin\alpha'} = \frac{n'}{n}$$

Note that the refraction is from a rarer to a denser medium, i.e., **n'>n** and the ray refracts toward the normal. In Figure 3.10, **n'<n** and the ray refracts away from the normal.

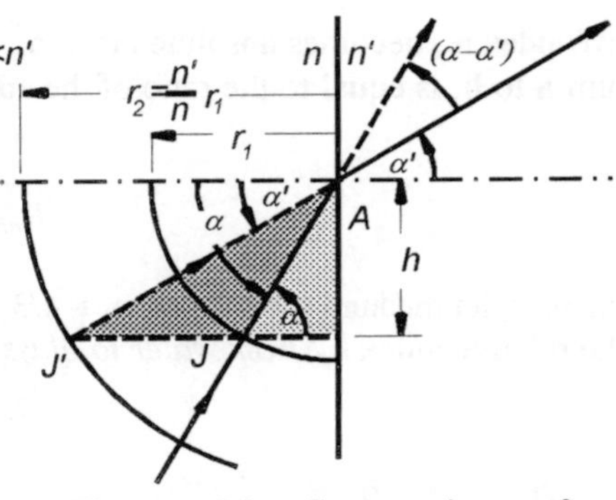

Figure 3.9 Construction of refracted ray (n'>n).

3.6 DEVIATION OF THE REFRACTED RAY

The angle of deviation of the refracted ray is the acute angle through which the refracted ray has to be turned to bring it into alignment with the incident ray.

$$\delta = \alpha - \alpha' \qquad 3.8$$

When the incident ray is normal to the surface, angle $\alpha = \alpha' = \delta = 0°$. As the angle of incidence is increased, so does the angle of deviation. In Figures 3.9 and 3.10, is δ positive or negative?

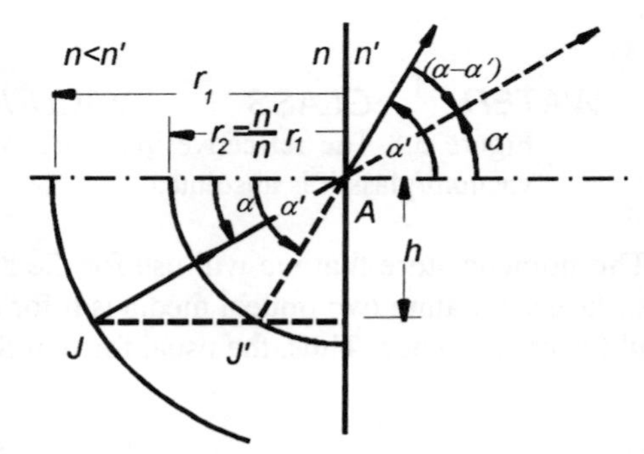

Figure 3.10 Construction of refracted ray (n'<n).

EXAMPLE: A ray makes an angle of incidence of 30° at a plane refracting surface separating air (n = 1) from glass (n' = 1.6). What is the angle of deviation?

$$\sin 30° = 1.6 \sin \alpha'$$

$$\sin \alpha' = 0.5/1.6 = 0.3125$$

$$\alpha' = 18.21°$$

therefore, $\delta = 30 - 18.21 = 11.79°$

3.7 DISPERSION

Except in vacuum and for normal incidence, white light, refracted by any medium, will be dispersed into colors or a spectrum. This is due to a variation of the velocity of light in the medium as a function of the wavelength. Consequently, the index of refraction of the medium varies with the wavelength. For example, the glass commonly used for spectacle lenses has an index in yellow light (λ = 589 nm) of n_d = 1.523; in blue light (λ = 486 nm) the index is n_F = 1.529; and in red light (λ = 656 nm) it is n_C = 1.520. When an index is stated without any qualification, it is understood to refer to yellow light, i.e., n = 1.523.

3.8 TOTAL INTERNAL REFLECTION

In going from a rarer to a denser medium the angle of incidence may reach a maximum of 90°, or *grazing incidence*. The refracted ray also will be a maximum, but less than 90°. This maximum angle α' is called the critical angle α_c. If now we reverse the light path, a ray incident at the critical angle will undergo *grazing emergence.* Rays incident at greater angles than α_c will undergo *total internal reflection.* We can find the critical angle from Snell's law: n sin 90° = n' sin α_c.

thus, $$\sin\alpha_c = n/n' \tag{3.9}$$

The critical angle for an air/glass interface is, $\sin\alpha_c = 1/1.5$

thus, $$\alpha_c = 41.81°$$

3.9 OPTICAL PATH LENGTH

Distinguishing between the geometric path length of light and the optical path length is important. The OPL allows for the slower velocity of light in any media through which the rays travel. For an optical system to form perfect images the optical path lengths of all the rays converging on the image must be equal.

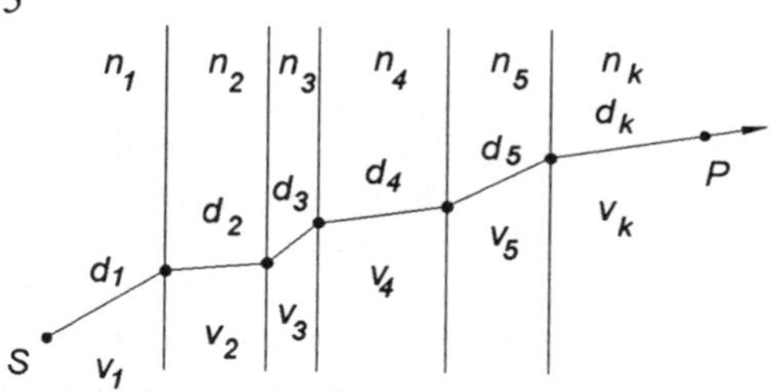

Figure 3.11 Optical path length.

Suppose light travels from point **S** to point **P** by traversing several layers of media of various indices, as shown in Figure 3.11. Individual segments of the path are indicated by **d**, and the velocity in each layer by **v**. The transit time will be

$$t = \frac{d_1}{v_1} + \frac{d_2}{v_2} + \cdots + \frac{d_k}{v_k}$$

This may be written as a summation:

$$t = \sum_{j=1}^{k} \frac{d_j}{v_j} \tag{3.10}$$

where subscript *j* represents a given layer.

The index of the *jth* layer is given by the velocity in a vacuum **C** divided by the velocity in the *jth* layer, therefore,

$$v_j = \frac{C}{n_j}$$

Substitution into Eq. 3.10 results in

$$t = \frac{1}{C}\sum_{j=1}^{k} n_j d_j = \frac{1}{C}\sum_{j=1}^{k} OPL \tag{3.11}$$

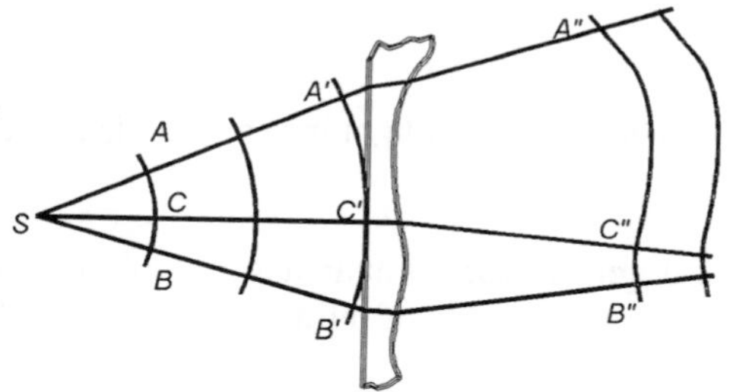

Figure 3.12 Wavefronts are deformed by an irregular plate.

The summation of the $n_j d_j$ terms is known as the *optical path* length (OPL). It is the product of the index of the medium and the path length in the medium. Equal OPLs occur for all ray paths between wavefronts. In Figure 3.12, spherical wavefronts proceed from source S. Wavefronts are separated from each other by equal optical path lengths. Thus, the ray paths AA', BB' and CC' are equal. When these rays traverse the irregular optical medium their OPLs are unequal, but each of the total OPLs A'A", B'B" and C'C" are equal (contain the same number of waves) and the emergent wavefront is deformed. All the light waves are in phase at the wavefront.

3.10 FERMAT'S PRINCIPLE

In Figure 2.6 clearly the path of the reflected ray MBR' is equal to the straight line path M'BR'. Since the straight line is the shortest distance between two points, it would appear that, in reflecting from M to R', light takes the shortest path. It is obvious in Figure 3.1 that light does not take the straight line path in refracting from point A to point C. Fermat's Principle originally said that a ray of light in going from one point to another takes the path that requires the least time. A complete statement of Fermat's Law is that the traversal time is an extreme, i.e., a minimum or a maximum, or has a stationary value with respect to other paths.

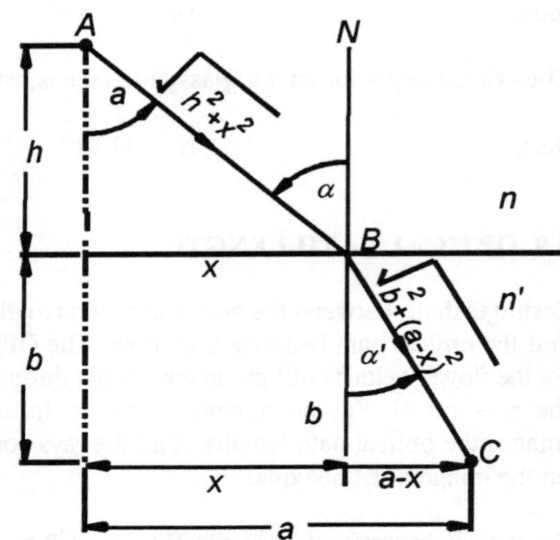

Figure 3.13 Fermat's principle subsumes Snell's law.

REVIEW QUESTIONS

1. State the law of refraction.
2. How may the reversibility of light be shown?
3. Does wavelength or frequency change when light refracts from one medium to another?
4. A glass fiber pipes light down its length. Would a fine tunnel of air in an ice cube act as a light pipe?

PROBLEMS

1. The velocity of light in sea water is 224,551 km/sec. What is the index of refraction of the water?

Ans.: 1.336

2. Light is incident at an angle of 30° at a plane refracting surface separating air from glass of index 1.5. Find: a) the angle of refraction? (b) the angle of deviation? Construct the refracted ray.

Ans.: a. 19.471°; b. 10.529°

3. Light is incident at an angle of 30° at a plane refracting surface separating glass of index 1.6 from water of index 1.333. Find: (a) the angle of refraction; (b) the angle of deviation; the critical angle; d) the relative index of refraction. Construct the refracted ray.

Ans.: (a) 36.881°; (b) -6.881°; (c) 56.421°; d) 0.833

4. The angles of incidence and refraction at a plane interface, separating air from a second medium, are 50° and 35°, respectively. The velocity of the light in air is 300,000 km/sec. Find (a) the relative index; (b) the absolute index of the second medium; (c) find the velocity of the light in the second medium; d) find

the relative index when the media are reversed. Construct the ray path.

Ans.: (a) 1.3356; (b) 1.3356; (c) 224,625 km/sec; d) 0.7488.

5. A ray makes an angle of incidence of 30° with a plane interface that separates water (index = 1.3333) from glass (index = 1.7500). Find: (a) the angle of refraction; (b) the angle of deviation; (c) the critical angle. Construct the ray path.

Ans.: (a) 22.3927°; (b) 7.6073°; (c) 49.6324°

6. A ray makes an angle of incidence of 45° with a plane interface that separates glass (index = 1.7500) from glass (index = 1.5000). Find: (a) the angle of refraction; (b) the angle of deviation; (c) the maximum angle of incidence that results in a refracted ray; d) the angle of refraction of this ray. Construct the ray path.

Ans.: (a) 55.5842°; (b) -10.5842°; (c) 58.9973°; d) 90°.

7. A beaker contains a 1 inch thick layer of water (index 1.333) A 0.5 inch layer of oil (index 1.45) floats on the water. A ray of light makes an angle of incidence of 60° with the upper surface of the oil. Find the angles of refraction at the air/oil and oil/water interfaces.

Ans.: 36.674°; 40.518°

8. What is (a) the actual path length and b) the optical path length of the ray from the point of incidence at the oil to the bottom of the beaker?

Ans.: a. 1.938 in.; b. 2.657 in.

9. You are 3 feet under water and looking straight up. Within what total angle will the heavens be contained?

Ans.: 97.18°

10. In Problem 9, what is the diameter on the water's surface through which light from the sky reaches your eye?

Ans.: 6.8 ft.

11. White light is incident at an angle of 45° at a plane refracting surface separating air from glass. The index of refraction of the glass for red light is 1.500; for blue light it is 1.505. (a) What is the angle between the refracted red and blue rays? (Note: this is called angular dispersion. b) which color is deviated more?

Ans.: (a) 6.1 minutes of arc. (b) blue.

NOTES:

CHAPTER 4

REFRACTION BY PLANES, PLATES AND PRISMS

4.1 EXACT RAY TRACING THROUGH A PLANE REFRACTING SURFACE

Snell's law exactly describes the change in the direction taken by rays as they cross interfaces separating optical media. Ideally, *all* rays from a point object at some distance **u** from the interface should meet, after refraction, to form a point image at some predictable image distance **u'** from the interface. Will, in fact, a point image be formed by rays refracted through a plane surface?

$$\tan\alpha = -\frac{h}{u}$$

$$\tan\alpha' = -\frac{h}{u'}$$

In Figure 4.1 an object **M** is at a distance **u = AM** from the vertex **A** of a plane surface that separates two media. The indices of object and image space are **n** and **n'**, respectively, where **n'>n**. A ray with a slope θ strikes the surface at height **AB** = h, and makes an angle of incidence of **α**, where **α = θ**. The ray is refracted toward the normal, according to Snell's law. A backward projection of the ray appears to cross the axis at $\mathbf{M_\theta'}$. The ray slope is **θ' = α'**. According to the figure, the image distance is given by $\mathbf{u_\theta' = AM_\theta'}$.

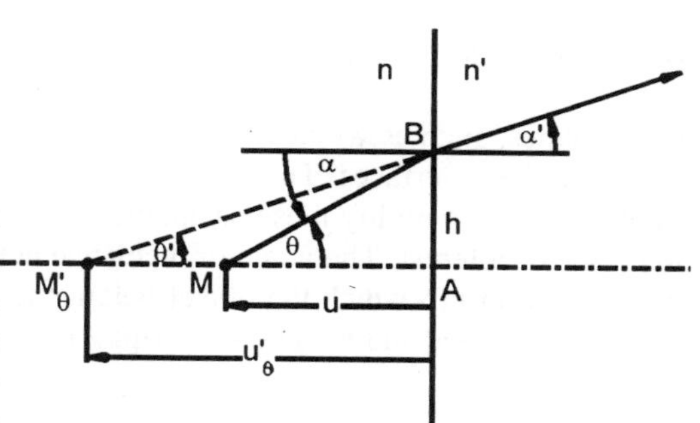

Figure 4.1 Refraction at a plane surface, (n'>n).

$$u_\theta' = \frac{\tan\alpha}{\tan\alpha'}u \qquad 4.1$$

Equation 4.1 says that the image distance **u'** depends on the ratio of the tangents of the angles of incidence and refraction. However, this ratio does not vary linearly as rays of increasing slopes (or incidence) are refracted. That is, the refracted rays cross the axis at varying distances from the surface, and a point image is *not* formed. The image is said to be *spherically aberrated.*

For example, if the object point is in air, (**n** = 1), the image in glass, (**n'** = 1.5) and the object distance u = -10 cm, we will obtain the following values for $\mathbf{u_\theta'}$ as a function of α.

α (deg)	0	15	30	45	60	75	87
u_θ' (cm)	-15	-15.30	-16.33	-18.71	-24.49	-44.34	-213.86

In Figure 4.2, rays diverge from the point **M**, in air. The rays refract into the higher index medium. They are bent toward the normal but still diverge. Their projection backwards results in virtual images. The Figure shows the difference in the image distances, measured along the optical axis, for a pair of nearly paraxial rays ($\mathbf{u_p' = AM_p'}$) and a pair of marginal rays ($\mathbf{u_m' = AM_m'}$). Rays lying in the plane of the diagram are called *meridional* rays. If the diagram is rotated about the optical axis, a *caustic* is produced. This three dimensional cusp shaped form is the envelope of the spherically aberrated rays. As the eye scans across the refracted rays, the image "point" will appear to move along the caustic surface. This swimming motion causes the image to appear distorted when viewed from different positions. An eye viewing the paraxial rays will see the image at M'_P. As the eye moves toward the marginal rays, the image appears to slide along the caustic to **B.**

Figure 4.3 is an illustration of the caustic for a point is immersed in a block of glass. The image slides along the caustic from M'_P to B. When the line of sight nearly grazes the surface, the image is seen at position B. The images are virtual.

4.2 EXACT RAY TRACING THROUGH A PARALLEL PLATE

Glass plates, such as windows, or layers of immiscible liquids in a beaker, comprise systems of two or more parallel interfaces. The paths of rays through these media are found by the iterative application of Snell's law. To simplify these calculations, a series of subscripts are added to the formula to identify the interface of interest. The angle of incidence at the first interface of the ray from the object and the index of the medium to which the object belongs is α_1 and n_1, respectively. The refracted angle and index in image space are α_1' and n_1'. The subscript 2 is used at the second interface, and so on. This is expressed as follows:

$$n_j \sin \alpha_j = n_j' \sin \alpha_j'$$

where **j** is the current surface.

The image medium or space following refraction at surface **j** becomes the object medium or space for the next surface, **j+1.** Thus, $\mathbf{n_j' = n_{j+1}}$. At the first surface, **j** = 1, and so, $\mathbf{n_1' = n_2}$.

For plates with parallel surfaces, surrounded by a uniform medium, the angle of refraction at surface **j** equals the angle of incidence at surface **j+1.** That is, $\alpha_1' = \alpha_2$. Consequently, the equations for refraction at the two surfaces of a parallel plate are:

Surface 1: $n_1 \sin \alpha_1 = n_1' \sin \alpha_1'$

Surface 2: $n_2 \sin \alpha_2 = n_2' \sin \alpha_2'$

but, $n_1' \sin \alpha_1' = n_2 \sin \alpha_2$

therefore, $n_1 \sin \alpha_1 = n_2' \sin \alpha_2'$

That is, $\alpha_1 = \alpha_2'$ the angle of emergence at the second surface is equal to the angle of incidence at the first surface. Although the ray emerges undeviated, it is displaced. See Figure 4.4a.

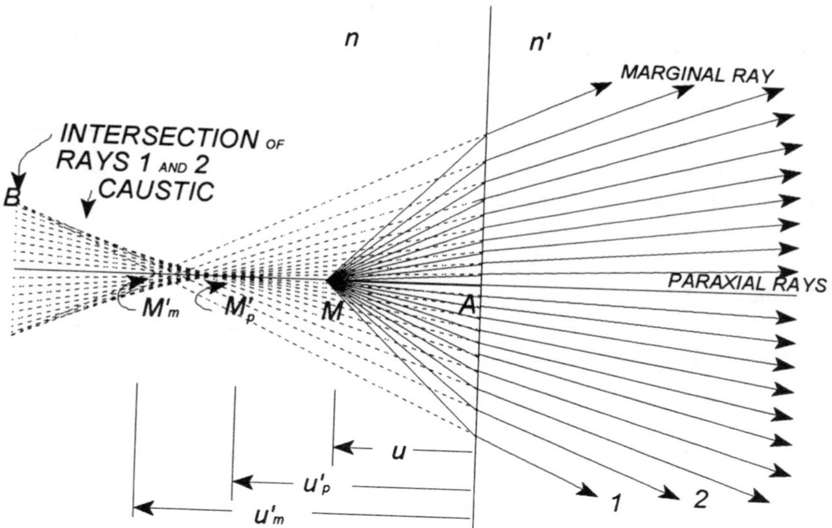

Figure 4.2 Spherical aberration by a plane refracting surface (n'>n). Paraxial rays from M are imaged at M_p'; marginal rays at M_m'. The image of point M moves along the caustic surface from M'_P to B, as the eye moves from the paraxial to the marginal zones of rays.

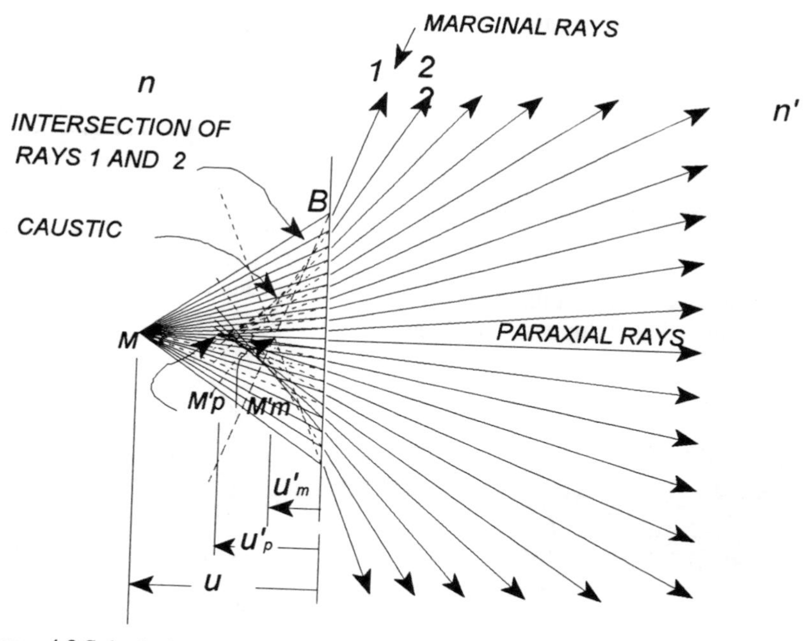

Figure 4.3 Spherical aberration by a plane refracting surface (n'<n). As the eye scans from the paraxial to the marginal rays the image slides along the caustic from M'_P to B.

4.3 DISPLACEMENT OF RAYS OBLIQUELY INCIDENT ON A TILTED PLATE

The path of a ray, incident at an angle **α**, through a plate of thickness **d** is shown in Figure 4.4b. The emergent ray is displaced through a distance ϵ. Ray path **s** is the common hypotenuse of two right triangles. In the upper triangle

$$\epsilon = s \sin(\alpha - \alpha') \tag{4.2}$$

In the lower triangle $s = d/\cos\alpha'$ 4.3

The displacement ϵ is, therefore,

$$\epsilon = \frac{d \sin(\alpha - \alpha')}{\cos\alpha'} \tag{4.4}$$

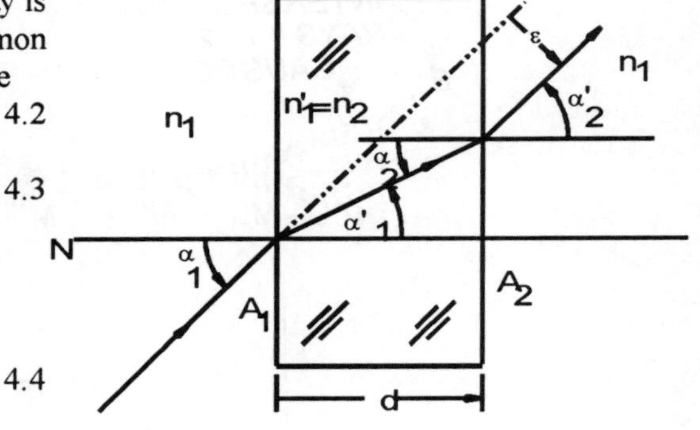

EXAMPLE 1: What will be the displacement of a ray through a 100-mm thick plate of index 1.5, which is tilted 40°?

1. Solve for **α'** $1.5 \sin\alpha' = 1 \sin 40°$

$\alpha' = 25.3740°$

2. From Eq. 4.4,

$$\epsilon = \frac{100 \sin(40 - 25.3740)}{\cos 25.3740} = 27.9469 mm$$

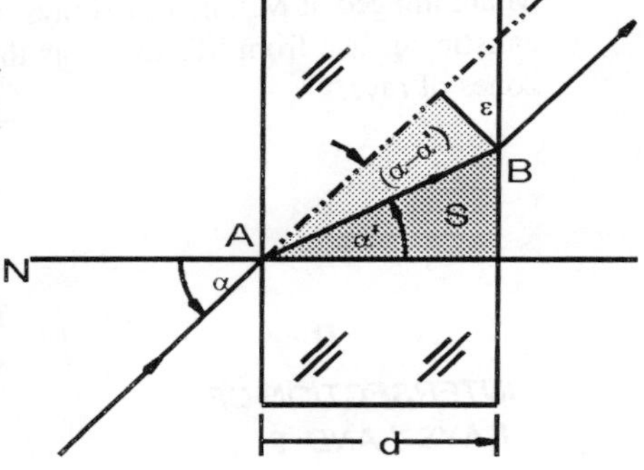

Figure 4.4 a) Path of a ray through a parallel plate in a uniform medium. (b) Displacement of the ray.

Will a distant object observed through a thick glass plate appear to move as the plate is tilted? Suppose the object is near? See Figure 4.5.

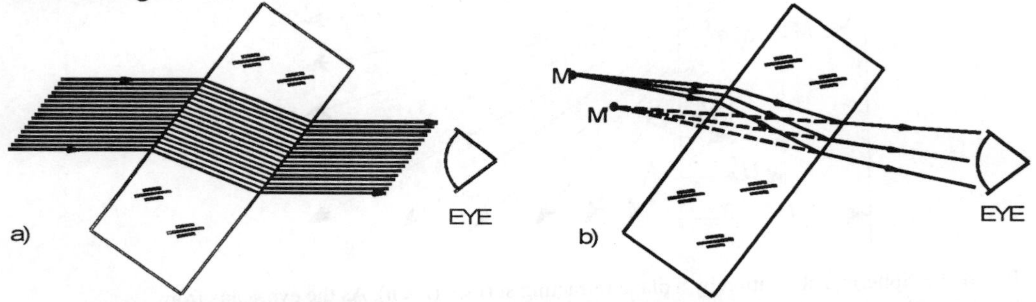

Figure 4.5 a) The direction of the image of a distant object is unchanged by a tilted plate. (b) A near object **M**, however, appears to be shifted to position M'.

4.4 REFRACTION THROUGH PRISMS

A *prism* comprises two inclined plane surfaces that meet to form an *edge*. The base of the prism is opposite the edge. Rotating the first face about the edge to make it coincide with the second face, produces a dihedral angle between the faces called the *refracting angle* ß. If the rotation is counterclockwise, ß is a positive angle. The path of a ray through a meridional or principal section of a prism is shown in Figure 4.6a.

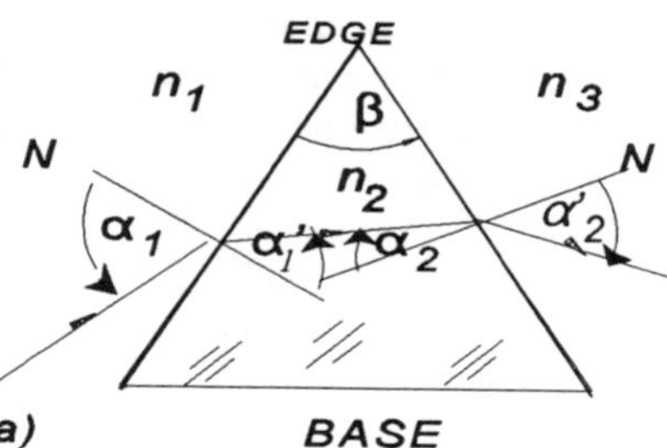

Figure 4.6b shows the ray path constructed with the method described in Chapter 3. A ray is drawn to the first surface of the prism. It makes an angle of incidence of α_1. Two arcs with radii proportional to $\mathbf{n}_1$ and $\mathbf{n}'_1$ are drawn at the point of incidence, $\mathbf{A}_1$. Let $\mathbf{n}'_1 > \mathbf{n}_1$. The normal to the surface is translated to point **P** where the incident ray intersects the arc corresponding the object space medium $\mathbf{n}_1$. Mark the point **Q** on the other arc, intersected by the normal. The line joining $\mathbf{QA}_1$ determines the direction of the refracted ray, and the angle of refraction, α'_1. The traversing ray makes an angle of incidence of α_2 at the point of incidence $\mathbf{A}_2$ at the second surface of the prism. Again draw arcs with radii proportional to $\mathbf{n}_2$ and $\mathbf{n}'_2$. Let $\mathbf{n}_2 > \mathbf{n}'_2$. Project the ray to its intersection **R** with the new object space arc. Drop a normal from **R** to the second surface. Mark its intersection **S** with the final image space arc. $\mathbf{A}_2\mathbf{S}$ is the direction of the emergent ray, and α'_2 is the angle of refraction. Refraction at the first surface deviates the ray through angle δ_1. It is deviated through angle δ_2 at the second surface. *A prism of a higher index than its surrounds will deviate light toward the base.*

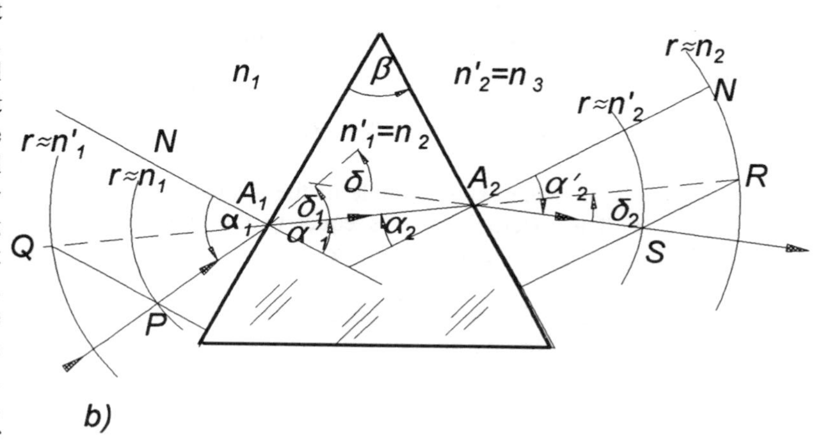

Figure 4.6 (a) Path of a ray through a principal section of a prism (b) Graphical construction of the path of a ray through a prism.

The following ***sign convention*** for angles where, ***counterclockwise rotations denote positive angles,*** will be adopted.

1. The refracting angle ß is produced by rotating the first face of the prism to bring it into coincidence with the second face.
2. Angles of incidence and refraction (α and α') are produced by rotating the normal into the ray.
3. Angles of deviation (δ, δ_1 and δ_2) are produced by rotating the refracted ray into coincidence with the incident ray.

In Figure 4.6b angles α_1, α'_1, δ, δ_1 and δ_2 are *positive;* α_2 and α'_2 are negative.

4.5 PRISM GEOMETRY

It will prove useful in tracing the path of a ray through a prism to know how the various angles relate to each other. According to the shaded triangle in Figure 4.7a, and noting that α_2 is negative, we can write

$$180° = \beta + (90°-\alpha_1') + (90°+\alpha_2)$$

thus,

$$\beta = \alpha_1' - \alpha_2 \qquad 4.5$$

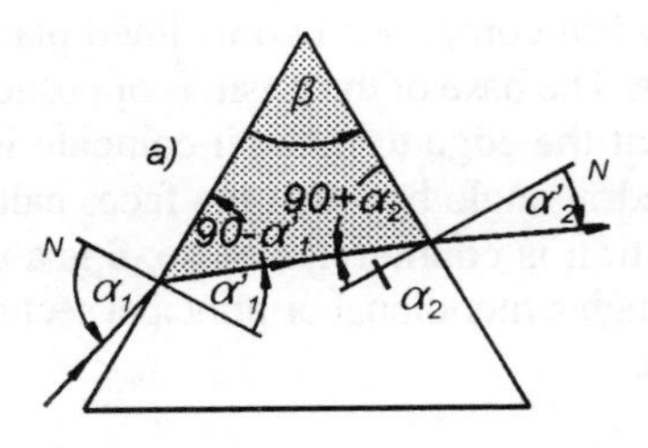

Alternatively, because the sum of the angles of the shaded quadrilateral in Figure 4.7b equals 360°, the angle opposite β is equal to 180°-β.

$$360° = 90° + 90° + \beta + 180° - \beta$$

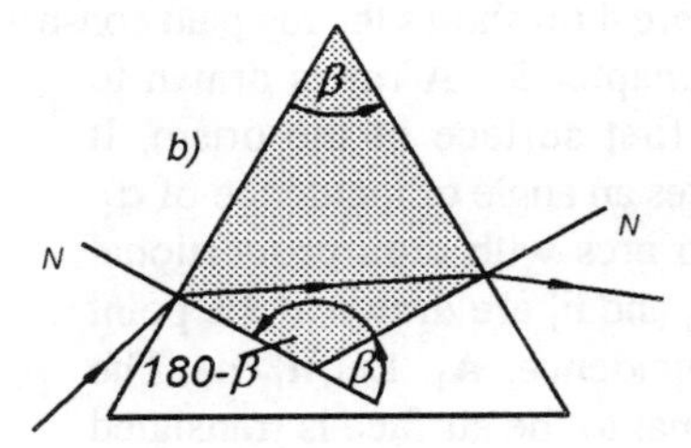

Consequently, β is the external angle of the shaded triangle in Figure 4.7c. The external angle is equal to the sum of two opposite interior angles, thus, $\beta = \alpha_1' - \alpha_2$.

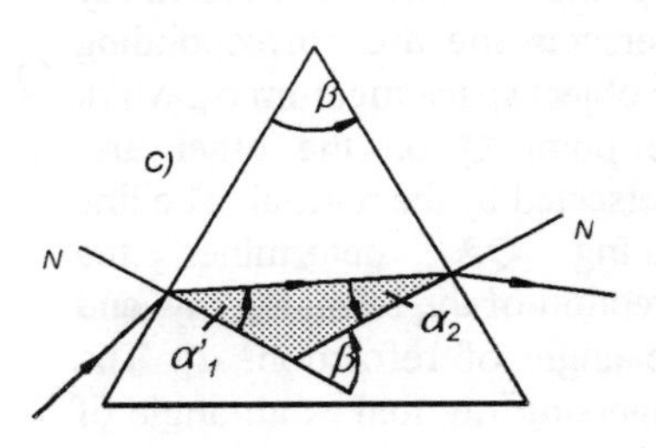

The total deviation of the ray δ is the external angle of the shaded triangle in Figure 4.7d, and so

$$\delta = \delta_1 + \delta_2 \qquad 4.6$$

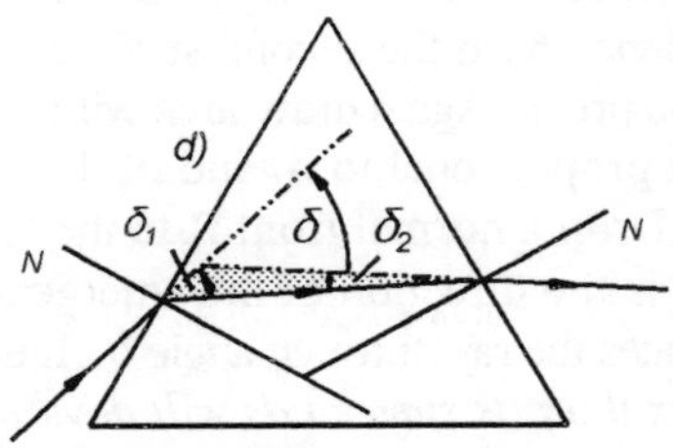

Figure 4.7 Prism geometry to show that $\beta = \alpha_1' - \alpha_2$. See text.

4.6 LIMITS TO PRISM TRANSMISSION: GRAZING INCIDENCE AND EMERGENCE

A ray will be transmitted through a prism of a higher index than its surrounds provided the angle of incidence is within a certain sector. A ray may have a maximum angle of incidence equal to +90°. The ray has *a grazing incidence.* It refracts into the prism with angle α_1' equal to the critical angle α_c , and emerges from the prism as shown in Figure 4.8a. The minimum angle of incidence α_{lim} is given by the ray that takes the reverse path. That is, it has grazing emergence. See Figure 4.8b.

$$\sin\alpha_{lim} = \frac{n'}{n}\sin(\beta - \alpha_c) \qquad 4.7$$

Only rays making angles of incidence between α_{lim} and $+90_o$ will be transmitted. The total deviation δ of the ray is identical for the two ray paths.

4.7 MINIMUM DEVIATION BY A PRISM

Figure 4.9 shows the total deviation $\boldsymbol{\delta}$ versus the angle of incidence $\boldsymbol{\alpha}_1$ of a 60° prism (n=1.5). Two possible angles of incidence will result in a particular angle of deviation, according to the principle of the reversibility of light. In other words, a ray incident at angle **a** and emergent at an angle **b**, will be deviated the same amount as the ray that is incident at angle **b** and emerges at angle **a**. The table in Figure 4.9 contains a series of pairs of incidence angles that result in the indicated deviation angles. For example, a deviation of $\boldsymbol{\delta}$ = 40° will result when $\boldsymbol{\alpha}_1$ = 36.54° or 63.46°. A horizontal line through $\boldsymbol{\delta}$ = 40° intersects the deviation curve at the two points corresponding to these values of $\boldsymbol{\alpha}_1$. The horizontal line through the point of minimum deviation ($\boldsymbol{\delta}_{min}$ = 37.18°) is *tangent* to the deviation curve at its lowest point. Consequently, just *one* angle of incidence (48.59°) corresponds to $\boldsymbol{\delta}_{\mathbf{min}}$. Reversibility of light requires those angles of incidence and emergence be equal, i.e., according to our sign convention, $\boldsymbol{\alpha}_1 = -\boldsymbol{\alpha}'_2$. It follows that the internal angles are also equal, i.e., $\boldsymbol{\alpha}'_1 = -\boldsymbol{\alpha}_2$. Thus, the ray that undergoes minimum deviation will *symmetrically traverse the prism*. In symmetric traverse $\boldsymbol{\alpha}'_1 = \boldsymbol{\beta}/2$.

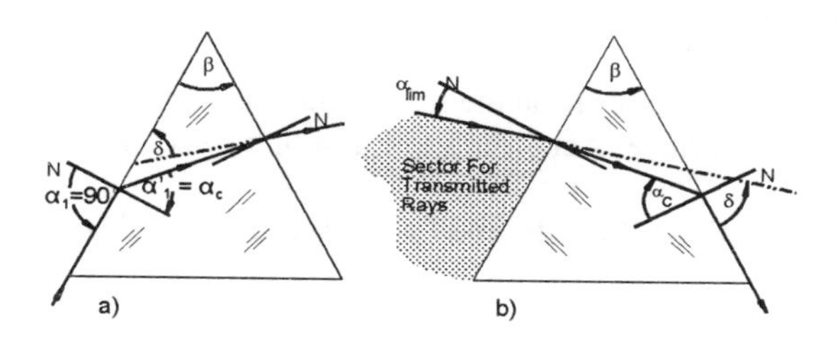

Figure 4.8 a) Grazing incidence represents the maximum angle of incidence, $\boldsymbol{\alpha}_1$ = 90°. (b) The minimum angle incidence $\boldsymbol{\alpha}_{\mathbf{lim}}$ corresponds to grazing emergence.

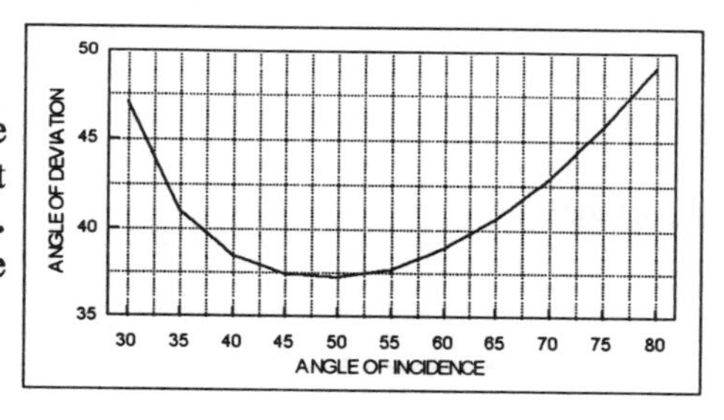

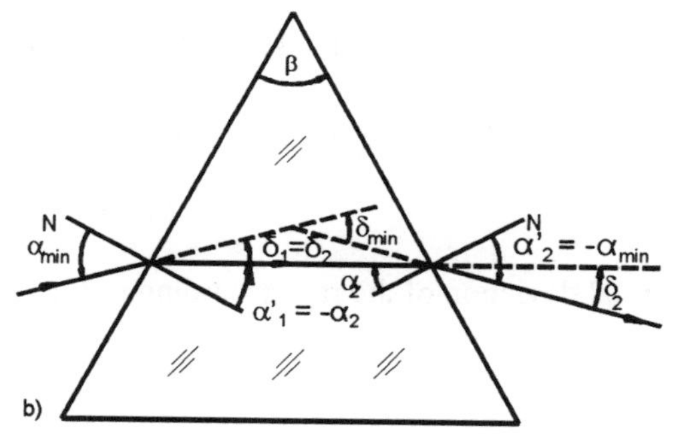

Figure 4.9 a) Deviation as a function of angle of incidence for a 60° prism. (b) The minimum deviation ray traverses the prism symmetrically.

Angle of Incidence		Angle of Deviation
α or	α	δ
30.000	77.096	47.096
35.000	65.997	40.997
40.000	58.466	38.466
45.000	52.381	37.381
48.591	48.591	37.181*
50.000	47.209	37.209
55.000	42.738	37.738
60.000	38.877	38.877
65.000	35.588	40.588
70.000	32.867	42.867
75.000	30.723	45.723
80.000	29.273	49.173

* Angle of minimum deviation.

4.8 CALCULATION OF THE PATH OF A RAY THROUGH A PRISM

Snell's law is used to calculate the path of a ray through a prism, given β, n_1, n_2, n_3 and α_1.

1. Find α_1' $$n_1' \sin \alpha_1' = n_1 \sin \alpha_1$$

To apply Snell's law at the second surface of the prism, we need to find the angle of incidence α_2 of the ray at that surface. According to Eq. 4.6, $\alpha_2 = \alpha_1' - \beta$.

2. Substitute this into Snell's law for the second surface and solve for the angle of emergence.

$$n_2' \sin \alpha_2' = n_2 \sin (\alpha_1' - \beta)$$

Example 2: A ray makes an angle of incidence of 30° at the first face of a 60° prism made of index 1.5, surrounded by air. Find the angle of emergence.

First surface:
$$\sin \alpha_1' = \frac{1}{1.5} \sin 30° = 0.3333$$

$$\alpha_1' = 19.4712°$$

From Eq. 4.5
$$\alpha_2 = 19.4712 - 60 = -40.5288°$$

Second surface:
$$\sin \alpha_2' = \frac{1.5}{1} \sin -40.5288°$$

$$\alpha_2' = -77.0958°$$

4.9 CALCULATION OF THE TOTAL DEVIATION BY A PRISM

The deviations of the ray by refraction at the first and second surfaces of the prism are

$$\delta_1 = \alpha_1 - \alpha_1', \text{ and } \delta_2 = \alpha_2 - \alpha_2'$$

The total deviation of the ray, according to Eq. 4.6, is

$$\delta = \delta_1 + \delta_2 = \alpha_1 - \alpha_1' + \alpha_2 - \alpha_2'$$

but,
$$\beta = \alpha_1' - \alpha_2$$

thus, the total deviation is
$$\delta = \alpha_1 - \alpha_2' - \beta \qquad 4.8$$

The deviation of the ray in the previous example is

$$\delta = 30° + 77.0958° - 60° = 47.0958°$$

The angle of *minimum deviation* δ_{min} occurs when the ray undergoes symmetric traverse. Accordingly, $\alpha_1 = -\alpha_2'$, and so

$$\delta_{min} = 2\alpha_1 - \beta \qquad 4.9$$

EXAMPLE 3: Find the angle of minimum deviation for the 60° prism.
According to Eq. 4.6a

$$\alpha_1' = 60/2 = 30°$$

From Snell's law, $\alpha_1 = 48.5904°$

thus,

$$\delta_{min} = 37.181°$$

REVIEW QUESTIONS

1. Are all rays from a point object refracted according to Snell's law focused at a point?
2. Why do the slopes of the rays from an object point on an axis perpendicular to a plane refracting surface equal the angles of incidence?
3. Define: meridional rays, the caustic, a prism, displacement of rays.
4. Is the apparent position of a star displaced, when viewed through a tilted parallel plate? What of a near object?
5. Define the angles of incidence and refraction at each surface, and the refracting angle of a prism.
6. In which direction is light deviated by a prism?
7. What is the path taken by the ray that undergoes minimum deviation?
8. What condition results in a ray having a minimum angle of incidence at a prism?

PROBLEMS

1. An object is 24 cm in front of a plane refracting surface separating water (n = 4/3) from glass (n' = 1.5). How far from the surface will a ray with a slope of 60° cross the axis, after refraction? Repeat for slopes of 45°, 30° and 0.001°. Draw a ray diagram to scale.
Ans.: -34.467 cm; -29.928 cm; -27.928 cm; -27.000 cm.

2. Repeat Problem 1 with the indices reversed.
Ans.: -9.615 cm; -18.282 cm; -20.367 cm; -21.333 cm.

3. What will be the displacement of a ray through a 50-mm plate of index 1.75, tilted 60°?
Ans.: 29.064 mm.

4. In Problem 3, what is the displacement when the plate is tilted 90°? Draw the ray path.

5. A ray makes an angle of incidence of 30° on a prism of index 1.75, with a 45° refracting angle: Find the angles of: (a) refraction at the first surface, (b) incidence and (c) emergence at the second surface, (d) deviation; (e) critical angle; f) minimum angle of incidence; g) angle of incidence for minimum deviation; and h) the minimum deviation angle. Construct the paths of the 30° ray and the minimum deviation ray.
Ans.: (a) 16.602°; (b) -28.398°; (c) -56.336°; (d) 41.336°; e, 34.850°; (f) 17.963°; (g) 42.044°; (h) 39.087°.

6. Repeat parts a-d, g and h, of Problem 5, except interchange the indices, i.e., make it an air prism in glass. Construct the path of the ray.

Ans.: (a) 61.045°; (b) 16.045°; (c) 9.087°; (d) -24.087°; (g) 12.631°; (h) -19.737°.

7. A ray, normally incident on a prism of index 1.5, has an angle of emergence of -48.590°. Find: (a) the prism angle, and (b) the deviation. Construct the path of the ray. Ans.: (a) 30°; (b) 18.590°

8. The index of the prism in Problem 7, for red light is 1.49, and for blue light 1.52. Find the angle between the red and blue light (this is called the dispersion). Ans.: 1.305°

9. A ray through a 30° prism undergoes a minimum deviation of 32.348°. (a) What is the index of the prism? (b) the angle of incidence (c) the limiting angle of incidence? Ans.: (a) 2; (b) 31.174°; (c) 0°

10. What must be the angle of incidence on a 45° prism of index 1.6, to obtain grazing emergence?

Ans.: 10.14°

11. What is the minimum index of refraction of a 60° prism so that no rays will be transmitted?

12. What is the minimum refracting angle of a prism of index 1.75 so that no rays will be transmitted?

Ans.: 69.7°

13. White light makes an angle of incidence of 30° at the first face of a 60° prism. Its index is 1.500 for red light and 1.505 for blue light. Find the angular dispersion. Ans.: 1.2364°

14. Rays traveling in air strike the first face of a 45° prism of index 1.5 with an incident angle of 30°. The rays emerge into air. Find: (a) the angle of deviation at the first surface; (b) the angle of incidence at the second surface; (c) the angle of emergence and (d) the total deviation. Construct ray path.

Ans.: (a) $\alpha_1' = 10.529°$; (b) $\alpha_2 = -25.529°$; (c) $\alpha_2' = -40.274°$; (d) $\delta = 25.274°$

15. For the prism in Problem 14, find: (a) the angle of incidence for minimum deviation and (b) the angle of minimum deviation. Ans.: (a) $\alpha_1 = 35.031°$; (b) $\delta_o = 25.063°$

16. Rays travelling in air strike the first face of a 60° prism of index 1.5 with an incident angle of 45°. The rays emerge into air. Find: (a) the angle of emergence and (b) the total deviation. Construct ray path.

Ans.: (a) -52.381°; (b) $\delta = 37.381°$

17. Repeat Problem 16 for the prism surrounded by water of index 4/3.

Ans.: (a) -23.842°; (b) $\delta = 8.842°$

18. Repeat Problem 17, but make the prism of water and the surrounds glass.

Ans.: (a) -6.483°; (b) $\delta = -8.517°$

19. Rays travelling in water strike the first face of a 60° prism of index 1.5 with an incident angle of 45°. The rays emerge into *air*. Find: (a) the angle of emergence and (b) the total deviation. Construct ray path.

Ans.: (a) -32.613°; b) $\delta = 17.613°$

CHAPTER 5

PARAXIAL REFRACTION AT PLANES, PLATES AND PRISMS

5.1 PARAXIAL REFRACTION AT A PLANE SURFACE

When angles are very small, the ratios of the angles in degrees, the ratios of their sines and their tangents are very nearly equal. *Paraxial rays* strike a surface in a small zone around the normal. They have small angles of incidence and refraction. Their paraxial ratios are

$$\frac{\tan\alpha}{\tan\alpha'} = \frac{\sin\alpha}{\sin\alpha'} = \frac{\alpha}{\alpha'} \qquad 5.1$$

By eliminating the sines of the angles we obtain the paraxial equivalent to Snell's law, $\dfrac{\alpha}{\alpha'} = \dfrac{n'}{n}$ 5.2

According to Eq. 4.1, $\dfrac{\tan\alpha}{\tan\alpha'} = \dfrac{u'}{u}$. The paraxial equivalent to this equation is $\dfrac{\alpha}{\alpha'} = \dfrac{u'}{u}$ 5.3

The right sides of Eqs. 5.2 and 5.3 are equal, therefore, $\dfrac{u'}{u} = \dfrac{n'}{n}$, or the *paraxial* image equation is

$$\frac{n'}{u'} = \frac{n}{u} \qquad 5.4$$

The angle α has been eliminated from the image equation. *All* rays from the object point **M** will be refracted to the real or virtual image point **M'**. The object and image points, **M** and **M'**, are *conjugate* points. For the sake of clarity, in *paraxial diagrams,* such as Figure 5.1, rays will be shown making exaggeratedly large angles of incidence and refraction.

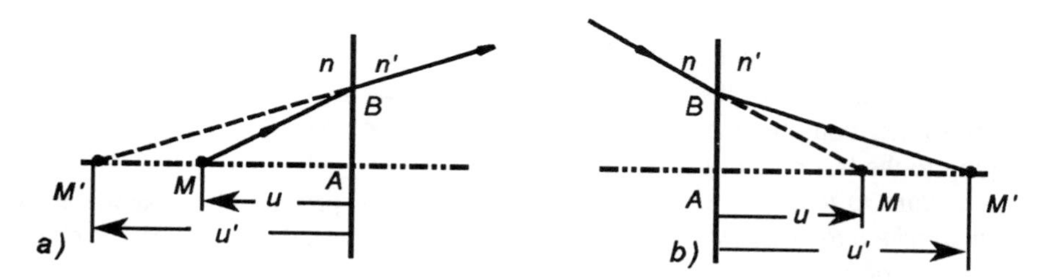

Figure 5.1 Paraxial object and image distances for refraction at a plane surface, (n'>n).
a) Real object, virtual image. b) Virtual object, real image.

5.2 PARAXIAL IMAGE FORMATION BY PARALLEL PLATES

Iterative application of Eq. 5.4 will find the paraxial image of a point object, produced by refraction at a series of parallel plane interfaces. The *optical axis* is the straight line, normal to the refracting surfaces and through the object point M. It intersects the surface at the *vertex* **A**, and contains the image point **M'**.

Although, it is essential to adopt a **sign convention** for optical calculations, no one sign convention is used universally. The sign convention used here is:

1. *Light from the object initially travels from* ***left*** *to* ***right****.*

2. *Distances measured in the direction that light initially travels (from left to right) are positive.*

3. *Distances measured opposite the direction that light initially travels are negative.*

The object distance **u** is the step from **A** to **M,** that is, **u = AM**; the image distance **u' = AM'** is the step from **A to M'.** In Figure 5.1a the object is real and the image is virtual. Both the object and image distances are negative. In Figure 5.1b the object is virtual, the image is real, and both object and image distances are positive.

In a system of several surfaces the vertices are numbered $\mathbf{A}_1$ to A_K, where **k** is the last refracting surface. The distance between vertices is **d.** The image produced by refraction at a given surface becomes the object for the next surface. As with Snell's law, it is useful to append a series of subscripts to the paraxial image equation to systematize surface by surface calculations.

Where will the image be seen by an observer, if an object is placed at a distance **EM** from the observer, (Figure 5.2a) and a plate of glass is interposed? As shown in Figure 5.2b, there are three indices of refraction: $\mathbf{n}_1$, $\mathbf{n}_2$ and $\mathbf{n}_3$. The plate is surrounded by the same medium, therefore, $\mathbf{n}_1 = \mathbf{n}_3$. Each surface will refract rays from the object. For the first surface the object point is $\mathbf{M}_1$. Its image $\mathbf{M}'_1$ belongs to the medium of the plate. Image $\mathbf{M}'_1$ becomes the object $\mathbf{M}_2$ for the second surface. After refraction, the final image $\mathbf{M}'_2$, belonging to the last medium, is formed. *It is important to understand that it is the medium in which the rays are traveling that determines object or image space, not where the object or image is located on your diagram.*

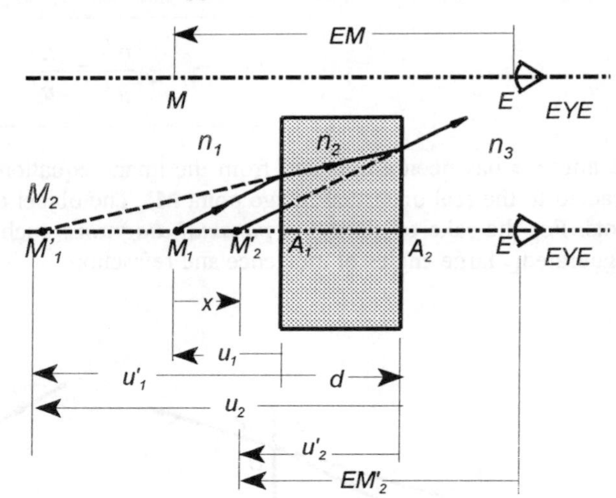

Figure 5.2 The image produced by a parallel plate ($n_1 = n_3$) is shifted through distance x. a) The object is at distance EM. b) Seen through the plate the image appears to be at distance EM_2'.

Interposition of the plate ($\mathbf{d = A_1A_2}$) will cause the observer to see image $\mathbf{M_2'}$ displaced from object $\mathbf{M_1}$. The image is at a distance $\mathbf{EM_2'}$ The displacement $\mathbf{x = M_1M_2'}$ is found as follows:

At the 1st Surface: $u_1=A_1M_1$, $u_1'=A_1M_1'$, $d=A_1A_2$, and the image equation is $\frac{n_1'}{u_1'} = \frac{n_1}{u_1}$ 5.5

The image distance after refraction at the first surface is $u_1' = \frac{n_1'}{n_1}u_1 = A_1M_1' = A_1M_2.$ 5.5a

The object distance to the second surface is: $u_2 = A_2M_2 = A_2A_1 + A_1M_2 = u_1' - d$ 5.6

At the 2nd Surface: $\frac{n_2'}{u_2'} = \frac{n_2}{u_2}$

Substitute equation 5.5a into 5.6 and solve for u_2'. Note: to simplify the nomenclature let $n_1 = n_2' = \mathbf{n}$, and $n_1' = n_2 = \mathbf{n'}$.

$$u_2' = u_1 - \frac{n}{n'}d = A_2M_2'$$

Now,

$$x = M_1M_2' = M_1A_1 + A_1A_2 + A_2M_2'$$

Thus, the equation for displacement is

$$x = \frac{n'-n}{n'}d$$ 5.7

The shift in the apparent position of an object seen through a parallel plate depends on the thickness and index of refraction of the plate and the index of the surrounding medium. It does not matter how far the object or the observer is from the plate. The above equation suggests that the index of the plate may be determined by measuring the displacement **x** and the plate thickness **d**, since,

$$n' = \frac{nd}{d-x}$$ 5.8

For example, suppose a microscope is focused on the surface of a table. If a 100-mm thick block of glass is interposed, it is necessary to refocus on the image of the table top by raising the microscope tube 35 mm, i.e., the image displacement is 35 mm. See Figure 5.3. Find the index of the block if n = 1, d = 100, and x = 35. According to Eq. 5.8, n' = 100/65 = 1.5385.

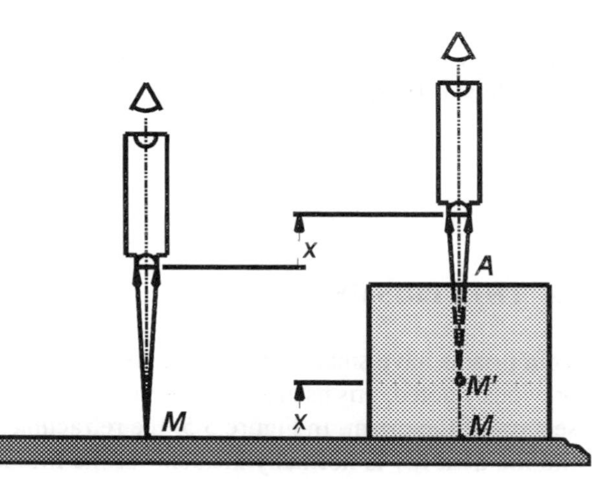

Figure 5.3 Experimental determination of the index of refraction of a transparent plate.

5.3 REDUCED THICKNESS

If the object distance $\mathbf{u_1} = 0$, i.e., the first surface of the plate is the object, an observer will see this surface displaced through distance $\mathbf{x}$. See Figure 5.4. The plate will appear to be as thick as the image distance $\mathbf{u'_2}$. If the plate is of higher refractive index than the surrounds, it will seem thinner than it is. The apparent thickness of the plate is called the *reduced thickness.* It is a positive quantity equal to $\mathbf{M'_2A_2} = \mathbf{-u'_2}$. According to Eq. 5.7,

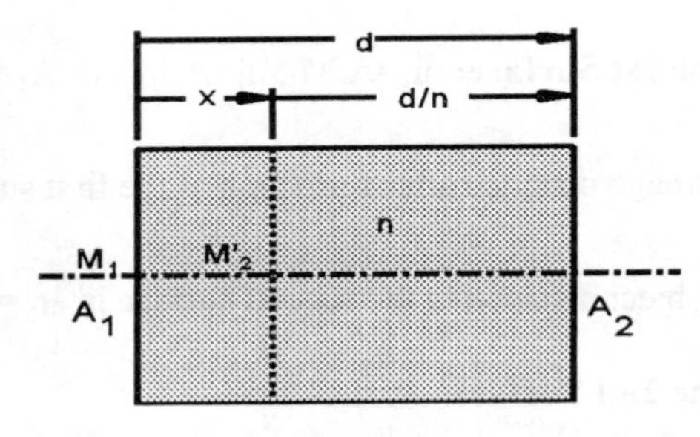

Figure 5.4 The reduced thickness of a plate.

$$u'_2 = u_1 - \frac{n}{n'}d$$

But, as noted, $u_1 = 0$. Therefore, the reduced thickness $M'_2A_2 = \frac{n}{n'}d$ 5.9

If, as is common, the plate is surrounded by air, $\mathbf{n} = 1$, then the index $\mathbf{n'}$ of the plate is the only index of concern and it is *renamed* $\mathbf{n}$. Therefore,

$$\textit{Reduced thickness} = c = \frac{d}{n} \qquad 5.10$$

Finally, given air as the surrounding medium, we can restate the displacement as the thickness minus the reduced thickness, namely,

$$x = d - \frac{d}{n} \qquad 5.11$$

5.4 THIN PRISMS

Thin prisms have small refracting angles and are used at very nearly minimum deviation angles. As previously noted, prisms in air will deviate light toward the base. The *base-apex* line lies in a principal section of the prism. In Figure 5.5, the refracting angles are exaggerated. To simplify the analysis, the incident ray is normally incident at the first face of the prism, thus $\boldsymbol{\alpha_1} = \boldsymbol{\alpha'_1} = \boldsymbol{\delta_1} = \mathbf{0}$. The ray makes an angle of incidence at the second face $\boldsymbol{\alpha_2}$ equal to the refracting angle $\mathbf{-\beta}$. The angle of refraction $\boldsymbol{\alpha'_2} = \mathbf{-(\beta+\delta)}$. According to Eq. 5.2, paraxial refraction at the second surface is given by

$$\frac{n_2}{n'_2} = \frac{\alpha'_2}{\alpha_2}$$

The prism index n_2 and the surrounding index n_2' may be renamed n' and n, respectively. With these substitutions we obtain

$$\frac{n'}{n} = \frac{\beta+\delta}{\beta}$$

Solve for **δ** to obtain the equation for deviation by a thin prism in a uniform medium.

$$\delta = \frac{n'-n}{n}\beta \qquad 5.12$$

Generally the prism is surrounded by air, consequently, **n** = 1 and **n'** is *renamed* **n**. The deviation of a thin prism in air is

$$\boxed{\delta = (n-1)\beta} \qquad 5.13$$

β and **δ** may be in degrees or radians.

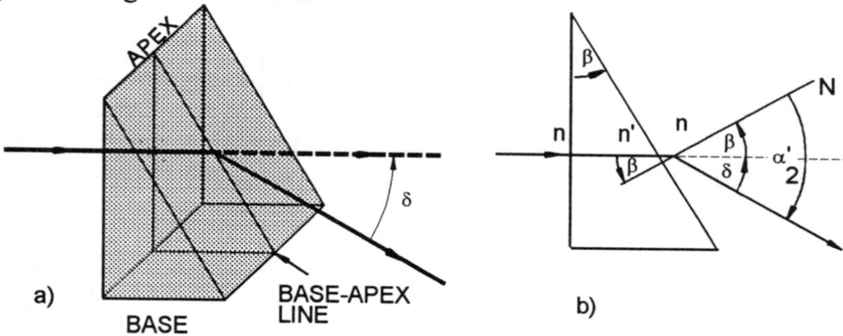

Figure 5.5 a) The base-apex line and deviation toward the base of a prism.b) Diagram for deriving the paraxial deviation formula.

5.5 OPHTHALMIC PRISMS, CENTRADS AND THE PRISM-DIOPTER

Ophthalmic prisms are thin prisms used in eye examinations and to treat certain binocular vision problems. The radian is too large a unit to express the small angular deviations of ophthalmic prisms, consequently, a unit called the centrad, equal to 1/100th of a radian, is used. The symbol for centrad is an inverted triangle $^{\triangledown}$. Since, 2π radians = 360°, an angle θ in degrees may be converted to radians as follows:

$$2\pi\theta/360° = \pi\theta/180° \text{ radians.}$$

Multiply the result by 100, to obtain centrads, i.e., $P^{\triangledown} = 100\pi\theta/180° = 1.74533\theta$

Example: Convert 2.5° to centrads: $P^{\triangledown} = (2.5)(1.74533) = 4.3633^{\triangledown}$.

However, the *prism-diopter* $^{\Delta}$ is more commonly used as an expression of the *power* of a prism. The prism-diopter is a deviation of one unit at a distance of 100 units. In Figure 5.6a, the prism has a power of 10 prism-diopters. Small angles of deviation **δ** in degrees may be converted to prism-diopters by multiplying their tangents by 100. Specifically, $\mathbf{P^{\Delta} = 100 \tan \delta}$. Eq. 5.13a yields a deviation in *prism-diopters* if the refracting angle **ß** is given in degrees.

$$P^{\Delta} = 100 \tan(n-1)\beta^{\circ} \qquad 5.13a$$

EXAMPLE: Directly solve for deviation in prism diopters, given ß =5° and n = 1.5.

$$P^{\Delta} = 100 \tan 2.5^{\circ} = 4.3661^{\Delta}$$

To convert prism-diopters to degrees, divide by 100 and find the arc tangent

$$\tan^{-1} (P^{\Delta}/100) = \theta^{\circ}$$

For example, to convert 1^{Δ} to degrees we have:

$$\tan^{-1} 0.01 = 0.5729^{\circ}.$$

Figure 5.6b illustrates the difference between centrads and prism-diopters. The angles in the diagram cut equal arc lengths on the circle, but progressively larger segments on the prism-diopter scale. In other words, the centrad is a linear unit, i.e., N centrals are exactly N times greater than one centrad, but N prism-diopters are not exactly N times greater than 1 prism-diopter. However, as the following table shows, the discrepancy in angular deviation between 20 centrads and 20 prism-diopters is 1.3%. This is not a significant error for ophthalmic applications.

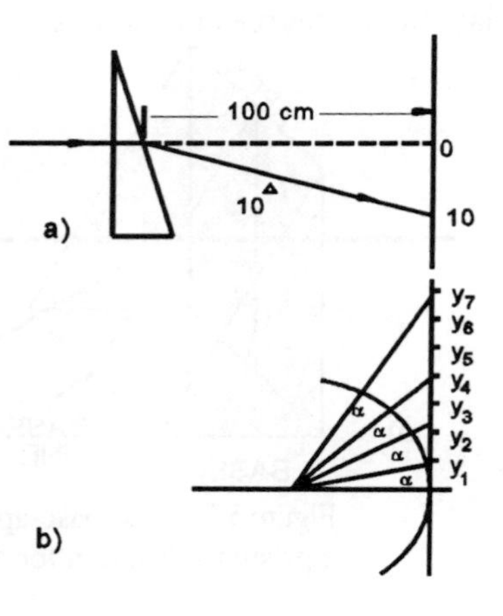

Figure 5.6 a) Ten prism-diopters. b) Centrads compared to prism-diopters.

CENTRADS vs. PRISM DIOPTERS

Centrads to Degrees	Prism diopters to Degrees	% Error
$1^{\nabla} = 0.572958^{\circ}$	$1^{\Delta} = 0.572939^{\circ}$	0.003
$10^{\nabla} = 5.729578^{\circ}$	$10^{\Delta} = 5.710593^{\circ}$	0.33
$20^{\nabla} = 11.459156^{\circ}$	$20^{\Delta} = 11.309932^{\circ}$	1.30
$40^{\nabla} = 22.918312^{\circ}$	$40^{\Delta} = 21.801409^{\circ}$	4.87

5.7 OBLIQUELY COMBINED PRISMS

Thin prisms are made in discrete powers. By combining two prisms it is possible to obtain a continuous range of powers. Rotating the pair of prisms will align the resultant base-apex line and deviate the rays along any desired meridian. *Meridian angles* with respect to the eyes of a patient are shown in Figure 5.7. The position of the base of the prism denotes the direction of deviation. Temporal placement is termed base-out; nasal placement is base-in. If the base-apex line coincides with the 45° meridian and the base is uppermost the prism is base-up at 45°. As shown in Figure 5.8 the prism in front of the right eye will deviate upwards and nasally (base-up-and-in at 45°); in front of the left eye, the prism will deviate upwards and temporally (base-up-and-out at 45°).

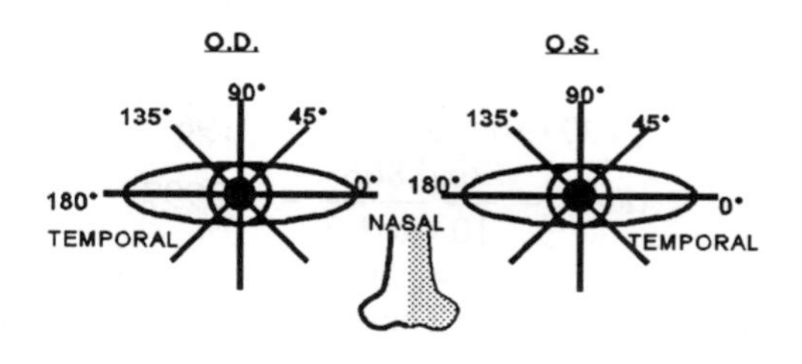

Figure 5.7 Convention for specifying meridians.

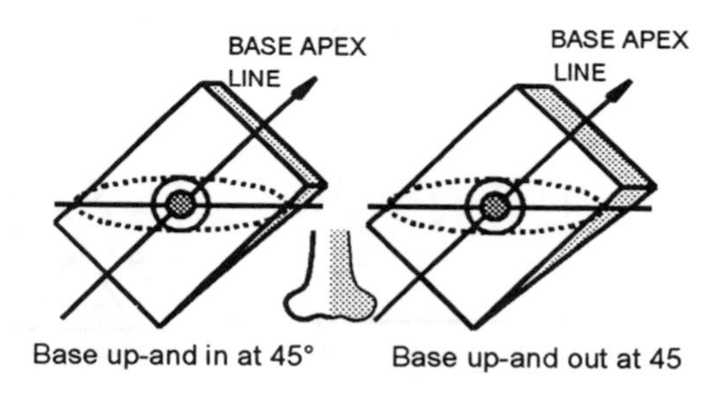

Figure 5.8 An example of obliquely positioned prisms.

Two obliquely combined prisms are shown in Figure 5.9. Their individual powers are $\mathbf{p}_1$ and $\mathbf{p}_2$. Prism $\mathbf{p}_1$ is positioned with its base-apex line to the right; the base-apex line of prism $\mathbf{p}_2$ is base-up at about 45°. The angle between the base-apex lines is θ. **R** is the power of the resultant or equivalent prism, and ϕ is its base-apex meridian, as shown in the vector diagram, Figure 5.10. The line segments **x** and **y** are produced to form right triangles. According to the figure,

$$y = p_2 \sin \theta, \quad x = p_2 \cos \theta$$

and

$$R^2 = y^2 + (p_1+x)^2$$

The resultant power is

$$R = \sqrt{p_1^2 + p_2^2 + 2p_1p_2\cos\theta} \tag{5.14}$$

The base-apex meridian is given by

$$\tan\phi = \frac{p_2\sin\theta}{p_1 + p_2\cos\theta} \tag{15.15}$$

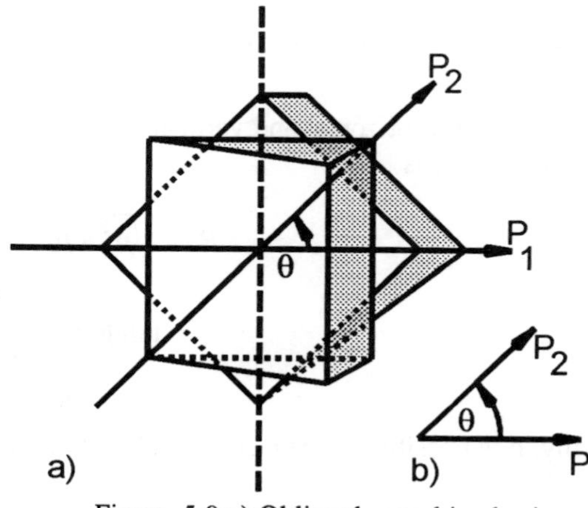

Figure 5.9 a) Obliquely combined prisms. b) Prism vectors.

Example: Two prisms are placed in front of the right eye of a patient. A 10 prism-diopter prism is set base-in and a 6 prism-diopter prism is set base-up-and-in at 60°. Find the power and position of the resultant prism. Given: $p_1 = 10^\Delta$; $p_2 = 6^\Delta$; $\theta = 60°$

$$R = \sqrt{100+36+120x0.5} = \sqrt{196} = 14^\Delta$$

$$\tan\phi = \frac{6x0.866}{10+6x0.5} = 0.3997$$

$$\phi = 21.79°$$

The resultant prism has a power of 14 prism-diopters and is base-up-and-in at 21.79°.

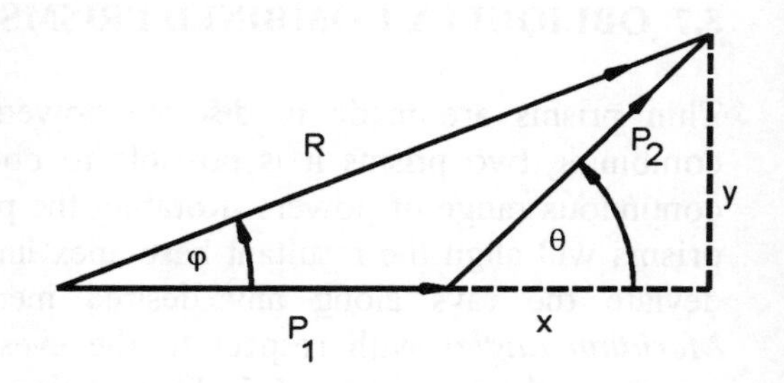

Figure 5.10 Vector diagram for finding the resultant prism.

An *alternate solution,* which obviates the need to memorize these equations, is to resolve each prism into its horizontal and vertical components and correspondingly add them. In the above example, the entire power of the first prism is horizontal, i.e., $p_{1H} = 10$. The vertical and horizontal vectors of the second prism are p_{2H} and p_{2V}. See Figure 5.11.

$$p_{2H} = p_2 \cos 60° = 6(0.5) = 3^\Delta, \text{ and } p_{2V} = p_2 \sin 60° = 6(0.866) = 5.196^\Delta$$

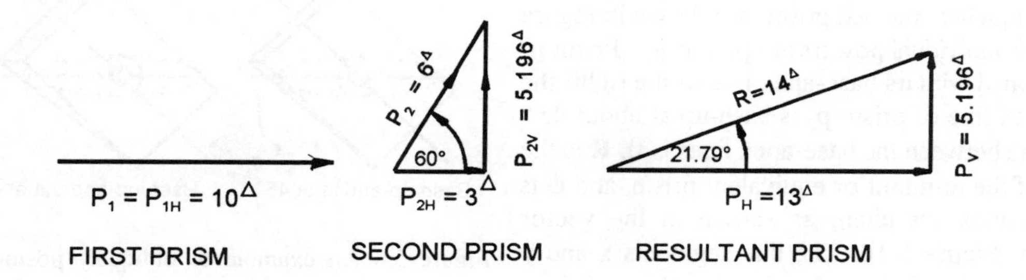

Figure 5.11 Compounding two obliquely combined prisms.

The sums of the horizontal and vertical components are $p_H = 10+3 = 13$, and $p_V = 0+5.196 = 5.196$. The resultant power is

$$R = \sqrt{p_H^2 + p_V^2} = \sqrt{169+27} = 14^\Delta$$

$$\tan\phi = \frac{5.196}{13} = 0.3997$$

$$\phi = 21.79°$$

Finally, it is possible to measure **R** and **ϕ** directly, simply by accurately drawing the vector diagrams to scale.

5.8 RISLEY PRISMS

The Risley prism is a device containing a pair of prisms of equal power that can be counter-rotated. See Figure 5.12. Risley prisms are part of the phoropter, an instrument used to measure the refractive error of the eyes. Because $p_1 = p_2 = p$, Eq. 5.14 can be simplified. Applying the Law of Sines to the vector diagram (Figure 5.13) we find

$$\frac{R}{\sin\theta} = \frac{p}{\sin\frac{\theta}{2}}$$

Solving for R, we obtain, $R = \frac{\sin\theta}{\sin\frac{\theta}{2}}p$ According to a trigonometry identity $\frac{\sin\theta}{\sin\frac{\theta}{2}} = 2\cos\frac{\theta}{2}$

Therefore,

$$R = 2p\cos\frac{\theta}{2} \qquad \phi = \frac{\theta}{2} \tag{5.16}$$

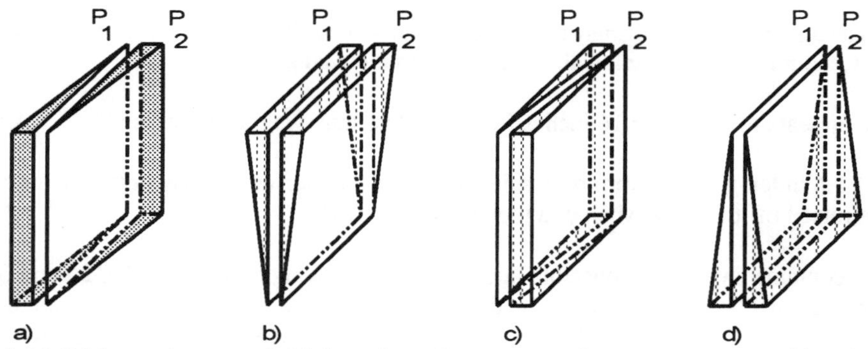

Figure 5.12 Risley prisms. a) Initial setting: Bases opposite - zero power. b) Prism P_1 is rotated 90° and prism P_2 is counter-rotated 90°; power = 2P, base-up. c) Prism P_1 is rotated 180° and prism P_2 is counter-rotated 180°; bases are again opposite - zero power. d) Prism P_1 is rotated 270° and prism P_2 is counter-rotated 270°; power = 2P, base-down.

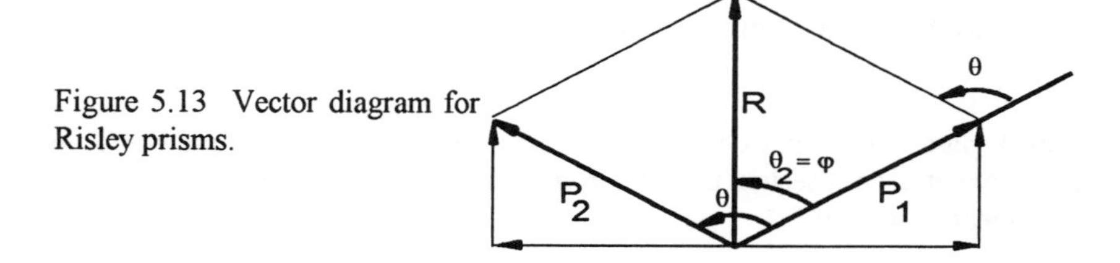

Figure 5.13 Vector diagram for Risley prisms.

REVIEW QUESTIONS

1. How does a paraxially calculated image of a point differ from an image calculated with Snell's law?
2. Define conjugate points.
3. Review our sign convention.
4. What distinguishes object space from image space?
5. Is it possible to have an object to the right of a plate of glass?
6. A glass plate is surrounded by water, where **n** is the water index and **n'** the glass index. When does **n** represent the glass index?
7. Define reduced thickness.
8. Define base-apex line for a thin prism.
9. Differentiate between centrad and prism-diopter. Which is a linear unit?
10. What is meant by the meridian angle? How are meridian angles specified in optometry?

PROBLEMS

1. A bird is 30 feet above a swimming pool. You are submerged and looking straight up at the bird. How high above the surface of the water does it appear to be? Ans.: 40 ft.

2. You are 6 ft under water. How far under water do you appear to be to the bird? Ans.: 4.5 ft.

3. A shark in an aquarium tank longingly looks at the visitors. The shark spots you at eye level. You appear to the shark to be 5 ft from the tank wall. How far from the wall are you? Ans.: 3.75 ft

4. The shark appears to you to be 16 inches from the glass tank wall. How far is it? Ans.: 21.33 in.

5. A cube of 1.6 index glass, 80 mm on a side, is used as a paper weight. By how much does the paper appear to be raised off the table when you look straight down? Ans.: 30 mm.

6. A beaker contains layers of immiscible liquids of the following thicknesses 10, 20 and 30 mm, with respective indices of 1.3, 1.4 and 1.45. How deep do the liquids appear to be, on looking down into the beaker? Ans.: 7.69+14.29+20.69=42.67 mm.

7. Repeat Problem 6 using Eq. 5.4.

8. A bank teller sits 18 inches behind a 3-inch thick Plexiglas window of index 1.5. You are 12 inches from the near side of the window. How far from you does the teller appear to be? Ans.: -32 in.

9. A microscope is focused on a table top. A parallel plate of glass 24 mm thick is placed on the table and the microscope must be raised 10 mm, to refocus on the top of the table. What is the index of the glass? Ans.: 1.714

10. A 24-cm thick glass plate of index 1.5 divides a tank, filled with water of index 4/3, into two sections. A submerged swimmer in each section is 24 cm from the nearer face of the plate. How far apart from each other do they appear to be now? Ans.: 69.333 cm

11. In Problem 10, water is drained from the first section. All other conditions are the same. How far apart do they appear to be: a) to the submerged swimmer, b) to the one in air?

Ans.: a) 77.333 cm; b) 58 cm.

12. A thin prism of index 1.65 has a power of 3 prism diopters. What are its deviation and refracting angles?

Ans.: 1.718°; 2.644°

13. An ophthalmic prism of index 1.523, deviates light by 2°. Find its power in prism diopters.

Ans.: 3.49^{Δ}

14. A thin prism of index 1.65 has a refracting angle of 2.644°. What is its deviation in prism diopters?

Ans.: 3^{Δ}

15. A thin prism has a refracting angle and a deviation angle of 5°. Find a) its index; b) the deviation in $^{\Delta}$; the deviation in $^{\nabla}$

Ans.: a) 2.0; b)8.749^{Δ}; c) 8.727^{∇}

16. What is the power along the horizontal meridian of a 15-diopter prism with a base-apex line at 15°?

Ans.: 14.49^{Δ}

17. To cause a turned right eye to look straight ahead you must use a 3-diopter prism base-in and a 4-diopter prism base-up. Find the resultant prism.

Ans.: 5^{Δ} base-up at 53°

18. To straighten a left eye you must use an 8^{Δ} prism base-in at 120° and a 5^{Δ} prism base-down at 210°. Find the resultant prism. Draw the vector diagram.

Ans.: 9.434^{Δ}. base-up at 152° .

19. A 15^{Δ} prism base-out and a 20^{Δ} prism base-up-and-in at 120 °are placed in front of the left eye to straighten it. Find the resultant prism.

Ans.: 18.03^{Δ}, base-up at 73.9°

20. In Problem 19 the angle is increased to 180°. Find the resultant prism.

Ans.: 5^{Δ} base-in.

21. A Risley prism device is set so that each prism has been counter-rotated from the zero power setting by 30°. The resultant power is 30^{Δ}. What are the powers of the component prisms? Draw the vector diagram.

Ans.: 30^{Δ}

22. The following prisms are placed in front of the right eye, 8^{Δ} base-in at 45° and 12^{Δ} base-up-and-out at 120°. Find the resultant prism.

Ans.: 16.05^{Δ} base-up at 91.2°

NOTES:

CHAPTER 6

REFRACTION AND REFLECTION AT SPHERICAL SURFACES

6.1 SIGN CONVENTION

To be consistent in solving optical problems and to obtain correct solutions, adopting a convention for measuring distances and angles is essential. In the section dealing with thick prisms we specified how to measure certain angles, we now will define the pertinent quantities for refraction at a single spherical refracting surface.

1. Light initially travels from left to right.
2. Distances measured from left -to-right are positive.
3. Distances measured from right-to-left are negative.
4. Distances measured up from the axis are positive.
5. Distances measured down from the axis are negative.
6. All angles are acute.

In Figure 6.1, **C** is the center of curvature of a convex spherical surface. **C** lies on the optical axis (x-axis) which intersects the surface at the vertex **A**. Point **A** is the origin of a coordinate system. The tangent to the surface at **A** is the **y** axis. An object **M** lies on the axis and sends a ray to the point **B** on the surface. **BC** is a radius to the surface and, therefore, the normal. The slope of radius **BC** is the central angle **ϕ.** After refraction, the ray crosses the axis at the image point, **M'**. The plane of the drawing is a meridional section. A perpendicular dropped from point **B** to the optical axis at point **D** is the height of incidence **h** of the ray. Angles that are measured counter clockwise (CCW) are positive, clockwise (CW) angles are negative:

	Angles		Symbol
Incidence:	NBR	=	α
Refraction:	N'BR'	=	α'
Slope of incident rays:	AMB	=	θ
Slope of refracted rays:	AM'B	=	θ'
Central angle:	BCA	=	ϕ

Distances and lengths are defined as follows:

Radius of curvature:	AC	=	r
Object distance:	AM	=	u
Image distance:	AM'	=	u'
Object height:	MQ	=	y
Image height	M'Q'	=	y'
First focal length:	FA	=	f
Second focal length:	F'A	=	f'
Newtonian object distance:	FM	=	x
Newtonian image distance:	F'M	=	x'

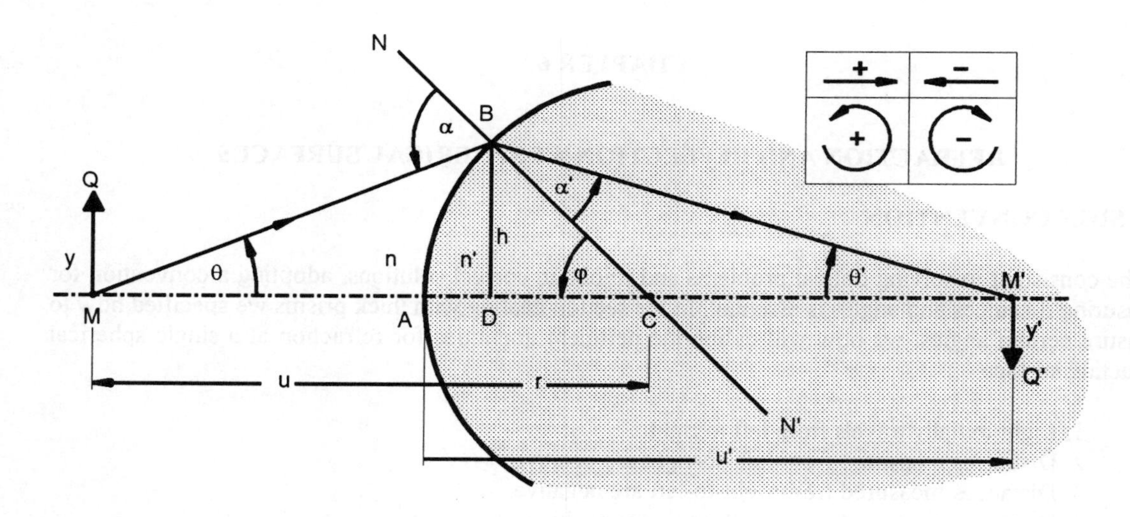

Figure 6.1 The nomenclature and sign convention for the single spherical refracting surface.

6.2 REFRACTION AT A SINGLE SPHERICAL REFRACTING SURFACE

6.2.1 THE PARAXIAL IMAGE EQUATION

Figure 6.1 illustrates the refraction at a spherical refracting surface of a ray from object M to image M'. In triangle MBC, $\alpha = \theta + \phi$ (6.1). In triangle M'BC, $\phi = \alpha' - \theta'$, or, $\alpha' = \phi + \theta'$ 6.2

The paraxial equation corresponding to Snell's law is: $n\alpha = n'\alpha'$ 6.3

Substitute Eqs. 6.1 and 6.2 into 6.3. $n(\theta + \phi) = n'(\theta' + \phi)$

$$n\theta + n\phi = n'\theta' + n'\phi$$

or,

$$\mathbf{n'\theta' = n\theta - (n'-n)\phi} \qquad 6.4$$

The paraxial condition says that as the ray slope becomes very small, the points **B** and **D** approach the vertex A. In the limit, BD = BA = h, BM = AM = u, BM' = AM'= u', and BC = AC = r. Thus,

$$\theta = \tan\theta = \frac{h}{MA} = -\frac{h}{u}$$

$$\theta' = \tan\theta' = \frac{h}{M'A} = -\frac{h}{u'}$$

$$\varphi = \tan\varphi = \frac{h}{AC} = \frac{h}{r}$$

Substitute the above angles into Eq. 6.4 and divide by -h, to obtain the *refraction equation.*. It is the fundamental equation for paraxial calculations.

$$\frac{n'}{u'} = \frac{n}{u} + \frac{n'-n}{r} \qquad 6.5$$

This equation eliminated the angles of incidence and refraction. The paraxial region of a spherical surface is a small zone about the vertex **A**. Paraxial diagrams represent the surface as a plane tangent to **A**. It is as if the diagram were made on a thin sheet of rubber and stretched only in the vertical direction. Ray slopes are exaggerated for the sake of clarity.

EXAMPLE 1: An object is located 100 mm in front of a convex spherical refracting surface separating air from glass of index 1.5. The radius of curvature of the surface is 20 mm. Find the position of the image. Given: n = 1, n' = 1.5, r = AC = +20 mm, and u = AM = -100 mm. Substitute into the refraction equation, and solve for **u'**.

$$\frac{1.5}{u'} = \frac{1}{-100} = \frac{1.5-1}{20}$$
$$u' = AM' = +100\,mm$$

6.2.2 Focal Points and Focal Lengths

The image of an axial object at infinity (u = ∞) is formed at the *second focal point* **F'**. The *second focal length* **f'= F'A**. See Figure 6.2. To find **f'** substitute u = -∞ into the refraction equation (Eq. 6.5) and solve for u'. The image point **M'** coincides with **F'**, thus, **u' = AM' = AF' = -f'**, therefore,

$$-\frac{n'}{f'} = \frac{n'-n}{r} \qquad 6.6$$

Figure 6.2 Focal points and focal lengths of a convex spherical refracting surface.

An object that produces an image at ∞ is at the *first focal point* **F.** The *first focal length* **f = FA, but FA = - AM = -u**. Substitution in Eq. 6.5 gives, $\frac{n'}{\infty}=\frac{n}{-f}+\frac{n'-n}{r}$

thus,

$$\frac{n}{f}=\frac{n'-n}{r} \qquad 6.7$$

From Equations 6.6 and 6.7 we find

$$\frac{f'}{f}=-\frac{n'}{n} \qquad 6.8$$

The two focal points **F** and **F'** of a spherical refracting surface lie on *opposite* sides of the vertex **A**, and at distances from it that are in the ratio of n to n'. A surface is *convergent* if it is convex and separates a rare medium from a dense medium (n'>n), or if it is concave and separates a dense from a rare medium (n'<n). Conversely, a surface is *divergent* if it is concave and separates a rare from a dense medium (n'>n), or if it is convex and separates a dense from a rare medium (n'<n). Shown in Figure 6.3, are converging (a-c) and diverging (d-f) refractive surfaces, using both ray and wavefront diagrams.

Equations 6.6 and 6.7 provide alternate forms of the refraction equation in terms of the focal lengths rather than the radius of curvature.

$$\frac{n'}{u'}=\frac{n}{u}+\frac{n}{f}, \quad and \quad \frac{n'}{u'}=\frac{n}{u}-\frac{n'}{f'}$$

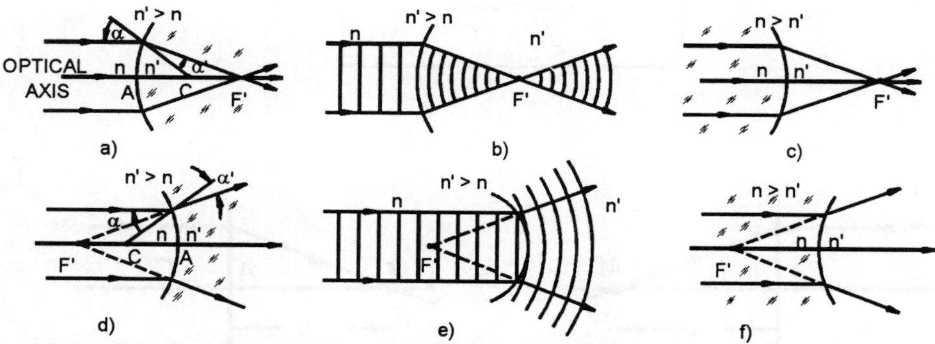

Figure 6.3 Convergent (a-c) and divergent (d-f) spherical refracting surfaces. (a) Ray diagram, (n'>n). (b) Wavefront diagram, (n'>n). (c) Ray diagram, (n>n'). (d) Ray diagram, (n'>n). (e) Wavefront diagram, (n'>n). (f) Ray diagram, (n>n').

EXAMPLE 2: Find the focal lengths of the spherical refracting surface in the previous example.
Eq. 6.7 gives f:

$$\frac{1}{f} = \frac{0.5}{20}$$

thus, f = FA = 40 mm. From Eq. 6.6 we find, $\frac{1.5}{-f'} = \frac{0.5}{20}$

Thus, f ' = FA = - 60 mm. Of course, once we have found f, we immediately can find **f'** with Eq. 6.8.

6.2.3 Refracting Power

The refracting power of the surface is denoted by F where

$$F = \frac{n'-n}{r} = \frac{n}{f} = -\frac{n'}{f'}$$

When r, f, and f ' are specified in meters the unit of power is *diopters*. The power of the surface in the above example is, F = n/f = 1/0.040 = 25 diopters. See Sect. 6.9.

6.2.4 Construction of On-axis Image Points

Figure 6.4 is a construction of the axial image point M' (for a convex surface, where n'>n). A perpendicular is dropped to the axis at point C, and points **P** and **P'** are marked on it according to the ratio,

$$\frac{CP}{CP'} = \frac{n'}{n}$$ or, CP = n' and CP' = n

The procedure for constructing the image point **M'** is to draw a ray MB where B lies on the straight line from M to P. The line BP' corresponds to the direction of the refracted ray. It crosses the axis at **M'.** Here the ray actually crosses the axis, therefore, M' is a real image. If the ray crosses the axis by a projection backwards, the image is virtual. (It may appear odd that **P** is associated with **n'**, and **P'** with **n**. However, according to the construction, unprimed points M and P are connected to form the object ray, and points P' and M' form the image ray. Unprimed characters denote object space and primed characters image space}.

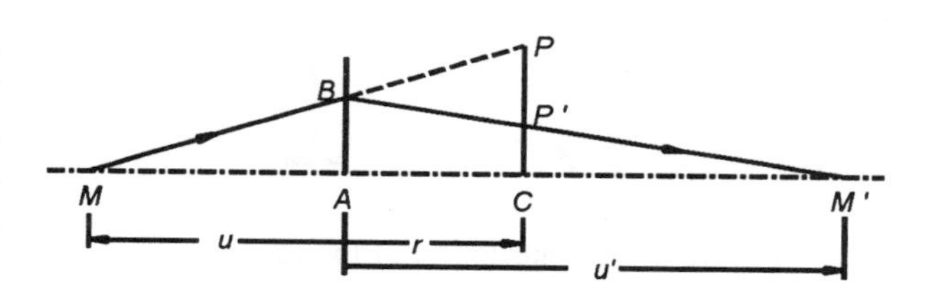

Figure 6.4 Paraxial construction of axial image point M'.

We will show that this construction satisfies the refraction equation. From the similar triangles MCP and MAB we can obtain the ratios

$$\frac{CP}{MC} = \frac{AB}{MA}$$

where, CP = n', MC = MA + AC = -u + r, and MA = -u, consequently,

$$AB = \frac{un'}{u-r} \tag{6.9}$$

In similar triangles M'CP' and M'AB, $\frac{CP'}{CM'} = \frac{AB}{AM'}$

and will obtain

$$AB = \frac{u'n}{u'-r} \tag{6.10}$$

If we set Eq. 6.9 equal to 6.10 we will find that the construction satisfies the refraction equation.

$$\frac{n'}{u'} = \frac{n}{u} + \frac{n'-n}{r}$$

This construction is a valuable way to check calculations and it illustrates some very important relationships concerning the focal lengths and the radius of curvature, to be developed in the next section.

6.2.5 Construction of the Focal Points

Figure 6.5 shows the construction of the paths of a ray from negative infinity that goes to **F'**, and a ray from F that goes to infinity, where, FA = f, F'A = f' and AC = r.

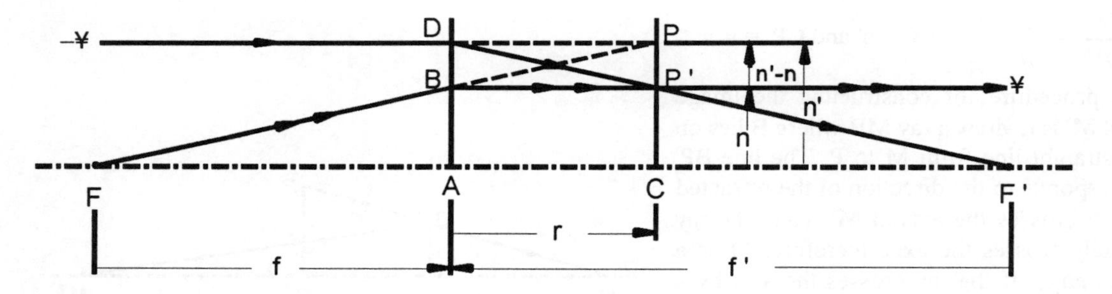

Figure 6.5 Construction of the focal points of a convergent spherical refracting surface.

According to the construction: CP'= AB = n, CP = AD = n', P'P= BD = n'-n

FAB and BP'P are similar triangles that give us the ratios

$$\frac{n'-n}{r} = \frac{n}{FA} \tag{6.11}$$

also, DBP' and P'CF' are similar triangles that give us the ratios

$$\frac{n'-n}{r} = \frac{n}{CF'} \tag{6.12}$$

Consequently, FA = CF'= f, *or the step from the center of curvature to the second focal point is equal to the first focal length.*

Similarly from the two sets of similar triangles FCP and FAB, and DAF' and P'CF' we obtain the ratios

$$\frac{n'}{FC} = \frac{n}{f}; \; also, \; \frac{n'}{AF'} = \frac{n}{f}$$

Consequently, FC = AF'= -f', *or the step from the first focal point to the center of curvature equals minus the second focal length.* This last equation reveals another relationship.

$$FC = FA + AC = f + r = -f'.$$

Thus,

$$f + f' + r = 0 \qquad 6.13$$

6.2.6 Construction of Off-axis Image Points

The arrow **MQ** lies in the object plane and has its end **M** resting on the axis. **Q** is the *off-axis* point at the other end of the arrow. To locate the image of **Q** it is necessary to select *two* rays through **Q** whose paths are known, and determine where they intersect. Figure 6.6 illustrates the paths of three predictable rays through a positive spherical refracting surface. Ray 1, parallel to the axis (as if from infinity), goes through point Q. After refraction, it must go through the *second focal point*. Ray 2, through the *first focal point*, after refraction, must travel parallel to the axis (to infinity). For good measure, ray 3, through the center of curvature will be undeviated after refraction. The three rays intersect at **Q'** in the conjugate image plane. A perpendicular dropped to the axis will locate **M'**.

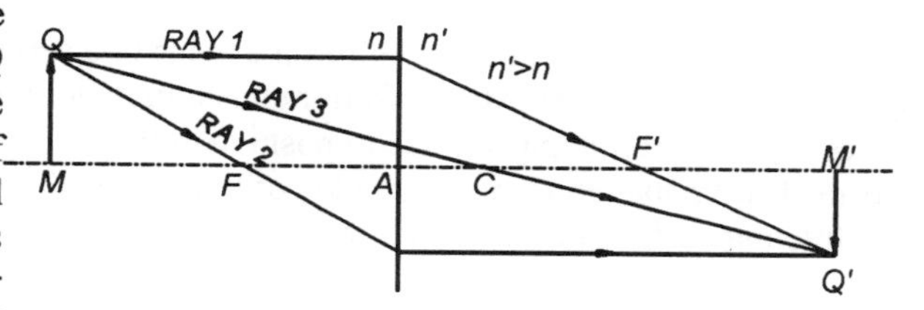

Figure 6.6 Construction of off-axis image point Q' produced by a convergent refracting surface.

A similar construction for a negative refracting surface is shown in Figure 6.7. Ray 1 must be projected back to **F'**. Ray 2, headed toward **F**, will travel parallel to the axis after refraction. It must be projected back to intersect ray 1. Ray 3 is undeviated by the surface. Is image **Q'** real or virtual? If virtual, does ray 3 enable you to form an image of **Q'** on a screen?

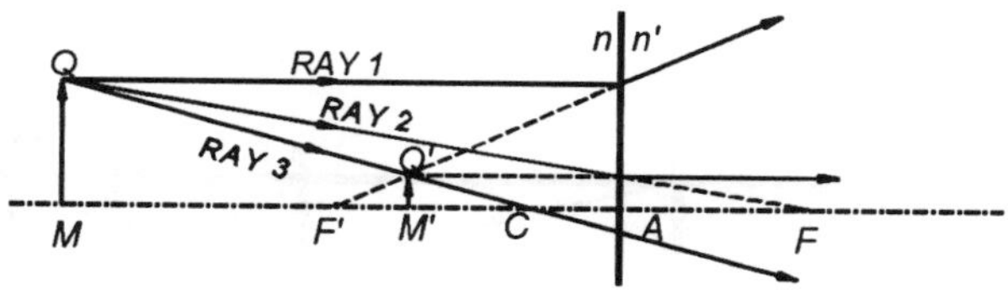

Figure 6.7 Construction of off-axis image point Q' produced by a divergent refracting surface.

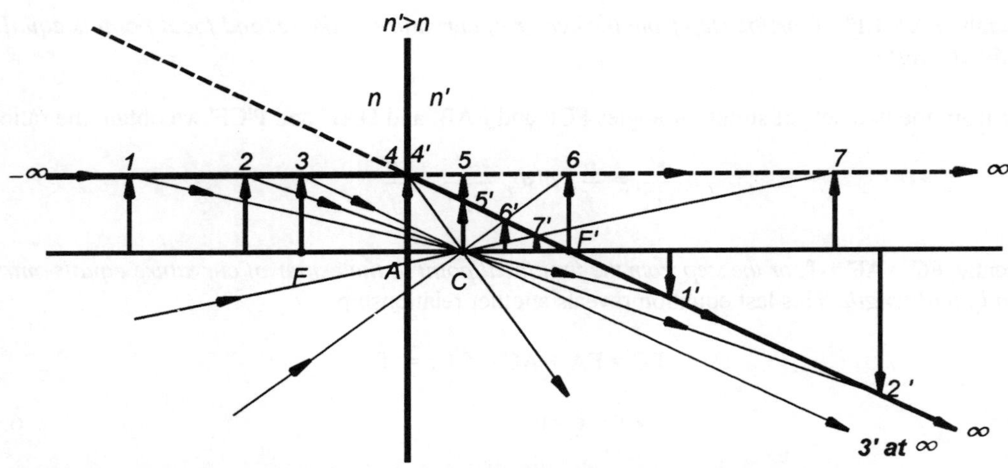

Figure 6.8 Object/image conjugates for a convergent spherical refracting surface.

6.2.7 Image Positions as an Object Approaches the Surface

In Figure 6.8 a ray from infinity defines the locus of a series of object and image points. The real part of the object ray (unbroken line) extends from minus infinity to the refracting surface. The virtual part (broken line) extends from the refracting surface to positive infinity. The real part of the image ray (unbroken line through **F'**) extends from the refracting surface to positive infinity. The virtual part (broken line) is projected back from the refracting surface to minus infinity.

Initially, the position of the object is at minus infinity. Its conjugate image is at **F'**. As the object is moved toward the surface to position 1, its *real inverted image* moves to 1'. On reaching *F*, the first focal plane (position 3), the object is imaged at infinity (3'). Advancing the object from F to A (through position 3 and to position 4), will move the *virtual erect image* from minus infinity to position 4'. The object is *virtual* when it moves past the surface. As it moves from position 4 to positive infinity the *real erect image* moves from position 4' to **F'**. Figure 6.9 is a similar construction for a negative refracting surface.

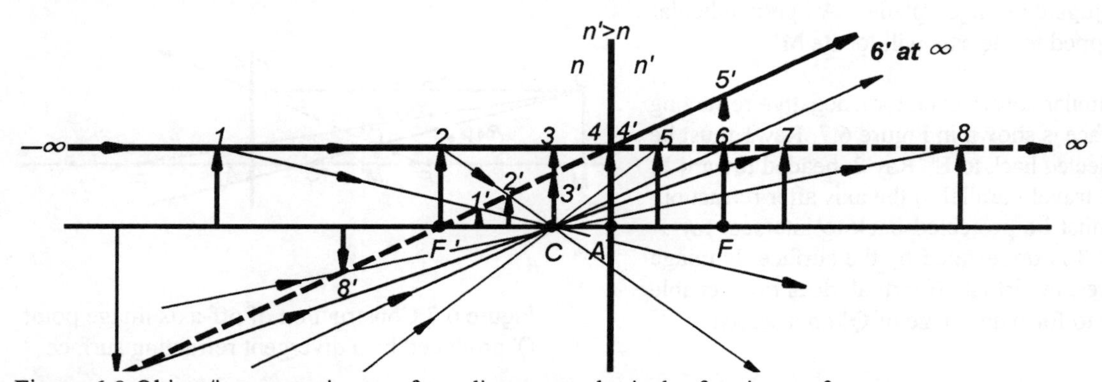

Figure 6.9 Object/image conjugates for a divergent spherical refracting surface.

6.2.8 POSSIBLE POSITIONS OF THE IMAGE

Checking for computational errors, such as keying in the wrong values or failing to observe the sign convention, is critical. Computations can be checked with graphical constructions (See above), or by using certain rules. The following rules are applicable to images formed by spherical refracting surfaces: *If the object point M does* ***not*** *lie within the radius AC, the image point M' will* ***not*** *fall within the radius AC. If the object point does lie within the radius AC, the image point M' will also lie within AC.* There are sixteen possible sequences of these four points, involving real and virtual objects (RO, VO) and real and virtual images (RI, VI), for spherical refracting surfaces. Four sequences apply to a convex refracting surface, where n'>n. See Figure 6.10.

SEQUENCE	OBJECT/IMAGE
M, A, C, M'	RO/RI Inverted
M', M, A, C	RO/VI Erect
A, M, M', C	VO/RI Erect
A, C, M', M	VO/RI Erect

The other twelve sequences correspond to:
b. Convex refracting surface, (n>n').
c. Concave refracting surface, (n'>n).
d. Concave refracting surface, (n>n').

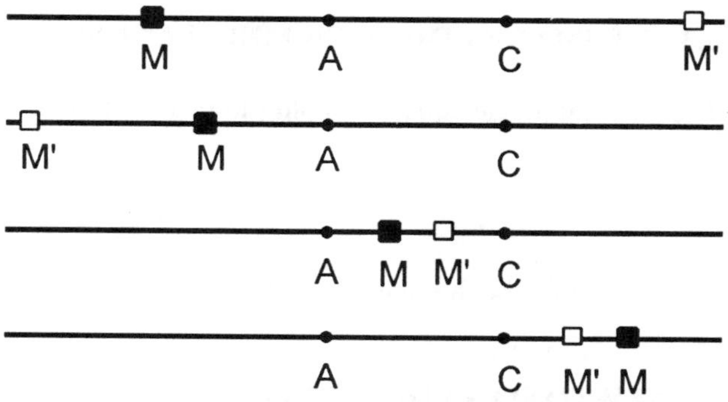

Figure 6.10 Possible sequences of points M, M', A and C, for a convex refracting surface, where (n'>n).

6.2.9 LATERAL MAGNIFICATION

The axial object M and image M' are conjugate points. Reversibility of light means that the object and image may be interchanged and will remain conjugate to each other. Furthermore, the transverse planes through M and M' are conjugate object and image planes. An off-axis *paraxial* point **Q** in the object plane will have its image **Q'** in the image plane. The size of the object is **y = MQ**, and the image size is **y' = M'Q'**. *Lateral magnification Y is the ratio between the size of the image to that of the object.*

$$Y = \frac{y'}{y} \tag{6.14}$$

The equation for the lateral magnification produced by a spherical refracting surface follows from the paraxial version of Snell's law.

$$\frac{\sin\alpha'}{\sin\alpha} = \frac{\alpha'}{\alpha} = \frac{n}{n'}$$

According to Figure 6.11, for paraxial angles, $\alpha = y/u$, and $\alpha' = y'/u'$. (radians), therefore,

$$\frac{y'}{u'}\cdot\frac{u}{y} = \frac{n}{n'}$$

and the lateral magnification is

$$\boxed{Y = \frac{y'}{y} = \frac{n}{n'}\cdot\frac{u'}{u}} \tag{6.15}$$

EXAMPLE 3: Find the lateral magnification and the size of the image for the example on Page 6.3, if the object is 5 mm high. Given: y = MQ = +5 mm.

$$Y = \frac{1}{1.5} \cdot \frac{100}{-100} = -0.67$$

y' = -0.67(5) = -3.33 mm.= M'Q'. A real inverted image.

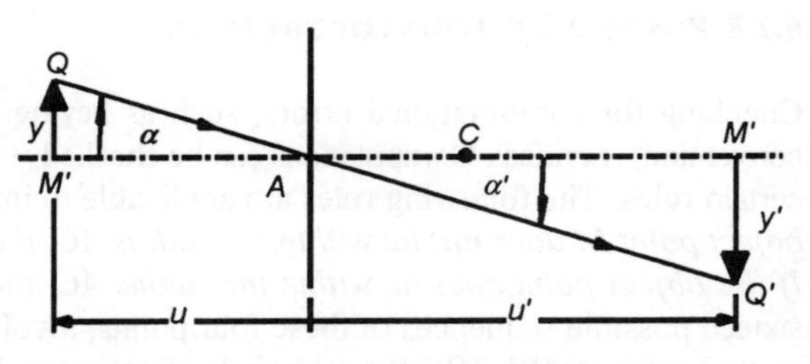

Figure 6.11 Ray diagram for lateral magnification.

6.3 THE IMAGE EQUATION FOR PLANE SURFACES

The paraxial image equation applies to plane surfaces, for which the radius of curvature **r** = ∞. Thus,

$$\frac{n'}{u'} = \frac{n}{u} + \frac{n'-n}{\infty}$$

and, we obtain the familiar equation,

$$\frac{n'}{u'} = \frac{n}{u} \qquad 6.16$$

6.4 NEWTONIAN EQUATIONS

So far, we have treated the vertex of the surface as the origin of a coordinate system. Equations attributed to Newton are based on designating **F** and **F'** as the origins of an object and image space coordinate system. The object and image distances, **u** and **u'**, are replaced with **x** and **x'**, where,

x = FM and **x' = F'M'**

Figure 6.12, based on the construction of the point **Q'**, illustrates two sets of similar triangles that directly lead to the Newtonian equation for lateral magnification.

$$Y = \frac{y'}{y} = \frac{f}{x} = \frac{x'}{f'} \qquad 6.17$$

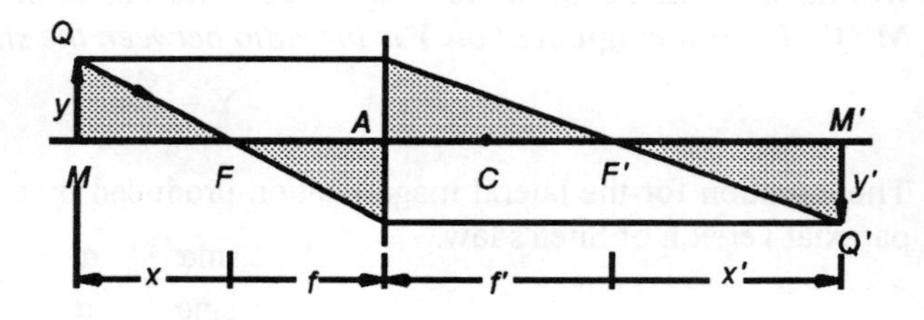

Figure 6.12 Newtonian diagram for a refracting surface.

From Equation 6.17 we obtain,

$$xx' = ff' \qquad 6.18$$

EXAMPLE 4: Continuing with the previous spherical refracting surface, find x' if an object is 60 mm in front of the first focal point. Given: x = FM = -60 mm, f = 40 mm, and f' = -60 mm. Substitute into Eq. 6.18 to obtain x' = F'M' = +40 mm.

6.5 THE FOCAL PLANES AND FOCAL IMAGES OF A SPHERICAL REFRACTING SURFACE

Focal points lie on transverse planes called the focal planes. The first or primary focal plane through F has a conjugate image plane at infinity. The infinity object plane is conjugate to the second focal plane at F'. An infinitely distant off-axis object **S**, will send collimated rays to the refracting surface. We will find the height (or size) of the image **S'** in the second focal plane with Figure 6.13. Ray 1, through the center **C**, is undeviated. It intersects the second focal plane at **S'**, as does parallel Ray 2. The size of the image is $\mathbf{y'_s} = \mathbf{F'S'}$. The slope of Ray 1 = θ_1. According to triangle CF'S',

$$\tan\theta_1 = \frac{F'S'}{CF'} = \frac{y'_s}{f}$$

or

$$y'_s = f\tan\theta_1 \qquad 6.19$$

For small angles, $\mathbf{y'_s} = \mathbf{f}\,\theta_1$.

This equation says that the size of the image of an infinitely distant extended object, formed by a spherical refracting surface, is equal to the product of the ray slope and the focal length.

The point **P** in the first focal plane represents an off-axis object. Ray 3, from **P** through **C,** will be undeviated and will travel parallel to Ray 2 in image space. These rays will form the image **P'** at infinity.

$$\tan\theta_2 = \frac{FP}{CF} = \frac{y_p}{f'}$$

or

$$y_p = f'\tan\theta_2 \qquad 6.20$$

For small angles, $\mathbf{y_p} = \mathbf{f'}\theta_2$

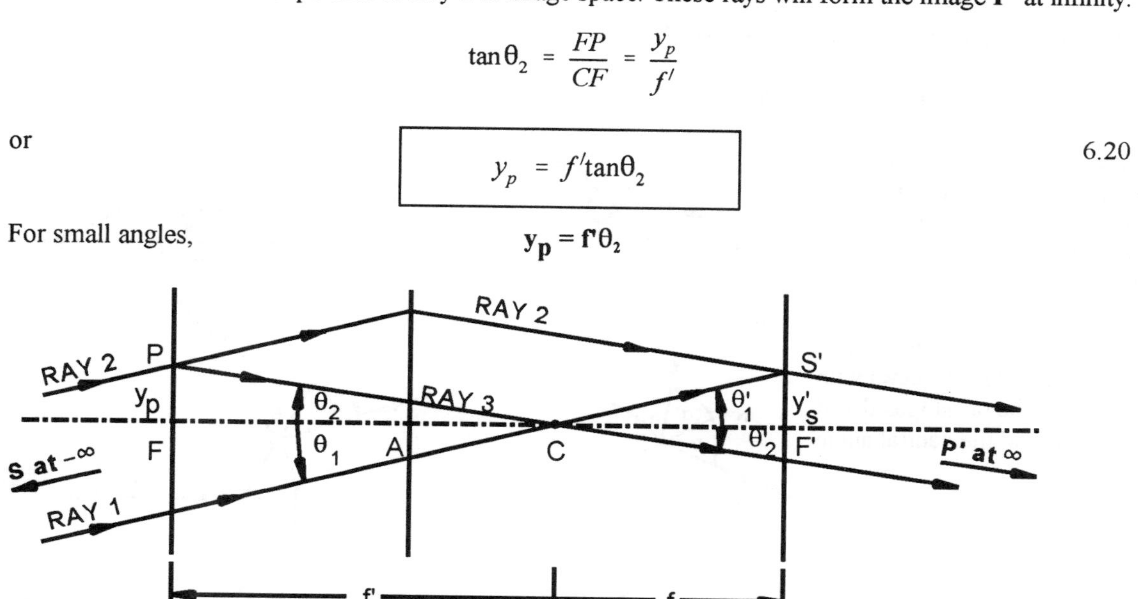

Figure 6.13 The size of focal plane objects and images.

6.6 THE SMITH-HELMHOLTZ FORMULA OR LAGRANGE INVARIANT

In Figure 6.14 the object ray **MB** becomes refracted ray **BM'** and, with the axis, forms triangle MBM'. According to the Law of Sines,

$$\frac{\sin\theta}{\sin\theta'} = \frac{BM'}{BM}$$

The paraxial equivalent is, $\dfrac{\theta}{\theta'} = \dfrac{u'}{u}$ 6.21

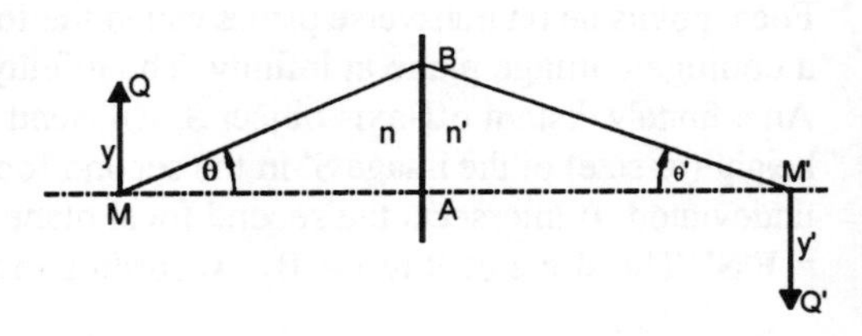

Figure 6.14 The Lagrange Invariant or Smith-Helmholtz equation.

According to the equation for lateral magnification, Eq. 6.15,

$$\frac{u'}{u} = \frac{n'y'}{ny} \qquad 6.22$$

By combining Equations 6.21 and 6.22, we obtain the Smith-Helmholtz or the Lagrange Invariant.

$$\boxed{n'y'\theta' = ny\theta} \qquad 6.23$$

According to this equation, no matter how many optical surfaces there are in a system, the product of the index of refraction, the object height and the axial ray slope will be a constant. This equation is an effective check on computations.

6.7 REFLECTION BY SPHERICAL MIRRORS

6.7.1 THE PARAXIAL IMAGE EQUATION

Figure 6.15 illustrates the two forms of spherical mirrors, namely, concave and convex. Concave mirrors converge light; convex mirrors diverge light. The object ray **MB** makes an angle of incidence α (cw) and an angle of reflection α' (ccw) with the normal. The ray slopes θ and θ' are ccw, and the central angle ϕ is cw.

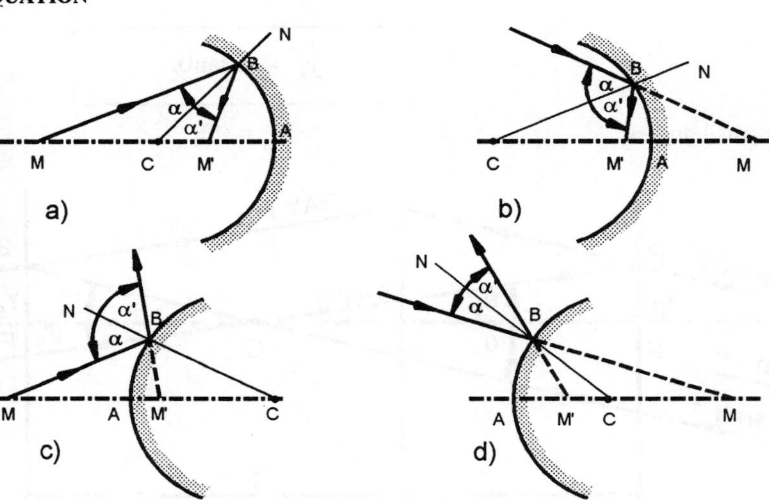

Figure 6.15 Real and virtual objects and images formed by concave and convex mirrors. a) RO/RI; b) VO/RI; c) RO/VI; d) VO/VI.

According to the law of reflection, $\alpha = -\alpha'$ 6.24

The image equation for a spherical mirror may be derived from Figure 6.16. In triangle **MCB** we find that

$$\phi = \alpha - \theta, \quad \text{or} \quad \alpha = \theta + \phi \qquad 6.25$$

In triangle **CM'B**

$$\theta' = \alpha' - \phi, \quad \text{or} \quad \alpha' = \theta' + \phi \qquad 6.26$$

Substitute Eqs. 6.25 and 6.26 into $\alpha = -\alpha'$,

$$\theta + \phi = -\theta' - \phi$$

$$\theta + \theta' = -2\phi \qquad 6.27$$

In the paraxial condition **B** and **D** simultaneously approach **A,** and the angles are small. Therefore,

$$\theta = -\frac{h}{u}; \quad \theta' = -\frac{h}{u'}; \quad \varphi = \frac{h}{r}$$

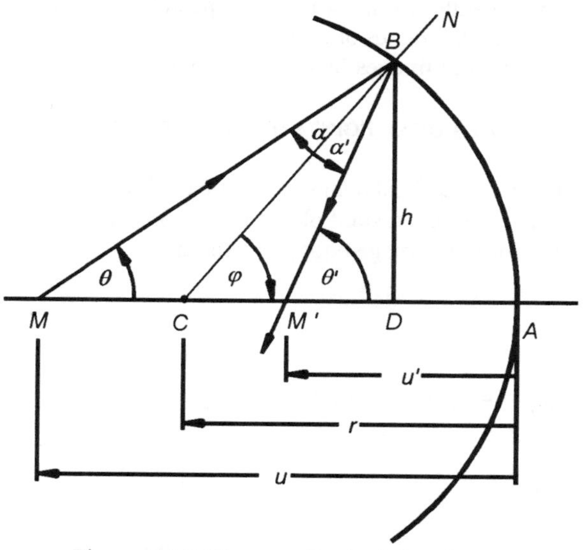

Figure 6.16 Diagram for deriving the image equation for a mirror.

Substitute these into Eq. 6.27, and divide by -h, to obtain the image equation for a spherical mirror that is frequently found in physics texts,

$$\frac{1}{u'} + \frac{1}{u} = \frac{2}{r} \qquad 6.28$$

Not needing to know a special equation for reflection is preferable. We may do so by a simple mathematical trick that, in effect, *treats reflection as a special case of refraction*. The index of refraction in object space is n; it is **n'** in image space. Since the medium in front of a mirror is the same for object and image, we ordinarily would say that **n = n'**. The trick is to say that *the index after each reflection is the negative of the index before each reflection.* Thus, for mirrors, $\mathbf{n_j' = -n_j}$

Substitution of -n for n', in the refraction equation for a spherical refracting surface, will make it applicable to mirrors, i.e.,

$$\frac{n'}{u'} = \frac{n}{u} + \frac{n'-n}{r}$$

becomes

$$-\frac{n}{u'} = \frac{n}{u} - \frac{2}{r}$$

If we divide by n and rearrange, we obtain the mirror equation (Eq. 6.28) from the refraction equation.

$$\boxed{\frac{1}{u'} + \frac{1}{u} = \frac{2}{r}} \qquad 6.29$$

Consequently, we do not need a special equation for mirrors. For *plane mirrors*, substitution of **n' = -n**, and ∞ for the radius of curvature, reduces the image equation to **u' = -u**; exactly what we know about the formation of images by plane mirrors.

6.7.2 The Focal Points of a Spherical Mirror

Definitions of focal points of spherical refracting surfaces also apply to mirrors. See Figure 6.17. The image of an infinitely distant object (**u = -∞**) is at the second focal point **F'**. The second focal length **f' = F'A**, therefore, the image equation becomes

$$\frac{n'}{u'} = \frac{n'-n}{r}$$

But, n' = -n, and u' = - f', therefore,

$$\frac{n}{f'} = -\frac{2n}{r}$$

and

$$f' = -\frac{r}{2} \qquad 6.30$$

Similarly, an object at the first focal point **F** will be imaged at infinity (**u' = ∞**). The first focal length **f = FA = -u.** Substitution in the image equation results in the following:

$$\frac{n'}{\infty} = \frac{n}{-f} + \frac{n'-n}{r}$$

and so

$$f = -\frac{r}{2} \qquad 6.31$$

Equations 6.30-31 show that **f = f'**. The first and second focal lengths are equal and, unlike the case for a spherical refracting surface, have the same sign. Consequently, *the focal points F and F' coincide.* However, **F** belongs to object space and **F'** to image space.

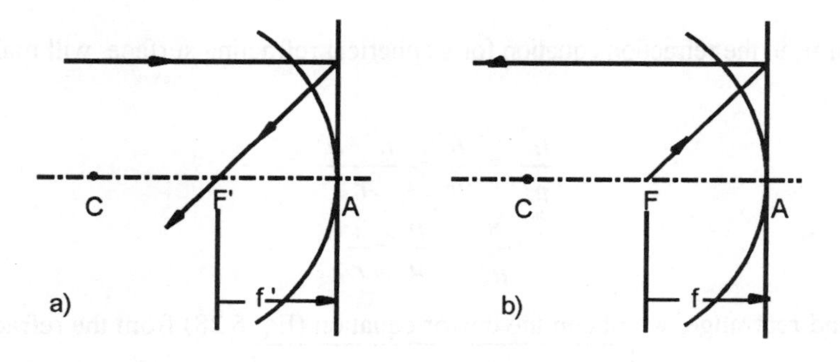

Figure 6.17 The focal points of a spherical mirror: a) second focal point; b) first focal point.

6.7.3 Construction of Axial Image Points for Spherical Mirrors

Based on treating reflection as a special case of refraction we may construct axial image points with the same construction as for a spherical refracting surface. Let,

$$\frac{CP}{CP'} = \frac{n'}{n} \quad \text{where, } n' = -n$$

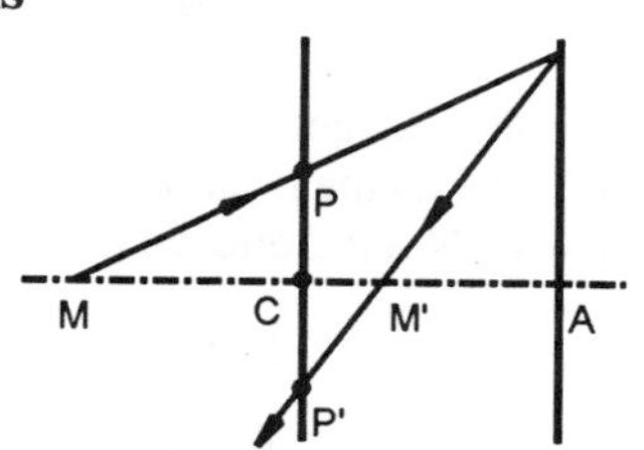

Figure 6.18 Construction of the axial image point for a spherical mirror.

Drop a perpendicular through the center of curvature **C** and mark equidistant points **P** and **P'** on opposite sides of the axis. A ray from **M** through **P** will, after reflection, go toward **P'**. The image point **M'** lies at the actual or projected intersection of the reflected ray with the axis, as shown in Figure 6.18.

6.7.4 Construction of Focal Points of a Spherical Mirror

In Figure 6.19 we have applied this construction to an axial object ray from infinity. After reflection the ray intersects the axis at **F'**. Two congruent triangles reveal that **CF' = F'A** and **CA = 2F'A** or,

$$f' = -\frac{r}{2}$$

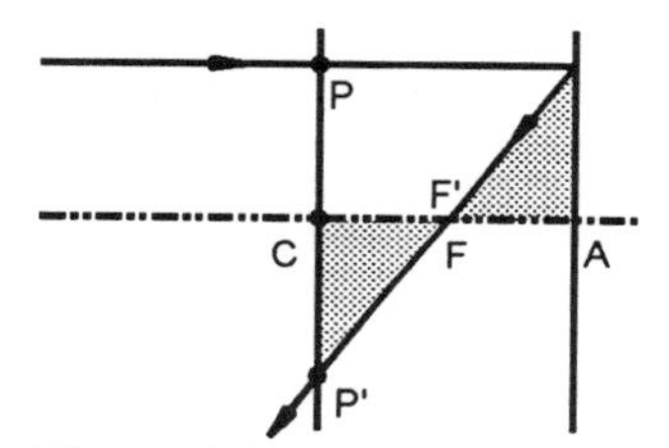

Figure 6.19 Construction of the focal points of a spherical mirror.

The image of an object at **F** may be constructed by reversing the path in Figure 6.19. Consequently, **CA = 2FA** or,

$$\mathbf{f} = -\frac{\mathbf{r}}{\mathbf{2}}$$

The construction agrees with the analytical results obtained above.

6.7.5 Construction of Off-axis Image Points Formed By a Mirror

The image **Q'** will be found at the intersection of any two rays from the object **Q**. As for a spherical refracting surface, the three predictable ray paths are the paths to the two focal points and the center of curvature. In Figure 6.20, off-axis image point **Q'** has been constructed for a concave and a convex mirror. Ray 1, from infinity, reflects to **F'**, Ray 2 through **F** reflects to infinity, and Ray 3 through **C** reflects back upon itself. All three rays intersect at **Q'**. The object plane is conjugate to the image plane. (Can you predict the path of a fourth ray from **Q** to **Q'**?)

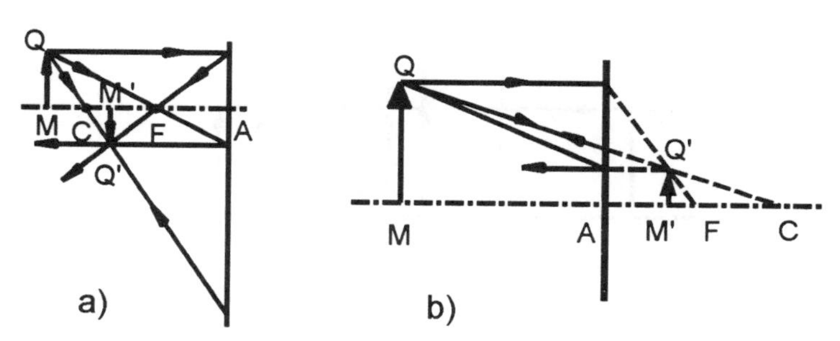

Figure 6.20 Construction of off-axis image points. a) concave mirror; b) convex mirror.

6.7.6 Image Positions as an Object Approaches a Spherical Mirror

Whether images are real or virtual, depends on the position of the object and the form of the mirror. Figure 6.21 shows the positions of the image as an object approaches a *concave mirror*. The results are tabulated below: Note: Concave mirrors, having positive power, *never form* VO/VI.

Object Travel	Image Travel	Real/Virtual	Orientation
From -∞ to C	F to C	RO/RI	Inverted
From C to F	C to -∞	RO/RI	Inverted
From F to A	∞ to A	RO/VI	Erect
From A TO ∞	A to F'	VO/RI	Erect

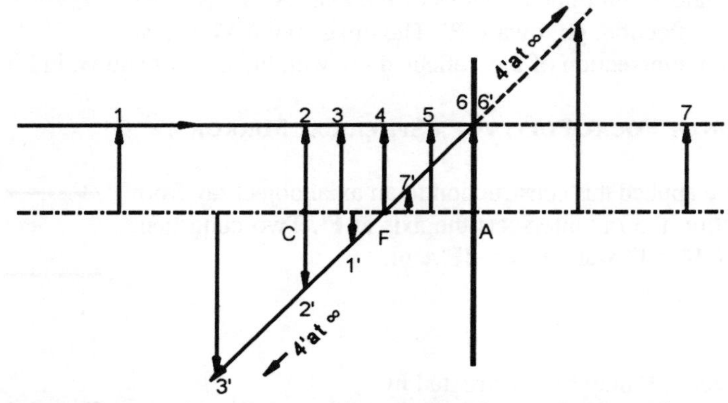

Figure 6.21 Image positions as an object approaches a concave mirror.

A similar diagram of object/image conjugates for a *convex mirror*, is shown in Figure 6.22. The student should work out an analogous tabulation. Is it possible to form RO/RI? Generalize your findings to spherical refracting surfaces.

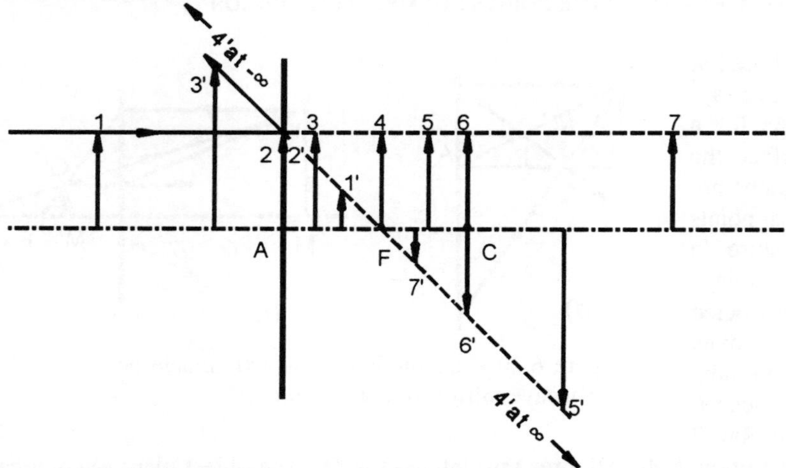

Figure 6.22 Image positions as an object approaches a convex mirror.

6.7.7 Possible Positions of the Image Formed By a Spherical Mirror

When calculating object/image problems for mirrors, the conditions to test are: *If the object point **M** does **not** lie within the radius AC, then the image point **M'** must lie within AC. If the object point **M** lies within AC then the image point **M'** cannot.*

Thus, the possible sequences of points for a concave mirror (See Figure 6.21) are: M C M' A, M' C M A, C M A M', C M' A M. They correspond to object positions 1, 3, 5 and 7, respectively. Determine the sequence for a convex mirror.

6.7.8 Lateral Magnification By Spherical Mirrors

In Figure 6.23 the image **Q'** has been constructed with a ray from **Q** parallel to the axis, and a ray to the vertex that forms a pair of similar triangles. These triangles directly show the lateral magnification relation.

$$Y = \frac{y'}{y} = -\frac{u'}{u}$$

However, the lateral magnification of a spherical refractingsurface (Equation 6.15) .

$$\mathbf{Y} = \frac{\mathbf{n\,u'}}{\mathbf{n'u}}$$

provides the same result when we substitute n' = -n.

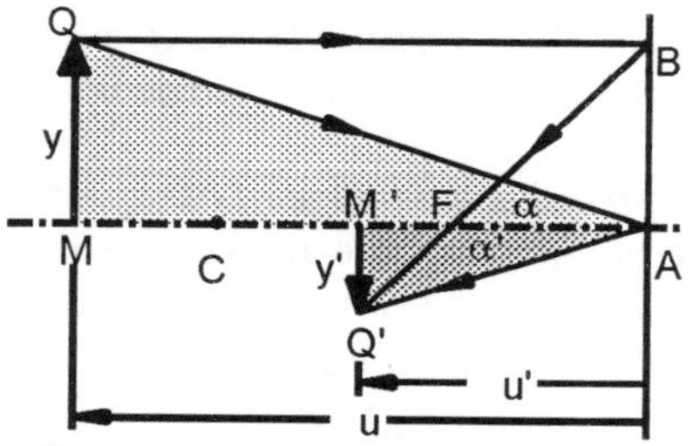

Figure 6.23 Lateral magnification diagram for a concave mirror.

6.7.9 Newtonian Equations for a Spherical Mirror

The Newtonian equations may be found simply by constructing **Q'** with the rays shown in Figure 6.24. Evidently, **y = MQ = AB,** and **y' = M'Q' = AD**. From similar triangles **MFQ** and **FAD** we obtain

$$\frac{AD}{MQ} = \frac{FA}{FM}, \text{ that is, } \frac{y'}{y} = \frac{f}{x}$$

From similar triangles **M'F'Q'** and **FAB** we obtain,

$$\frac{M'Q'}{AB} = \frac{F'M'}{FA}, \text{ that is } \frac{y'}{y} = -\frac{x'}{f}$$

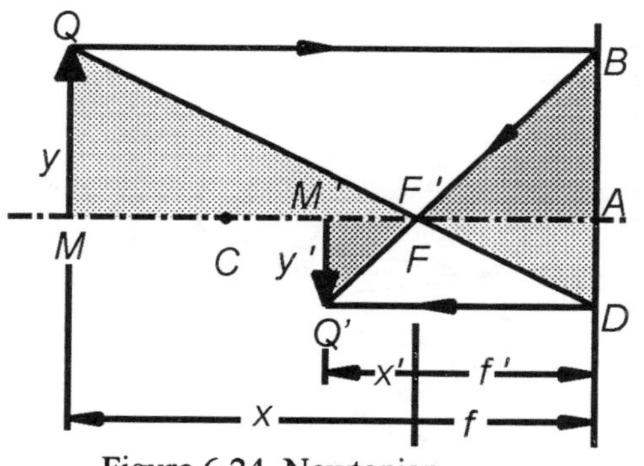

Figure 6.24 Newtonian diagram for a concave mirror.

Thus, the Newtonian equations for a spherical mirror are:

$$xx' = f^2$$
$$Y = \frac{y'}{y} = \frac{f}{x} = \frac{x'}{f} \qquad 6.32$$

6.7.10 THE FIELD OF VIEW OF A SPHERICAL MIRROR

We found that the real (object) and apparent (image) fields of view (FOV) of a plane mirror were equal. Spherical mirrors may have larger or smaller apparent fields than real fields. That is, they may magnify or minify. An eye is shown looking into a mirror in Figure 6.25. We will assume that the pupil of the eye (**B'O'C'**) is the *aperture stop* of this system. The mirror is the *first* element in the system (the light strikes the mirror before reflecting to the eye). To find the *entrance pupil* (**BOC**) we must find the image of the eye produced by the mirror, because *the entrance pupil is the image of the aperture stop formed by all elements in front of it.* The entrance pupil is a virtual and erect image of the aperture stop. Being the only other element in the system, the mirror is the *field stop* by default.

Rays 1 and 2, from the edge of the *field stop* toward the center **O** of the *entrance pupil*, delineate the angle θ, the *real FOV*. Because **O** and **O'** are conjugate points, any ray going to **O** must be reflected to **O'**. The *apparent FOV* is equal to the angle θ'.

The exit pupil is the image of the aperture stop formed by all elements that follow it. It coincides with the *aperture stop* because the eye is the final element of this system. The apparent FOV is smaller than the real FOV, that is, the virtual and erect image is compressed by the convex mirror. This is the *minification* observed in rear-view mirrors in automobiles. Object point **S** lies within the FOV.

In Figure 6.26 the mirror is **concave**. Here, the entrance pupil is a real inverted image (BOC) of the eye or aperture stop (B'O'C'). Rays 1 and 2, in object space, go through **O** to the edge of the field stop (the mirror). The real FOV equals angle θ, and the apparent FOV equals angle θ'. Again, the image field is compressed but the image is real and inverted. Object **S** lies within the FOV.

Draw the diagram for a magnifying mirror, such as a cosmetic mirror. Is it convex or concave, Is the image real? Is it erect?

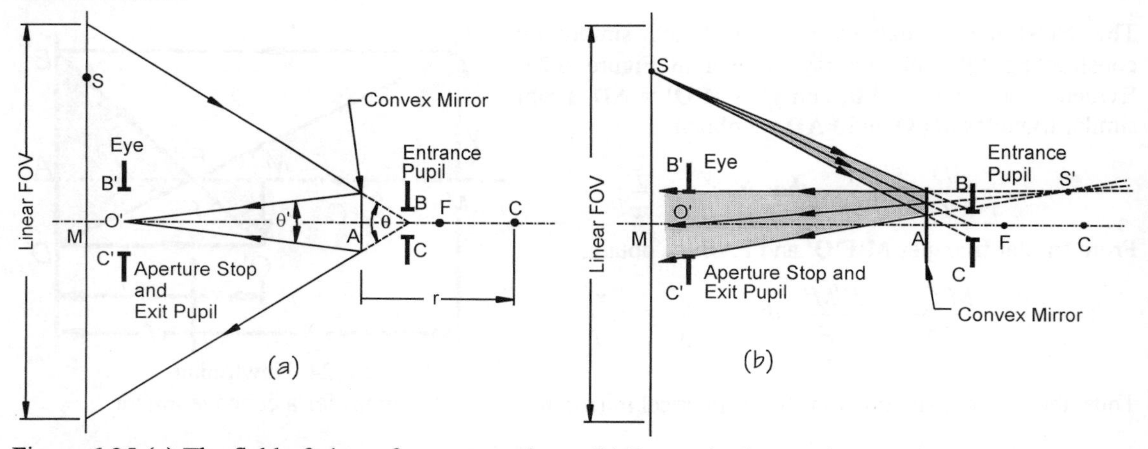

Figure 6.25 (a) The field of view of a convex mirror. (b) Ray paths from point S to the eye.

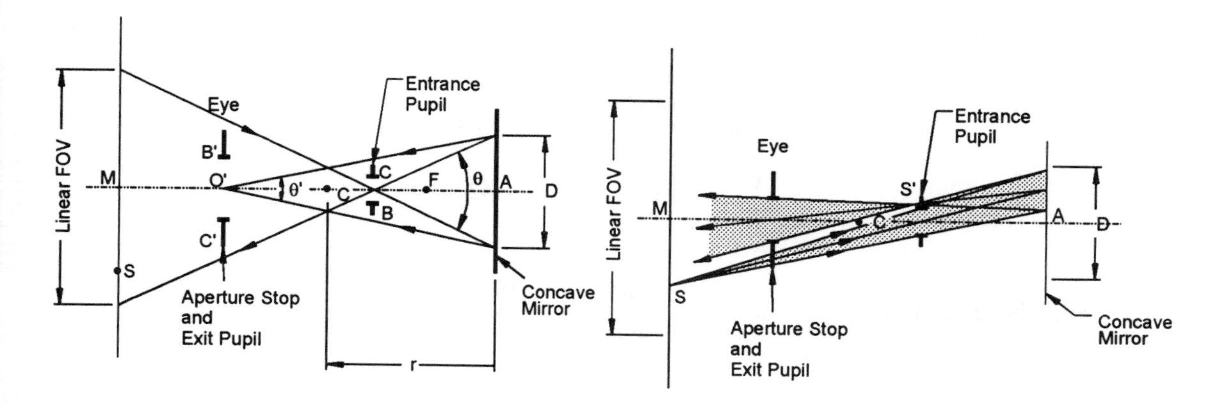

Figure 6.26 (a) The field of view of a concave mirror. (b) Ray Paths from point S to the eye.

6.8 ANGULAR MAGNIFICATION

The angular magnification of the mirror is the ratio between the apparent and real FOV,

$$M_a = \frac{\tan\theta'}{\tan\theta} = \frac{\theta'}{\theta} \quad \textit{for small angles.} \tag{6.33}$$

EXAMPLE 5: Determine the FOVs and angular magnification of a rear-view mirror. A convex rear-view mirror, four inches in diameter, has a radius of curvature of 10 inches. It is 20 inches from a driver. What will be the real and apparent FOVs of the mirror (absolute values)? Treat the eye as an aperture stop and place it on-axis.

Step 1. Find the entrance pupil. Given: u = -20", r = +10", n = 1, n' = -1, and D = 4".

$$\frac{-1}{u'} = \frac{1}{-20} - \frac{2}{10} = \frac{-5}{20}$$

Therefore, u' = +4". The entrance pupil is 4" to the right of the mirror.

Step 2. Find real FOV subtended by rays from the edge of the field stop (the mirror) to the center of the entrance pupil.

$$\tan\frac{\theta}{2} = \frac{D}{2u'} = \frac{2}{4} = 0.50$$

$$\frac{\theta}{2} = 26.57°, \therefore \theta = 53.13°$$

Step 3. Find the apparent FOV subtended by the rays from the edge of the field stop to the center of the exit pupil.

$$\tan\frac{\theta'}{2} = \frac{D}{2u} = \frac{2}{20} = 0.10$$

$$\frac{\theta'}{2} = 5.71°, \therefore \theta' = 11.42°$$

Step 4. Find the angular magnification, $M_a = \dfrac{\tan\theta'}{\tan\theta} = \dfrac{0.10}{0.50} = 0.20.$ (*minified*)

6.9 VERGENCE, POWER AND CURVATURE

6.9.1 VERGENCE AND POWER

Vergence is a measure of the convergence or divergence of light to or from a point in units called *diopters*. The diopter is the *reciprocal of the distance in meters*. Thus, light that converges to a point in air, 2 meters away, has a vergence of +0.5 D. Light that diverges from a point in air, after traveling 4 meters, will have a vergence of -0.25 D. The *reduced distance* **c** is used to account for media other than air and can be thought of as an air-equivalent thickness.

$$\mathbf{c = d/n}$$

Other reduced distances are: u/n, u'/n', f/n, f'/n'. The reciprocals of these reduced distances are the *reduced vergences*. In Equation 6.5, the terms n'/u' and n/u are *reduced vergences*. They are in *diopters* if u and u' are measured in *meters*. Thus, the reduced vergences of the object and image are **U** and **U'**, where

$$\mathbf{U} = \frac{\mathbf{n}}{\mathbf{u}}; \text{ and } \mathbf{U'} = \frac{\mathbf{n'}}{\mathbf{u'}}.$$

Equations 6.6 and 6.7 contain the terms **n/f** and **-n'/f'**. These are the reciprocals of the *reduced focal lengths* and correspond to *reduced vergences* when the focal lengths are in meters, and equal *refracting power F* in diopters.

$$F = \frac{n}{f} = -\frac{n'}{f'} = \frac{n'-n}{r}$$

The refraction equation can be rewritten in vergence form,

$$U' = U + F \tag{6.34}$$

Lateral magnification is given by

$$Y = \frac{U}{U'} \tag{6.35}$$

By setting n' = -n, the vergence equations are applicable to spherical mirrors.

The vergence form of the Newtonian Equation is obtained by substituting x = n/X, x' = n'/X', f = n/F and f' = -n'F into xx' = ff'.

$$XX' = -F^2$$

6.9.2 CURVATURE AND VERGENCE

Curvature is the rate of change in the direction of a curve. This rate of change is constant for a spherical surface, therefore, the curvature of a spherical surface **R** is the same anywhere on the surface. The curvature of paraboloids, ellipsoids and hyperboloids varies with the portion of the surface of interest. *Curvature is the reciprocal of the radius of curvature.*

$$\mathbf{R = 1/r} \qquad 6.36$$

The units of curvature are *reciprocal meters*, $\mathbf{m^{-1}}$. For example a spherical refracting surface, or a wavefront with a radius of +200 mm has a curvature R = +5 $\mathbf{m^{-1}}$. As spherical wavefronts diverge and converge, their curvature constantly changes. See Figure 6.27.

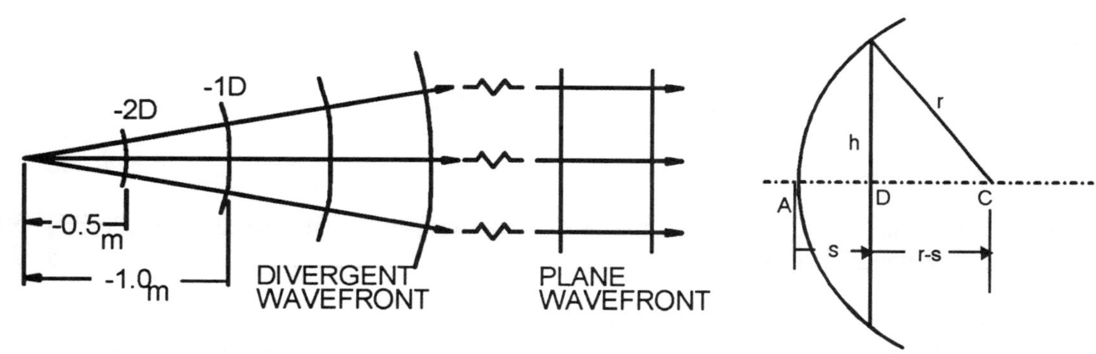

Figure 6.27 The vergence of wavefronts.

Figure 6.28 The sagitta.

Wavefront curvatures may be expressed as reduced vergences in *diopters*. The radius of the wavefront is the distance **u** or **u'**, and the reciprocals of their reduced distances are **U** and **U'**. A wavefront that has traveled 0.5 meters from its source, in air, has a vergence of -2 diopters. *If this wavefront had traveled 0.5 meters in index 1.5, its curvature would remain -2 D, but the vergence would be -3 diopters.* When the distance of the source is very great, the wavefronts are plane and the curvature and vergence are zero. At the instant a wavefront curvature is acted on by an optical surface it undergoes an abrupt change in vergence equal to the power of the surface as dictated by the vergence equation.

EXAMPLE 6: Given a source one meter from a spherical refracting surface with a power of +2 D that separates indices 1.0 and 1.5, find a) the image vergence; b) the image distance; and c) the curvature of the wavefront after refraction. Given: U = -1 D, F = +2 D.

U' = U + F = -1 + 2 = +1 D. (Ans. a)
u' = n'/U' = 1.5/1 = 1.5 m (Ans. b)
R' = 1/u' = 1/1.5 = 0.67 m^{-1} (Ans. c)

6.9.3 CURVATURE AND SAGITTA

Besides describing wavefronts, curvature can be related to the surface powers of lenses and mirrors. The curvature of these surfaces can be found by measuring their sagittas or sags. In Figure 6.28, **s** is the sag, **h** the semi-chord and **r** the radius of curvature of the surface. Applying the Pythagorean theorem to the right triangle, we obtain:

$$r^2 = h^2 + (r-s)^2 = h^2 + r^2 - 2rs + s^2$$

Canceling the r^2 terms, we find , $s^2 - 2rs + h^2 = 0$ 6.37

This quadratic equation has two solutions. Letting a=1, b=-2r and c= h^2, we will find, $s = +r \pm \sqrt{r^2 - h^2}$.

The minus solution is the sag; the positive solution is the distance from the chord to the opposite side of the spherical surface.

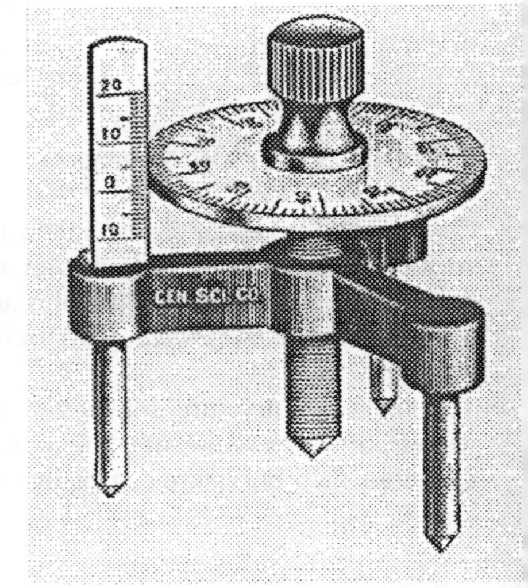

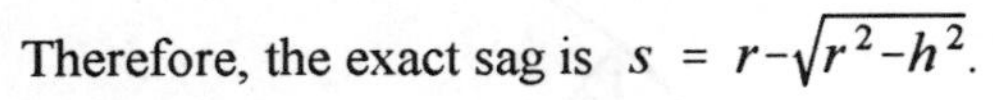

Therefore, the exact sag is $s = r - \sqrt{r^2 - h^2}$.

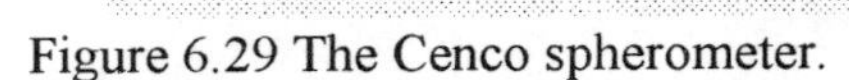

Figure 6.29 The Cenco spherometer.

Rearranging Eq. 6.37 we find the exact radius of curvature, $r = \dfrac{h^2 + s^2}{2s}$ 6.38a

Therefore, the exact curvature is $R = \dfrac{2s}{h^2 + s^2}$ 6.38b

Given **s** and **h**, we can find **R**. By "measuring" the curvature R, we can find the refracting or reflecting power of a spherical surface, because $F = (n'-n)R$ 6.39a

We can substitute Eq.6.38b into Eq. 6.39a to find the power of a surface. To obtain diopters, if s and h are in mm, multiply this equation by 1000 mm/meter. Thus, the exact power in terms of **s** and **h** is

$$F = \frac{2000(n'-n)s}{h^2 + s^2} \qquad 6.39b$$

An instrument that measures the sag of a surface can be calibrated to indicate its curvature or radius of curvature. One such instrument, the *spherometer*, has three legs arranged on a circle (corresponding to the chord) and a threaded center post that is adjustable in height. Zero calibration is obtained by placing the three fixed legs on a flat surface and adjusting the center leg until it just touches also. The sag is equal to the amount that the center post is extended or withdrawn to contact concave and convex surfaces, respectively. See Figure 6.29.

When the sag is very small compared with the radius of curvature, there is little error in dropping the s^2 term in Eq. 6.38b. A very useful approximate curvature equation is obtained.

$$\boxed{R \approx \frac{2s}{h^2}} \qquad 6.40$$

The approximate curvature is directly proportional to the sag **s**. The $2/h^2$ term is a proportionality constant that depends only on the separation of the legs of the instrument. According to this approximation, *the sag is proportional to the curvature*. This constant proportion is the basis for the design of the *lens measure*, used by optometrists and opticians. It provides a linear readout of the dioptric power of the surfaces of ophthalmic lenses. See Figure 6.30. The lens measure has two fixed legs and a spring-actuated center leg. It, thus, can read the power along a particular meridian of the surface. This is important when measuring non-rotationally symmetrical surfaces, such as those of cylindrical and toric lenses.

Figure 6.30 The Geneva lens measure.

The *approximate power equation* is obtained by substituting Eq. 6.40 into Eq. 6.39a.

$$F = \frac{2000(n'-n)s}{h^2} \qquad 6.41$$

A lens index of 1.53 and a 20-mm chord (separation of the fixed legs= 2h)is used to calibrate the lens measure. Substituting these values in Eq. 6.41 shows that **F = 10.6s**. This says that each mm of sag corresponds to 10.6 diopters of refractive power. The *instrument constant* is 10.6. Given curvature **R** and semi-chord **h**, and rearranging Eq. 6.40, we find the *approximate sag*

$$s = \frac{h^2R}{2} \qquad 6.42$$

6.9.4 The Correction of Lens Measure Readings

Refractive power varies with the index for any particular radius of curvature. The lens measure is calibrated to read power, however, it merely measures sags, which are proportional to curvature. The power for which it is calibrated, F_c, will be accurate only if the glass is made of the calibration index, n_c =1.53.

$$F_c = (n_c - 1)R \qquad 6.42a$$

Correcting the power reading for the actual refractive index $\mathbf{n_a}$ is easy. The actual power F_a is

$$F_a = (n_a - 1)R \qquad 6.42b$$

Solve Eq. 6.42a for R, and substitute into Eq. 6.42.b.

$$F_a = \frac{n_a - 1}{n_c - 1} F_r \qquad 6.43$$

EXAMPLE 7: A lens measure, calibrated for ophthalmic glass of index 1.523, reads +2.00 D when placed on the surface of a lens made of index 1.580 What is the actual power of the surface?

$$F_a = \frac{0.580}{0.523} 2.00 = +2.22\ D.$$

REVIEW QUESTIONS

1. Does light that initially travels from left-to-right travel in the positive direction?
2. Are distances measured from left to right positive?
3. List the line segments corresponding to **r, u, u', y** and **y'**.
4. If line segment MA is positive, is **u** positive?
5. If a ray slopes up to the right, is the slope angle positive?
6. Define the focal points.
7. Will the image of a distant object produced by a spherical refracting surface lie between the vertex and the center of curvature?
8. What does the Lagrange Invariant mean?
9. What is the difference between ***F*** and ***F'*** for a spherical mirror?
10. Will the image of a distant object produced by a spherical mirror lie between the vertex and the center of curvature?
11. Will a concave or convex mirror form a virtual image of a virtual object?
12. How is the field of view of a mirror found?
13. What is meant by reduced vergence, curvature and refractive power?
14. Why does the sagitta of an ophthalmic lens surface give the surface power?

PROBLEMS

1. Where are the focal points of a plane spherical refracting surface and a plane mirror?

2. A glass rod of index 1.5 has a convex radius of curvature of 50 mm on its front end. Its rear end is ground flat, but not polished. An image of the sun is formed on this end. How long is the rod?
Ans.: 150 mm.

3. The above rod is placed in a swimming pool. Find: a) the object distance that is conjugate to the ground end. b) the lateral magnification. Ans.: a) virtual object at 200 mm, b) 0.667.

4. An object 10 cm high, is 100 cm in front of a concave surface of 50-cm radius that separates air from glass of index 1.75. Find a) the image distance, b) the size of the image. Construct image Q'.
Ans.: u' = -70 cm, y' = +4 cm.

5. A glass ball of index 1.65 is 250 mm in diameter. How far is the image of the front vertex of the ball from the second vertex? Ans.: u' = -714.29 mm

6. Find the first and second focal lengths of a spherical refracting surface of radius +75 mm, separating air from index 1.85. Ans.: f = 88.235 mm, f'= -163.235mm

7. An object is 100 mm in front of a spherical refracting surface separating air from glass of index 2. A real inverted image is formed 200 mm from the surface. Find the radius of curvature. Ans.: +50 mm.

8. A virtual object is located 200 mm from a spherical refracting surface of +50 mm radius. A real image

is formed at 80 mm. The surface separates air from glass. Find the index of the glass. Ans.: 2

9. A virtual object 7.5 mm high is 75 mm from a concave spherical refracting surface of radius 25 mm. The surface separates air from glass of index 1.5. Find the position and size of the image. Construct Q'.
Ans.: Virtual, inverted image at u' = -225 mm, y' = - 15 mm

10. A simplified optical model of the eye (a reduced eye) treats the cornea as a spherical refracting surface separating air from aqueous humor of index 1.336. Let the radius of this surface be 8mm. a) How long is the eyeball if the retina is conjugate with infinity (emmetropic)? Find: b) the focal length, and c) the power of the eye.
Ans.: a) 31.81 mm, b) 23.81 mm, c) 42 D.

11. A reduced eye is 24 mm long. The radius of the cornea is 7 mm. a) How far is the object plane that is imaged on the retina (the far point)? b) Is the eye myopic or hyperopic? c) How many diopters is the error?
Ans.: a) 130.435 mm, b) hyperopic, c) 7.67 D.

12. When you look into the eye in Prob. 11, you see the pupil of the eye. It appears to be 3mm from the cornea, and appears 3 mm in diameter. a) How far is the iris from the cornea? b) How large is its aperture?
Ans.: a) 3.503 mm, b) 2.622 mm.

13. A galaxy subtends 5°. How large will its image be on the retina of the eye given in Problem 10?
Ans.: 2.083 mm.

14. A fish is at the center of a spherical fish bowl, 12 inches in diameter, filled with water. You observe the fish from a distance of 10 inches from the vertex of the bowl. The fish appears to subtend 5°. How long is the fish?
Ans.: 1.05 in.

15. A spherical refracting surface separating air from glass of index 1.6, has a focal length of 100 mm. Find the position of an image of an object that is 200 mm in front of the surface. Ans.: 320 mm.

16. When water is the first medium in the above problem, the focal length becomes 300 mm. Find the image distance for the 200 mm distant object.
Ans.: -720 mm

17. A +64 mm radius spherical refracting surface separates air and glass. Where must an object be placed to obtain an inverted image of the same size as the object? Ans.: 256 mm in front of the surface.

18. Repeat Prob. 17, but obtain an erect image of the same size.

19. A distance of +420 cm separates a real object and its real image, produced by a spherical refracting surface of radius 35 cm. The surface separates air from glass of index 1.5. How far is the object?
Ans.: u = -105 cm or - 280 cm.

20. A 5-mm high object is 45 mm in front of a concave mirror of radius 20 mm. Find the position and size of the image. Construct Q'.
Ans.: u' = -12.857 mm, y' = -1.429 mm. The image is real and inverted.

21. Repeat Prob. 20 with a convex mirror. Construct Q'.
Ans.: u' = 8.182 mm, y' = 0.909 mm. The image is virtual and erect.

22. A spherical mirror projects an image of a flashlight bulb onto a wall 200 inches away. The bulb is at a distance of 0.667r from the mirror, and the image is 3 times as large as the object. Find the radius of curvature of the mirror. Ans.: -100 in.

23. A real object is midway between the center of curvature and focal point of a concave mirror. What is the lateral magnification of the image? Construct Q'. Ans.: -2

24. A virtual object is midway between the center of curvature and focal point of a convex mirror. What is the lateral magnification of the image? Construct Q'. Ans.: -2

25. A real object is midway between the focal point and vertex of a concave mirror. What is the lateral magnification of the image? Construct Q'. Ans.: +2

26. How large is the image produced by reflection from the cornea of a 6-ft high window, if the window is 12 ft away, and the radius of the cornea is 7.25 mm? Ans.: 1.80 mm.

27. A light bulb is 5 meters from a projection screen. What is the radius of curvature of a mirror that will project an image of the bulb magnified 6 times? Ans.: 171.43 cm. concave.

28. A 4-inch diameter convex rear-view mirror has a radius of 12 inches and is 24 inches from a driver. Find the real and apparent fields of view. Ans.: 45.24°, 9.53°.

29. An object 10 cm high, is 100 cm in front of a concave surface of 50 cm radius that separates air from glass of index 1.75. Find a) the image vergence, b) the image distance.
Ans.: -2.50 D, -70.cm.

30. Find the refracting power of a spherical refracting surface of radius +75 mm, separating air from index 1.85. Ans.: F = 11.33 D

31. An object is 100 mm in front of a spherical refracting surface separating air from glass of index 2. A real inverted image is formed 200 mm from the surface. Find the refractive power of the surface.
Ans.: 20 D.

32. A 7.5 mm high virtual object is 75 mm from a concave spherical refracting surface of 20 D power. The surface separates air from glass of index 1.5. Find a) the vergence, b) the position, and c) the size of the image. Ans.: a) -6.67 D, b) Virtual, inverted image at u' = -225 mm, c) y' = - 15 mm

33 A 5-mm high object has a vergence of -22.222 D at a concave mirror of 100 D reflective power. Find the position and size of the image. Ans.: u' = -12.857 mm, y' = -1.429 mm.

34. Repeat Prob. 33 with a convex mirror. Ans.: u' = 8.182 mm, y' = 0.909 mm.

35. A plane wavefront strikes a +10 D refracting surface separating air from index 1.5. What is the curvature of the wavefront immediately on entering the image space medium? Ans.: 6.67 m^{-1}

36. Two spherical refracting surfaces on an optical axis are separated by a distance of 30 mm. The surfaces form a thick lens containing glass of index 1.5 between the surfaces. Air surrounds the lens. With respect to the incident light, the first surface is convex. It has a radius of curvature of 100 mm. The second surface is concave and has a radius of 200 mm. A real object, 10 mm high, is 400 mm from the front surface. Find: a) the image distance $u_2' = A_2M_2'$, b) the size of the final image, c) the refracting power of surfaces 1 and 2. Ans.: a) +194.87 mm, b) -5.13 mm, c) +5.00 D and + 2.50 D.

37. Two spherical refracting surfaces on an optical axis are separated by a distance of 30 mm. The surfaces form a thick lens containing glass of index 1.5 between the surfaces. Air surrounds the lens. With respect to the incident light, the first surface is concave. It has a radius of curvature of 100 mm. The second surface is convex and has a radius of 200 mm. A 10 mm high real object is 400 mm from the front surface. Find: a) the image distance $u_2' = A_2M_2'$, b) the size of the final image, c) the refracting power of surfaces 1 and 2. Ans.: a) -110.84 mm, b) +2.41 mm, c) -5.00 D and -2.50 D.

38. Two spherical refracting surfaces on an optical axis are separated by a distance of 30 mm. The surfaces form a thick lens containing glass of index 1.5 between the surfaces. Air surrounds the lens. With respect to the incident light, the first surface is convex. It has a radius of curvature of 100 mm. The second surface is convex and has a radius of 200 mm. A 10 mm high real object is 400 mm from the front surface. Find: a) the image distance $u_2' = A_2M_2'$, b) the size of the final image, c) the refracting power of surfaces 1 and 2. Ans.: a) +7,600.00 mm, b) -200.00 mm, c) +5.00 D and -2.50 D.

39. Two spherical refracting surfaces on an optical axis are separated by a distance of 30 mm. The surfaces form a thick lens containing glass of index 1.5 between the surfaces. *Air is in front of the lens; water is behind the lens.* The first surface is convex. It has a radius of curvature of 100 mm. The second surface is concave and has a radius of 200 mm. A real object, 10 mm high, is 400 mm from the front surface. Find: a) the image distance $u_2' = A_2M_2'$, b) the size of the final image, c) the refracting power of surfaces 1 and 2. Ans.: a) +384.81 mm, b) -7.59 mm, c) +5.00 D and + 0. 833 D.

40. Two spherical mirrors are 50 mm apart. An object, 3 mm high, is 20 mm in front of one mirror. This mirror is convex and has a 10-mm focal length. After reflection the rays strike the other mirror. It is concave and has a focal length of 10 mm. Find the image distance and size after reflection by both mirrors. Construct image point Q_2'.

Ans.: the inverted image is 12.143mm to the right of the second mirror and 0.2143 mm high.

41. Repeat the previous problem, but make both mirrors concave.

Ans.: the erect image is 15 mm to the right of the second mirror and 1.5 mm high.

42. Two spherical mirrors are 20 mm apart. An object, 3 mm high, is 15 mm in front of one mirror. This mirror is concave and has 20 mm radius of curvature. After reflection the rays strike the other mirror. It is convex and has a radius of curvature of 40 mm. Find the image distance and size after reflection by both mirrors. Construct image point Q_2'

Ans.: the inverted image is 20 mm to the right of the second mirror and 12 mm high.

43. Two concave spherical mirrors with 20 mm radii of curvature are 10 mm apart. The second mirror serves as an object for the first mirror. Find the position of the image of the second mirror after two reflections. What is the lateral magnification of the image? Construct image point Q_2'

Ans.: the image of the second mirror coincides with the first mirror. Y = -1.

44. A spherical refracting surface separating air from index 1.523 has a radius of curvature of +100 mm. Find a) the exact and approximate sags and b) powers for a 20-mm chord.

Ans.: a) exact: 0.5013 mm, +5.230 D, b) approx.: 0.500 mm, +5.243 D.

45. A spherical refracting surface separating air from index 1.523 has a radius of curvature of +100 mm. Find a) the exact and approximate sags and b) powers for a 50-mm chord.

Ans.: a) exact: 3.175 mm, +5.230 D, b) approx.: 3.125 mm, +5.314 D.

46. A lens measure, calibrated for index 1.530, reads +4.50 D when placed on a surface separating air from glass of index 1.589. What is the true surface power? Ans.: +5.00 D.

47. A lens measure reads -5.00 D on a surface separating air from glass of index 1.523. The true power of the surface is -5.230 D. For what index is the lens measure calibrated? Ans.: 1.500

48. A lens measure calibrated for index 1.5 gives a true reading of +2 D. The center leg is displaced 0.450 mm from the zero position. Find a) the radius of curvature of the surface, and b) the separation between the outer legs of the lens measure. Ans.: a) 250 mm, b) 30 mm.

CHAPTER 7

THIN LENSES

7.1 CONVEX AND CONCAVE LENSES

A simple lens generally comprises two spherical refracting surfaces separated by a transparent medium. If the lens medium is of *higher* refractive index than the surrounding medium, *convex* lenses will cause light rays to become *more convergent* or less divergent. Conversely, *concave* lenses make the rays *more divergent* or less convergent. Convex and concave lenses are also called positive and negative lenses, respectively. Positive lenses may be produced in biconvex, plano-convex or convex meniscus forms. All convex lens forms are thicker at the center than at the edge. The corresponding forms for negative lenses are biconcave, plano-concave and concave meniscus. They are thinner at the center than at the edge. See Figure 7.1a.

The three lens forms comprise a continuum of *bendings*. Lenses of any form can produce a given amount of convergence or divergence, i.e., have a given focal length. The form of the lens determines the aberrations it will produce. Thin lens optics neglects aberrations, consequently, the form of thin lenses may be neglected.

When the refractive index of the medium surrounding the lens is *greater* than the lens index, convex lenses will diverge and concave lenses will converge light rays. See Figure 7.1b.

Although the images formed by lenses can be found by applying the equations for a spherical refracting surface at each surface of the lens, we will develop more direct solutions for thin lenses.

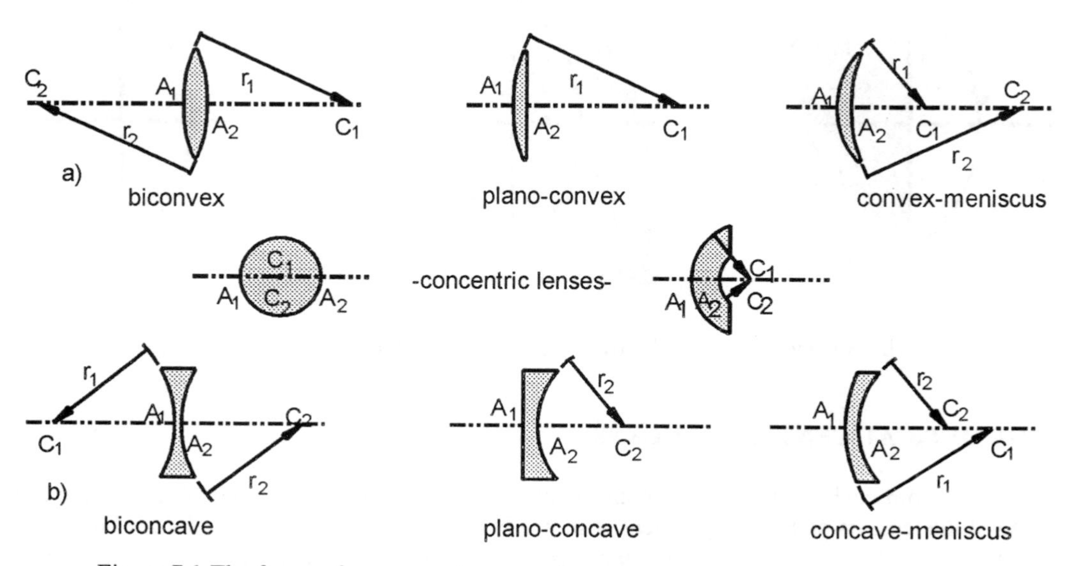

Figure 7.1 The forms of convex and concave lenses.

7.2 LENS NOMENCLATURE

The radii of curvature of the first and second surfaces of a lens are $\mathbf{r}_1$ and $\mathbf{r}_2$. Their corresponding centers of curvature are $\mathbf{C}_1$ and $\mathbf{C}_2$. A lens has a front vertex $\mathbf{A}_1$, and a rear vertex $\mathbf{A}_2$. Lens thickness is $\mathbf{d} = \mathbf{A}_1\mathbf{A}_2$. The radius of curvature is measured from the vertex **A** to the center **C**, i.e., $\mathbf{r}_1 = \mathbf{A}_1\mathbf{C}_1$; $\mathbf{r}_2 = \mathbf{A}_2\mathbf{C}_2$

According to our sign convention, a biconvex lens has a positive first surface radius, and a negative back surface radius. In an *equiconvex* lens, $\mathbf{r}_1 = -\mathbf{r}_2$. The *symmetrical* form of diverging lenses is called *equiconcave.* A sphere is a symmetrical lens with *concentric* surfaces, i.e., $\mathbf{C}_1$ and $\mathbf{C}_2$ coincide.

The *optical axis* is the straight line through the centers of curvature. The indices of refraction of the object space, the lens, and the image space media are $\mathbf{n}_1$, $\mathbf{n}_2$ and $\mathbf{n}_3$. Images formed by the lens may be found by using the refraction equation for a spherical refracting surface at each surface. At the first surface, the ray refracts from object space into the lens, i.e., from index $\mathbf{n}_1$ into $\mathbf{n}'_1$, where, $\mathbf{n}'_1 = \mathbf{n}_2$. The ray then refracts from the lens into image space, i.e., from $\mathbf{n}_2$ into $\mathbf{n}'_2$, where, $\mathbf{n}'_2 = \mathbf{n}_3$. See Figure 7.2.

When the lens is surrounded by a uniform medium the first and last indices are identical, i.e., $\mathbf{n}_1 = \mathbf{n}_3$. In this case we may simplify our nomenclature by calling the index of the surrounding medium **n**, and the index of the lens **n'**. A further simplification is made when the surrounding medium is air. Then $\mathbf{n}_1 = \mathbf{n}_3 = 1$, and the lens index is **n**. The following quantities are applicable to thin lenses:

Object distance:	AM	=	u	Image distance:	AM'	=	u'
Object height:	MQ	=	y	Image height	M'Q'	=	y'
First focal length	FA	=	f	Second focal length	F'A	=	f'

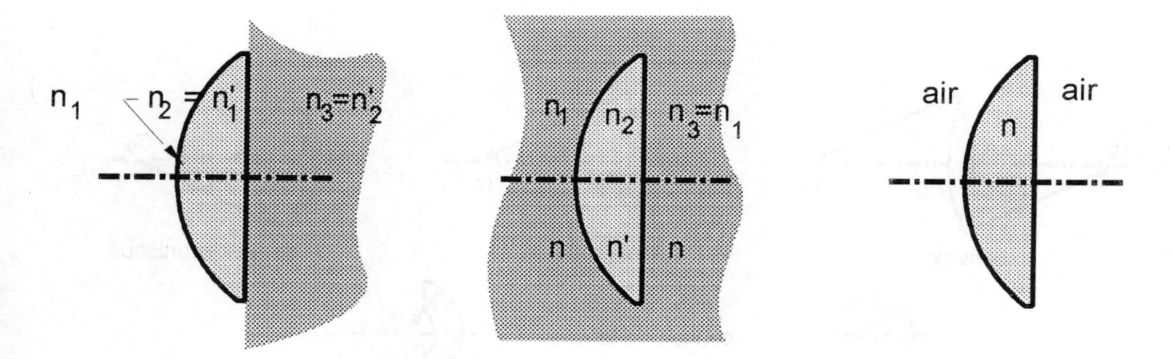

Figure 7.2 The nomenclature for refractive index varies with the media surrounding the lens.

7.3 THE OPTICAL CENTER OF A LENS

The *chief* ray from each off-axis object point, really or virtually, crosses the optical axis at the *optical center* **O** of the lens. Its path through a lens is shown in Figure 7.3. The incident chief ray heads toward the *first nodal* point **N**. After refracting into the lens the ray crosses the optical center, **O**. The ray refracts out of the lens as if leaving from the *second nodal* point **N'**, traveling in a direction parallel to the incident ray. By definition, rays that cross, or appear to cross the nodal points of a lens, will emerge undeviated. The position of the optical center of a lens, as a function of its bending and thickness, is given by

$$A_1O = \frac{r_1 d}{r_1 - r_2} \qquad 7.1$$

The path of a ray through the optical center is constructed by drawing parallel radii of curvature from the two centers of curvature. Within the lens, the path of the ray is found by connecting the points $\mathbf{I}_1$ and $\mathbf{I}_2$ at which the radii intersect the surfaces. If the ray path does not cross the axis, project it to the axis to find **O**. The direction of the incident and emergent rays depends on the indices of the lens and surrounding media. Figure 7.4 shows an arbitrarily chosen ray path to illustrate the positions of the nodal points. The incident ray crosses the axis at **N**, but the emergent ray must be projected back to the axis to find **N'**. Note how the points **O, N,** and **N'** of the meniscus lens are shifted, compared with their positions in the biconvex form.

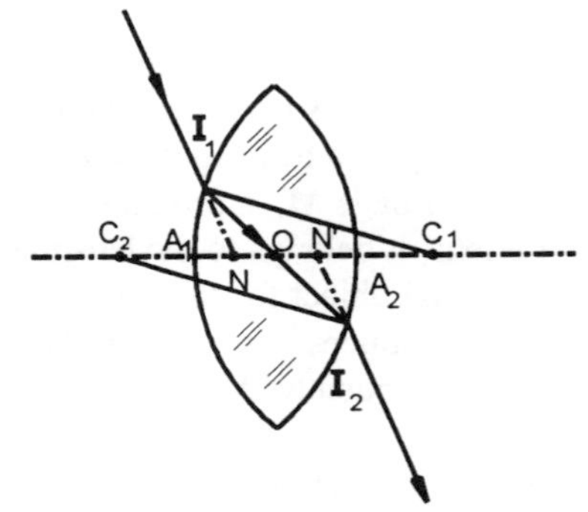

Figure 7.3 The optical center and nodal points of a biconvex lens.

EXAMPLE 1: Find the optical center of a lens with r_1 = +10 cm,
r_2 =+20 cm, and d = 2 cm. See Figure 7.4.

$\mathbf{A_1O}$ = 10(2)/10-20 = -2.00 cm.

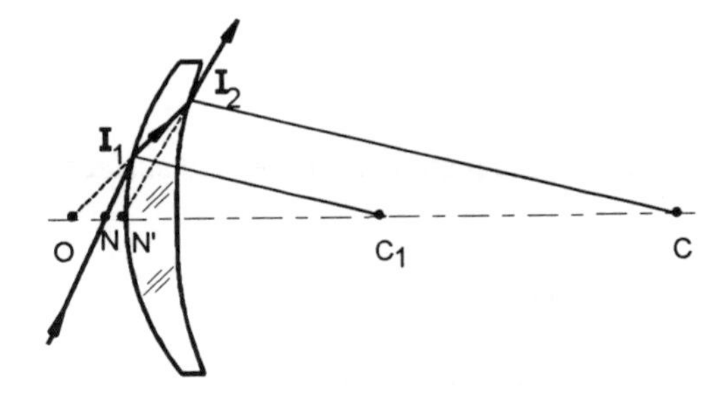

Figure 7.4 The optical center and nodal points of a positive meniscus lens.

7.4 THE THIN LENS

We may neglect the thickness of a lens if is very small compared with the radii of curvature. Consequently, d = 0, the vertices A_1 and A_2 coincide, and the lens vertex is simply denoted by **A**. The vertex **A** coincides with the optical center **0** and the nodal points **N** and **N'**. These assumptions enable us to find a simple thin lens equation.

7.5 THE THIN LENS EQUATION

The thin-lens equation for a lens in a uniform medium is obtained by applying the refraction equation for a spherical refracting surface at each of the two surfaces of the lens. Refraction at the first surface produces an image that serves as the object for the second surface. Although d = 0, the lens thickness is exaggerated in Figure 7.5 for the sake of clarity.

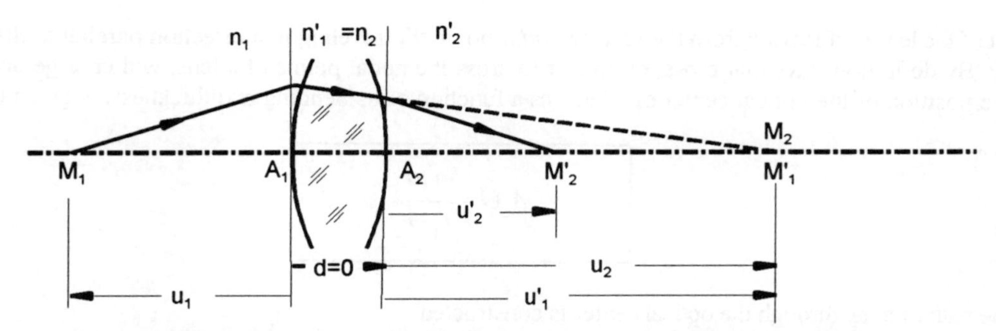

Figure 7.5 Refraction at each surface of a lens. M_1' is an intermediate image that becomes intermediate object M_2.

Point $\mathbf{M}_1$ is the object to be imaged by the lens. It is at object distance $\mathbf{u}_1 = AM_1$. After refraction at the first lens surface, an image $\mathbf{M}_1'$ is produced at a distance $\mathbf{u}_1' = \mathbf{AM}_1'$. Although the image appears to be to the right of the lens, it belongs to the lens medium into which the rays have refracted. The image point $\mathbf{M}_1'$ serves as a virtual object for the second lens surface, consequently, it is renamed $\mathbf{M}_2$. Refraction at the second surface produces a final image $\mathbf{M}_2'$. The figure shows that $\mathbf{M}_1$ and $\mathbf{M}_1'$ are conjugate points, as are $\mathbf{M}_2$ and $\mathbf{M}_2'$. Consequently, $\mathbf{M}_1$ and $\mathbf{M}_2'$ are conjugate to each other.

The equation for refraction at the first surface of the lens is
$$\frac{n_1'}{u_1'} = \frac{n_1}{u_1} + \frac{n_1'-n_1}{r_1} \qquad 7.2a$$

However, as noted in Section 7.2, $n_1' = n'$, and $n_1 = n$. Also, the image distance u_1' equals the object distance u_2. Thus, Eq. 7.2a becomes
$$\frac{n'}{u_2} = \frac{n}{u_1} + \frac{n'-n}{r_1} \qquad 7.2b$$

The refraction equation at the second surface is
$$\frac{n_2'}{u_2'} = \frac{n_2}{u_2} + \frac{n_2'-n_2}{r_2} \qquad 7.3a$$

This equation becomes
$$\frac{n}{u_2'} = \frac{n'}{u_2} - \frac{n'-n}{r_2} \qquad 7.3b$$

Upon substituting Eq. 7.2b into Eq. 7.3b, and substituting $u_1 = u$ and $u_2' = u'$, we obtain
$$\frac{1}{u'} - \frac{1}{u} = \frac{n'-n}{n}\left(\frac{1}{r_1} - \frac{1}{r_2}\right) \qquad 7.4$$

As defined for a spherical refracting surface, the image of an object at infinity ($u = \infty$) is termed the *second focal point* **F'** of a lens. An object at the *first focal point* **F** produces an image at ∞. That is, when the object distance AM = -FA or u = -f, the image distance u' = ∞. Substitution of these values of u and u' into Eq. 7.4 results in the *Lensmakers' Formula:*

$$\boxed{\frac{1}{f} = \frac{n'-n}{n}\left(\frac{1}{r_1} - \frac{1}{r_2}\right)} \qquad 7.5$$

In the Lensmakers' formula, n' is the index of refraction of the lens and n is the index of the surrounding medium. The *Thin Lens Equation* follows from Eqs. 7.4 and 7.5:

$$\frac{1}{u'} = \frac{1}{u} + \frac{1}{f} \qquad 7.6$$

EXAMPLE 2: An object is 100 mm in front of a thin negative lens of 50 mm focal length. Find the image distance. Given: u = -100 mm; f = -50 mm.

$$\frac{1}{u'} = -\frac{1}{100} - \frac{1}{50} = -\frac{3}{100}$$

$$u' = -33.33\,mm$$

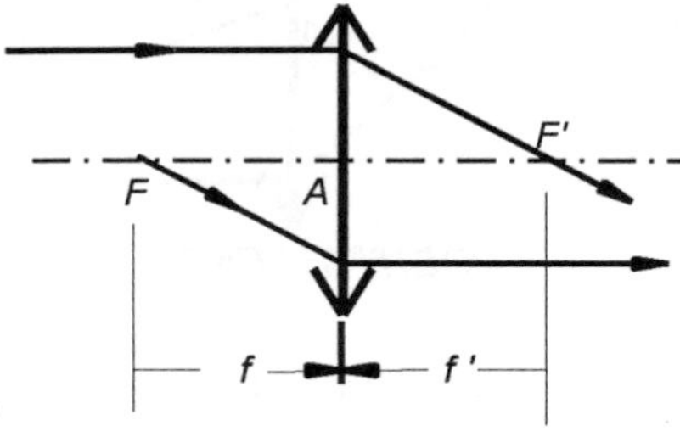

7.6 FOCAL LENGTHS

The second focal point **F'** lies at the point of intersection of the optical axis with the second focal plane of the lens. When **u** = ∞ the image point **M'** coincides with **F'**, therefore, AM' = AF'. According to Eq. 7.6, **u' = f**, or AM' = FA. Consequently,

$$FA = AF', \text{ but } AF' = -F'A = -f'$$

Thus,

$$f = -f' \qquad 7.7$$

This means that the focal points of a thin lens are equidistant but on opposite sides of the lens. For *positive* lenses, the first focal point **F** is in front of the lens and the first focal length **f** is *positive.* The second focal point **F'** is behind the lens, and **f'** is *negative.* For *negative* lenses, the sequence of focal points is reversed, the first focal length **f** is *negative* and **f'** is *positive.* See Figure 7.6.

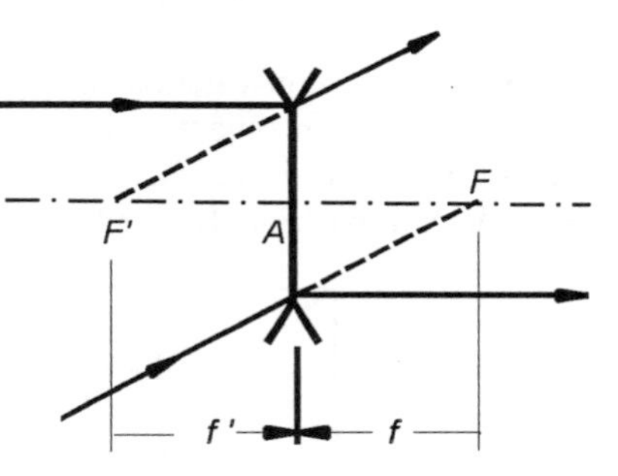

Figure 7.6 Focal points and lengths of a thin positive and negative lens.

7.7 CONSTRUCTION OF THE OFF-AXIS IMAGE POINT

The arrow **MQ** lies in the object plane. To find the image of **Q** it is necessary to select *two* rays through **Q** whose paths are known, and determine where they intersect. Figure 7.7a illustrates the paths of three predictable rays. Ray 1, parallel to the axis (as if from infinity), goes through point Q. After refraction, it must go through the *second focal point.* Ray 2, through the *first focal point*, travels parallel to the axis, after refraction. Ray 3, through the optical center, will be undeviated after refraction. The three rays intersect at **Q'** in the conjugate image plane. A perpendicular to the axis will locate **M'**.

A similar construction for a negative lens is shown in Figure 7.7b. Ray 1 must be projected back to **F'**; ray 2, headed toward **F**, will travel parallel to the axis after refraction. It must be projected back to intersect ray 1. Ray 3 through the optical center is undeviated by the lens.

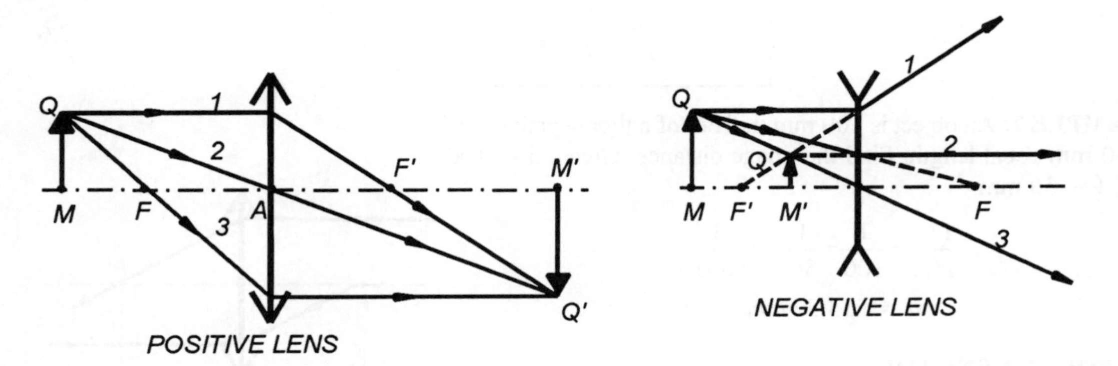

Figure 7.7 Construction of the image of an off-axis object point.

7.8 LATERAL MAGNIFICATION

According to the construction shown in Figure 7.7, the chief ray is a straight line that connects **Q** and **Q'**. The heights of an object and its image are, **y = MQ** and **y' = M'Q'**, as shown in Figure 7.8. Lateral magnification **Y** is the height of the image to that of the object.
From the similar triangles we find

$$Y = \frac{y'}{y} = \frac{u'}{u} \qquad 7.8$$

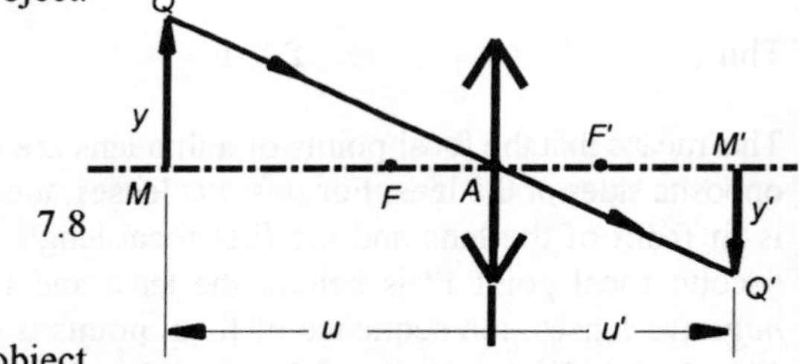

Figure 7.8 Lateral magnification.

Simply stated, the size of the image is to the size of the object as the image distance is to the object distance.

EXAMPLE 3: The size of the object in the previous thin lens example is 6 mm. Find the size of the image. Given: f = -50 mm; u = -100 mm; u' = -33.33 mm; and y = +6 mm.

$$Y = \frac{y'}{6} = \frac{-33.33}{-100} = \frac{1}{3}$$

$$y' = +2\ mm$$

7.9 VERGENCE EQUATIONS AND POWER OF A THIN LENS

The *vergence equation* is not merely another way of writing the thin lens equation. Because it depicts the change in vergence of the light, it offers useful insights into the formation of images.

$$U' = U + F \tag{7.9}$$

where, the reduced vergence of light from an object is **U = n/u**,
the reduced vergence of light to the image is **U' = n'/u'**,
and the power of the lens is $\boldsymbol{F} = \mathbf{n/f = -n'/f'}$

The object space medium is **n**, and **n'** represents the *image space* medium. When the lens is surrounded by a uniform medium, **n = n'**. If the lens is in air **n = n' = 1**. In either case, Eqs. 7.6 and 7.9 are identical.

Power and vergence are specified in *diopters*. The vergence equation graphically depicts the flow of light from the object to the image, with the power of the lens abruptly changing that flow. By requiring simple addition or subtraction, the equation lends itself to mental calculation.

EXAMPLE 4: In the previous example, f = -50 mm; u = -100 mm. Thus *F* = 1/-0.05 = -20 D and U = 1/-0.10 = -10 D. Therefore,

$$U' = -10 - 20 = -30 \text{ D}$$

and

$$u' = 1/-30 = -0.033 \text{ m} = -33.33 \text{ mm}$$

Lateral magnification, in terms of vergence, is equal to

$$Y = \frac{U}{U'} \tag{7.10}$$

Given an object 6 mm high, using vergence, find the size of the image.

$$Y = -10/-30 = 1/3 = y'/6$$

thus,

$$y' = +2 \text{ mm}$$

7.9.1 THIN LENS POWER IN NON-UNIFORM MEDIUM

When the optical media on either side of the lens differ the lens is equivalent to a single spherical refracting surface. Its power is given by the sum of the surface powers: $F = F_1 + F_2$, where $F_1 = (n_1' - n_1)/r_1$ and $F_2 = (n_2' - n_2)/r_2$. The object and image space indices **n** and **n'** must be substituted into the reduced vergence equations.

7.10 OBJECT AND IMAGE FORMATION BY A THIN LENS

Real objects are always in front of the lens, and virtual objects behind it. Virtual objects are produced by some other optical element in the system. Real images are always formed behind the lens, and virtual images lie in front of it. We now will examine image formation by lenses in greater detail. In this treatment, distinguishing between object and image space is important. No matter whether objects and images lie in front of or behind a lens, they belong to object and image space, respectively. Thus, both object and image spaces extend to $\pm\infty$. Light travels in object space before refraction, and in image space following refraction.

7.10.1 POSITIVE LENSES

In Figure 7.9, a ray from an infinitely distant on-axis object is drawn parallel to the optical axis. The real portion of the object ray is represented by the unbroken line drawn to the lens. That is, real objects lie in front of the lens along the real portion of this ray. The broken line, projected behind the lens, is the virtual portion of the object ray. Virtual objects lie behind the lens along this portion of the object ray.

As seen in the figure, after refraction occurs, the image ray crosses the axis at **F'**. The unbroken line segment corresponds to the real portion of the image ray. Real images lie behind the lens along this portion of the image ray. Virtual images lie along the broken backward projection of the image ray.

To survey the character of the imagery produced by the positive lens we will move object **MQ** from infinity toward and through the lens. Because two intersecting rays are necessary to find image **Q'**, we will send a chief ray from **Q** through the optical center of the lens and see where it intersects the previous image ray.

Starting with the object at $-\infty$, the image is a point at **F'**, (Fig. 7.9a). As the real object **MQ** advances toward **F**, a real, inverted image **M'Q'**, growing in size, moves toward $+\infty$, (Figures 7.9b-c). As the real object continues to advance from **F** to **A**, a virtual, erect, shrinking image moves from $-\infty$ to **A**, (Figs. 7.9d, e). The object and image reach the lens simultaneously, and coincide at **A**, (Figure 7.9f). Advancement of the object into the region between **A** and $+\infty$, requires that the object be virtual. The image, meanwhile, moves from **A** to **F'**. It is erect, real and smaller than the object, (Figures 7.9g-i).

In brief, where RO = real object, RI = real image, VO = virtual object and VI = virtual image, we find:

Object motion	Image motion	Characteristic
$-\infty$ to F	F' to $+\infty$	RO/RI Inverted image
F to A	$-\infty$ to A	RO/VI Erect image
A to $+\infty$	A to F'	VO/RI Erect image

A positive lens never satisfies the condition of VO/VI.

7.10.2 NEGATIVE LENSES

The results of a similar analysis of image formation by negative lenses are:

Object motion	Image motion	Characteristic
$-\infty$ to A	F' to A	RO/VI Erect image
A to F	A to $+\infty$	VO/RI Erect image
F to $+\infty$	$-\infty$ to F'	VO/VI Inverted image

A negative lens never satisfies the condition of RO/RI.

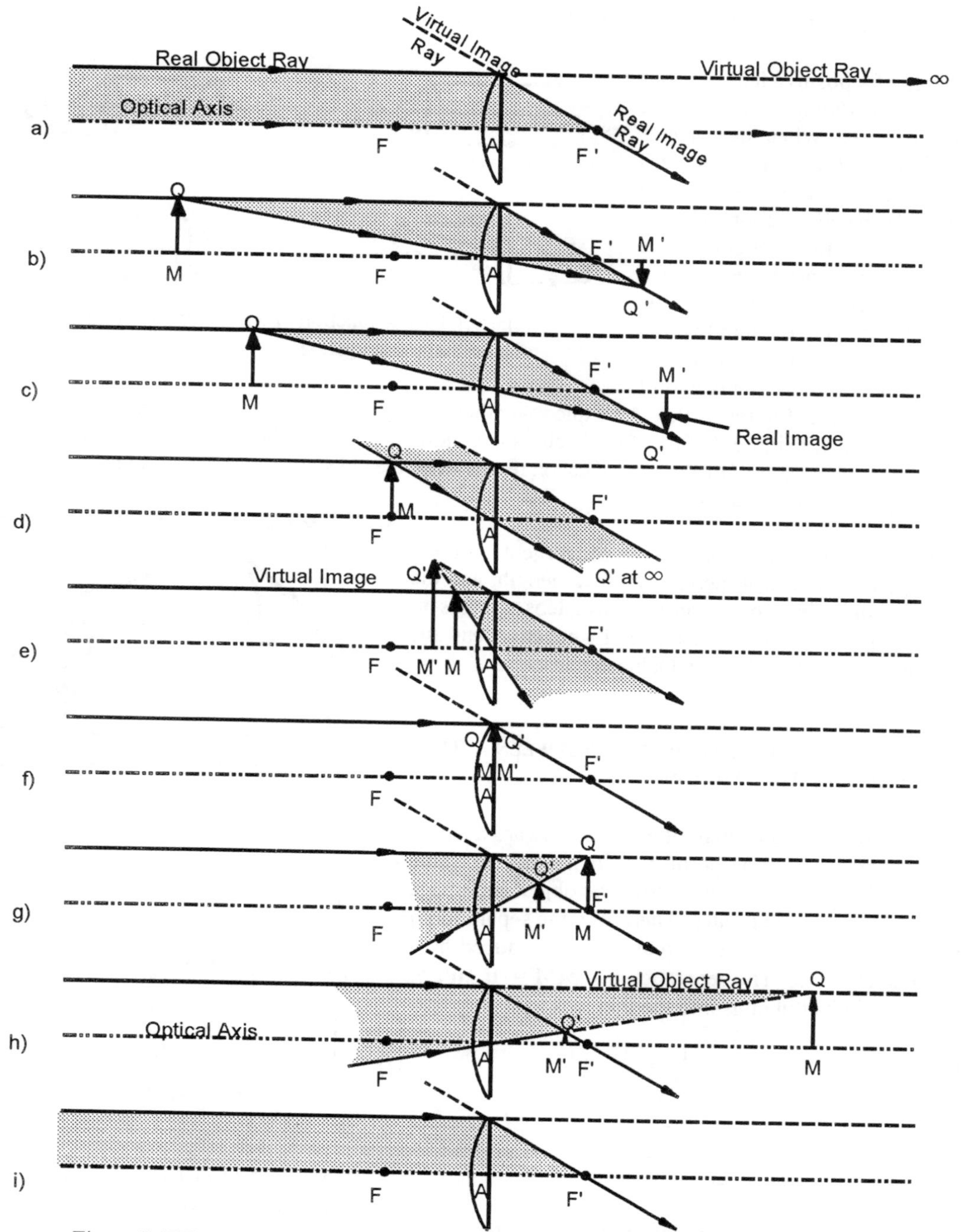

Figure 7.9 The total range of object/image conjugates of the positive lens.

7.10.3 APPLICATIONS OF LENSES WITH VARIOUS CONJUGATES

The role of a positive lens depends on the object/image conjugates at which it is used. When the object is very distant, it acts as a camera lens or telescope objective, (Figure 7.10a). As the object approaches to within several focal lengths of the "camera" lens, it produces minified images, (Figure 7.10b). When the object distance is just slightly more than a focal length, the image is magnified, and its action is that of a projection lens, (Figure 7.10c). A simple magnifier, or collimator lens results when the object is at the first focal plane, (Figure 7.10d). This is also the condition for using the positive lens as an ocular in a telescope or microscope. Greater magnification results when the object distance is just under a focal length. The images are erect and virtual.

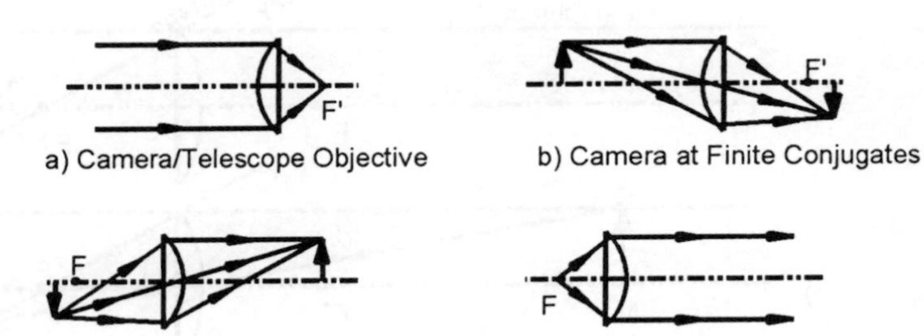

Figure 7.10 The function of a lens varies with the object/image conjugates.

A simple negative lens placed in the convergent light of a telescope objective will increase its focal length, and the power of the telescope. If an objective lens appears to converge light at the first focal plane of a negative lens, the latter serves as the ocular of a Galilean telescope.

7.10.4 THE MINIMUM SEPARATION BETWEEN A REAL OBJECT AND ITS REAL IMAGE

The left-to-right advancement of an object toward a lens is accompanied by a retreat of the image in the same direction. As the above tables show, the object and image move at far from equal rates. Therefore, the separation **q** between a real object and its real image, produced by a positive lens, varies. The separation **q = MM'= -u +u'**; or **u' = u + q**. Substitute this into the thin lens equation, and solve for u. $\frac{1}{u'} = \frac{1}{u} + \frac{1}{f}$

$$u = \frac{-q \pm \sqrt{q(q-4f)}}{2} \qquad 7.11$$

If we are to find two real solutions to the quadratic, (q-4f) must equal or be greater than zero. A (q-4f) term greater than zero means that there are two object distances that result in the same separation between object and image,

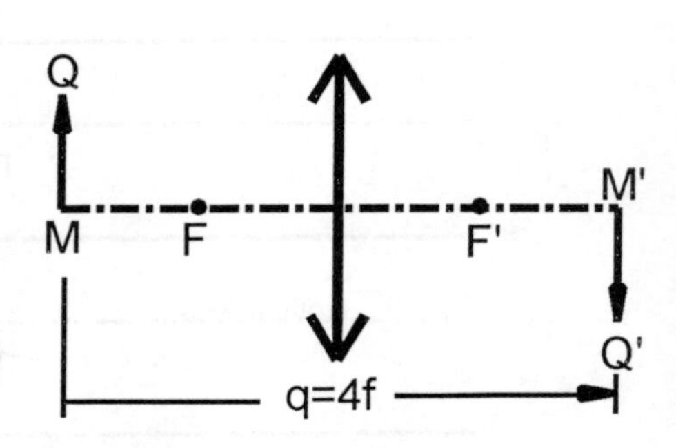

Figure 7.11 a) Two positions of a positive lens will satisfy the object/image conjugates for a fixed separation between M and M'. b) The minimum separation for real object and image is equal to 4f.

(Figure 7.11a). Minimum separation occurs when (q-4f) = 0. Then, according to Eq. 7.11, **u = -2f,** and **u' = 2f**, and the lateral magnification equals **-1.** See Figure 7.11b.

7.11 NEWTONIAN EQUATIONS

By designating the focal points **F** and **F'** as the origins of an object and image-space coordinate system we can write the Newtonian Equations for a lens. The object and image distances, **u** and **u'**, are replaced with **x** and **x'**, where,

$$\mathbf{x = FM} \text{ and } \mathbf{x' = F'M'}$$

Figure 7.12, based on the construction of **Q'**, illustrates two sets of similar triangles that directly lead to the Newtonian equation for lateral magnification.

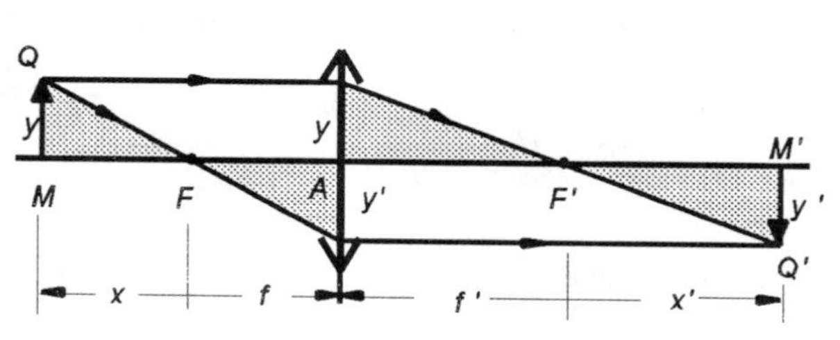

Figure 7.12 The Newtonian diagram for a thin positive lens.

$$Y = \frac{y'}{y} = \frac{f}{x} = \frac{x'}{f'} \qquad 7.12$$

We obtain **xx' = ff'** from this equation. However, for a lens, **f' = -f,** therefore,

$$xx' = -f^2 \qquad 7.13$$

7.12 THE FOCAL PLANES AND FOCAL IMAGES OF A THIN LENS

The focal points lie on transverse planes called the focal planes. See Figure 7.13. The first or primary focal plane through F has a conjugate image plane at infinity. The infinity object plane is conjugate to the second focal plane at F'. An infinitely distant off-axis object **S**, will send collimated rays to the lens. To determine the height (or size) of the image **S'** in the second focal plane, send Ray 1 through the center **A**. It will intersect the second focal plane at **S'**. Ray 2, parallel to Ray 1, intersects the point **P** in the first focal plane. Because Ray 2 originated from object **S**, it will be refracted to **S'** also. The size of the image $\mathbf{y'_S}$ = **F'S'.** The ray slope = θ. According to triangle AF'S',

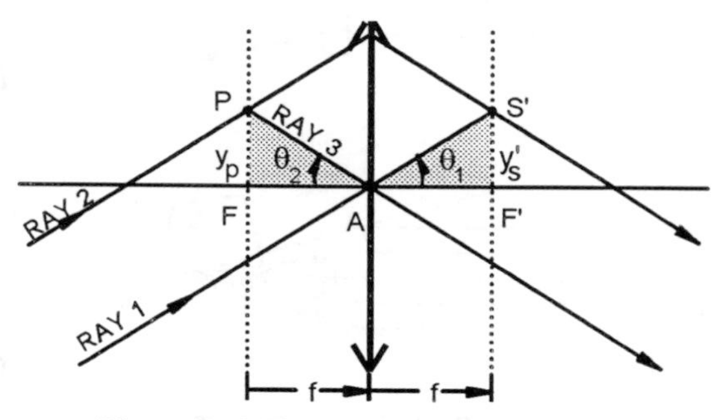

Figure 7.13 The size of objects and images in the focal plane of a lens.

$$\tan\theta_1 = \frac{F'S'}{AF'} = \frac{y'_s}{f} \quad \text{or} \quad \mathbf{y's = f \tan \theta_1} \tag{7.14a}$$

For small angles,

$$\mathbf{y'_s = f\theta_1}. \tag{7.14b}$$

The point **P** represents an off-axis object in the first focal plane of the lens. Rays from **P** will be collimated by the lens and form image **P'**, at infinity. The slope θ_2 of the chief ray (Ray 3) is constant in object and image space. Triangle PFA provides the object height $\mathbf{y_p} = \mathbf{FP}$.

$$\tan\theta_2 = \frac{FP}{AF} = \frac{y_P}{f'} \quad \text{or} \quad \mathbf{y_P = f' \tan \theta_2} \tag{7.15a}$$

For small angles,

$$\mathbf{y_P = f'\theta_2} \tag{7.15b}$$

7.13 PRISMATIC POWER OF A THIN LENS

A collimated pencil of rays is uniformly deviated by a thin prism whatever the height of the rays, as shown in Figure 7.14a. An observer would see the image up-to-the-left, along angle **δ.** Two base-to-base prisms produce opposite deviations. The emergent rays are split into two collimated pencils. See Figure 7.14b. The apparent direction of the image depends on which *pencil* the observer looks along. A lens converges the collimated pencil of rays to a focus. Each ray in the pencil is deviated as a function of the ray height and lens power. See Figure 7.14c. These deviations are called *prismatic effects*. When the observer's eye is on the optical axis, the prismatic effect is zero and the image is seen straight ahead. If the lens is decentered by a distance Δ, as in Figure 7.14d, the image appears to move upward.

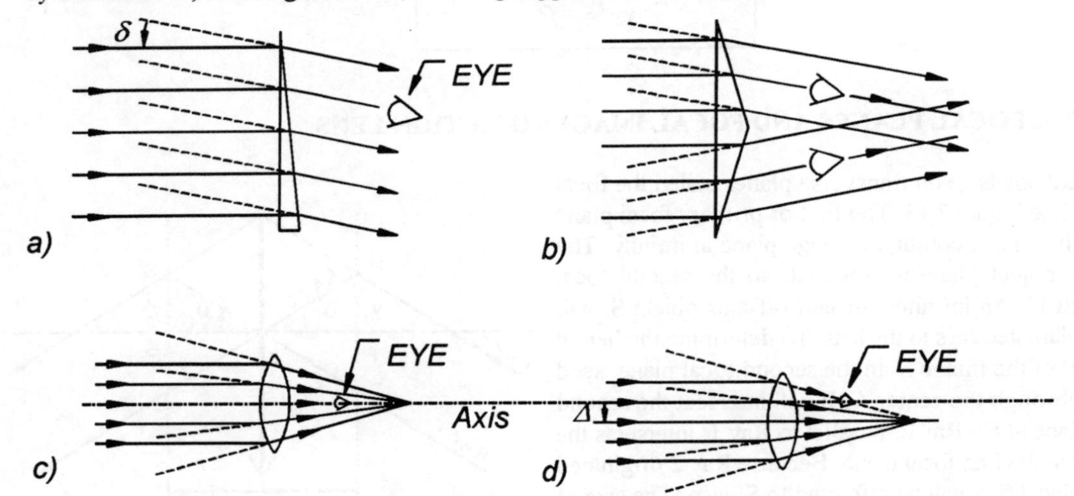

Figure 7.14 Comparison of prisms and the prismatic effect of a positive lens.

The prismatic effect of a ray at height **h**, or for a lens decentered through a distance **h**, is given by

$$\theta' = \frac{h}{f'} = -\frac{h}{f} = -hF$$

But, $\theta' = -\delta$, therefore, $\delta = \mathbf{h}F$ radians, where h is measured in meters.

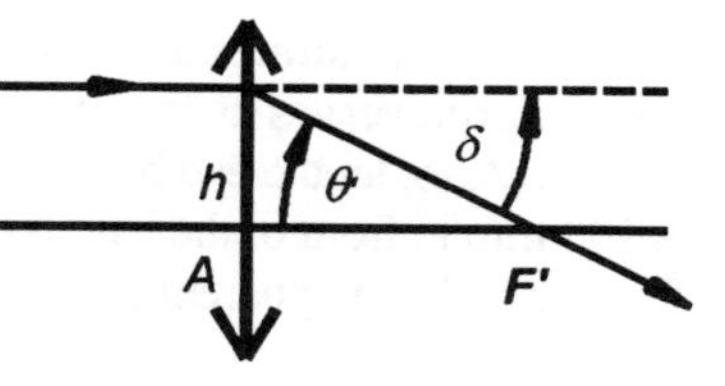

Radians and meters are inconveniently large units for expressing small decentrations and deviations. A more appropriate unit is the *centrad* $^\triangledown$. It is 1/100th of a radian. Prismatic effect in centrads $\mathbf{P}^\triangledown$ is

Figure 7.15 Primatic effect.

$$\mathbf{P}^\triangledown = 100\mathbf{h}F \text{ centrads, where } \mathbf{h} \text{ is in meters.}$$

However, 100*h in meters is equivalent to specifying **h** in cm.

$$\mathbf{P}^\triangledown = \mathbf{h}F \text{ centrads, where } \mathbf{h} \text{ is in cm.}$$

The final step toward specifying prismatic effect is to express it in *prism diopters* $^\Delta$. For small angles the centrad is equivalent to the prism diopter, thus,

$$\mathbf{P}^\Delta = \mathbf{h}F \qquad \mathbf{7.16}$$

EXAMPLE 5: A +5 D spectacle lens is decentered 4 mm down with respect to an eye. Find the prismatic effect, and the direction of motion of the image.

$P = 0.4 \times 5 = 2^\Delta$ Base-down. The image moved upwards.

7.14 STOPS AND PUPILS OF LENS SYSTEMS

An eye looking through a lens constitutes a very simple optical system. The components of this system, the eye and the lens, act as stops. One will be the *aperture stop* and the other will be the *field stop.* Aperture stops control the amount of light from each point in the object to reach the image. Field stops control the extent or size of the object to be imaged. Stops are tangible elements. Images of stops may be formed by other optical elements in a system. *Pupils* are images of the aperture stop. The *entrance pupil* is the image of the aperture stop produced by all optical elements in front of it. Alternatively, it is the smallest opening seen from the object position. The *exit pupil* is the image of the aperture stop formed by all optical elements behind it, and the smallest opening seen from the image position. *Entrance and exit ports* are images of the field stop formed by other optical elements in the system. We will consider the ports of an optical system in a subsequent chapter.

In the above simple optical system, finding the entrance pupil is necessary to decide whether the eye or the lens is the aperture stop. The field stop will be the other element, by default.

7.14.1 Procedure for Finding the Stops, Pupils and Field of View of a Thin Lens

Figure 7.16 illustrates an optical system comprising a thin positive lens and an eye looking through it. A page of print serves as the object. Let: the focal length of the lens, f= 50 mm, and the lens diameter D_L = 25 mm. Also, let the eye be 25 mm behind the lens, and its pupil diameter D_E = 10 mm. The object plane is 100 mm in front of the lens. a) Find the aperture stop, entrance and exit pupils, b) find the field stop, c) find the real and apparent angular fields of view, d) find the linear FOV (the diameter of the page of print that can be seen).

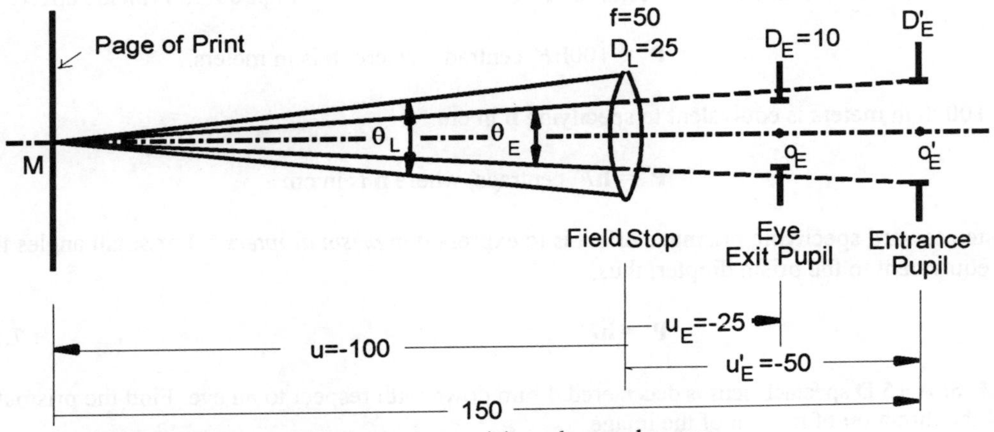

Figure 7.16 The stops of a system comprising a lens and an eye.

a. Find the entrance pupil.

Procedure: Look into the optical system from the object point **M**. Determine whether the lens or the eye appears to subtend the smallest angle.

The lens is directly seen. It subtends angle $\theta_L = D_L/AM = 25/100 = 0.25$

Because the eye is behind the lens, you will see an image of it, formed by the lens. Treat the eye as a real object that sends light to the lens from **right-to-left**. Therefore, the object distance u_E = -25. Solve for image distance.

$$\frac{1}{u_E'} = \frac{1}{-25} + \frac{1}{50}$$

thus, u_E' = -50 mm. The image of the eye is to the right of the lens (virtual). Find the size of the image, D_E'.

$$\frac{y'}{y} = \frac{D_E'}{10} = \frac{u_E'}{u_E} = \frac{-50}{-25} = +2$$

thus, $D_E' = +20$ mm

Find the angle θ_E subtended by the image at **M.**

$$\theta_E = D'_E/EM = 20/150 = 0.13$$

θ_E is less than θ_L, therefore, the image of the eye is the *entrance pupil*. The eye is the *aperture stop*. There are no lenses behind the eye, therefore, the *exit pupil* coincides with the aperture stop.

b. Find the field stop.
The lens is the *field stop* by default, because it is the only other element in the system.

c. Find the fields of view.
For this simple system, our definition of real FOV is the angle subtended by the field stop at the center of the entrance pupil. See Figure 7.17. To obtain a FOV in degrees, solve for the tangent of the angle subtended by half the diameter of the field stop (12.5 mm) at the entrance pupil. Then double the result for the total FOV.

$$\tan \frac{FOV}{2} = \frac{12.5}{50} = 0.25$$

The real FOV/2 = 14°; total real FOV = 28°

The apparent FOV is given by the angle subtended by the field stop at the exit pupil.

$$\tan \frac{FOV'}{2} = \frac{12.5}{25} = 0.5$$

The apparent FOV/2 = 26.6°; total apparent FOV = 53.1°

d. Find the linear field of view.
The linear FOV at the object is given by the real field angle and the distance from the object to the entrance pupil. Let y be half the diameter of the object, then 0.25 = y/150 and y = 37.5 mm. Therefore, a 75-mm diameter of the page is visible through the lens.

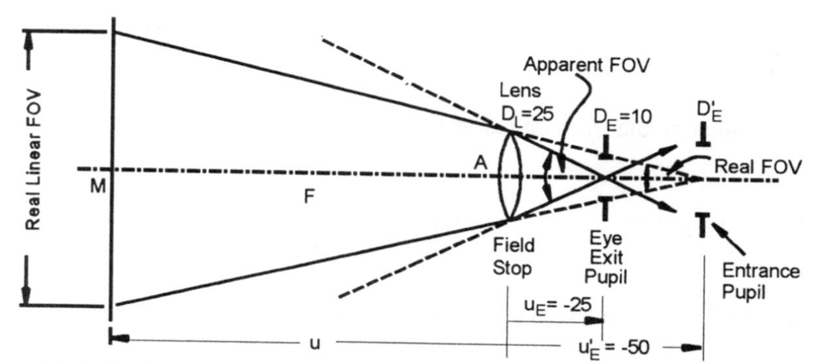

Figure 7.17 Real, apparent and linear fields of view.

REVIEW QUESTIONS

1. What conditions would cause a convex lens to diverge light?
2. List the distances measured from the vertex of a lens.
3. Define the nodal points, optical center and chief ray of a lens.
4. Define the diopter.
5. Compare what happens to the image as an object is moved through $\pm\infty$ with respect to plus and minus lenses.
6. What is meant by the prismatic effect of a lens?
7. A lens forms an image of a tree. How does this image change when the upper half of the lens is masked?

PROBLEMS

1. Find the position of the optical center of the following, 3 mm thick, lenses: (a) equiconvex, with radii of 10 mm, (b) plano-convex with a front surface radius of 10 mm, (c) concave meniscus with radii of 10 and 5 mm. Construct the optical centers.

Ans.: A_1O = (a) 1.5 mm, (b) 0 mm, (c) 6 mm

2. A biconvex lens of index 1.5 has radii of 10 and 15 cm. a) What is its focal length? (b) Find the image of an object located 24 cm in front of this lens, (c) Find the lateral magnification of the lens, (d) Construct the point Q'.

Ans.: (a) 12 cm, (b) u'=24 cm, (c) -1.

3. A positive meniscus lens, of index 1.5, has radii of 10 and 25 mm. a) What is its focal length? (b) Where is the image of an object, 10 mm high, located 25 mm in front of this lens? (c) how large is the image? (d) Construct the point Q'

Ans.: (a) 33.33 mm.(b) u'=-100 mm (c) y'=+40 mm.

4. A meniscus lens, of index 1.5, has radii of +50 and +25 mm. (a) What is its focal length? (b) Where is the image of a 6-mm high object, located 25 mm in front of this lens? (c) How large is the image?

Ans.: (a) f=-100 mm (b) u'=-20 mm (c) y'=4.8 mm.

5. An equiconvex lens of index 1.65, surrounded by water, has a focal length of 210.526 mm. Find its radii.

Ans.: 100 mm

6. An equiconvex lens surrounded by water has radii of 75 mm, and a focal length f = 157.8947 mm. Find its index of refraction.

Ans.: 1.65

7. Where should an object be placed with respect to a +18 cm focal length lens to produce a real image magnified 9x?

Ans.: -20 cm.

8. A lens with a focal length of -25 mm produces an erect image of an object magnified 4x. Find the position of the object.

Ans.: virtual object, 18.75 mm from the lens.

9. A real object is 16.67 cm from a lens of +8.33 cm focal length. What is the vergence of the light on leaving the lens?

Ans.: +6 D.

10. A lens (index=1.6), of 60 D power in air, is implanted in the eye. It is surrounded by aqueous humor of index 1.336. Find (a) its power in the eye, (b) its focal length.

Ans.: (a) 26.4 D. (b) 50.61 mm.

11. A 100mm focal length lens (index=1.5) projects an image of a 36-mm wide transparency onto a screen. The image fills the 4-ft wide screen. How far is the screen from the lens?

Ans.: 11.44 ft.

12. Three thin lenses are placed in contact. Their focal lengths are +800 mm, -250 mm and +625 mm. What is the focal length of the combination?

Ans.: -869.57 mm

13. A plano-cx lens of index 1.5 and a plano-cc lens of index 1.75 have identical radii of 75 mm. When put together they form a parallel plate. What are the (a) focal length and (b) power of the plate?

Ans.: (a) -300 mm, (b) -3.33 D.

14. An object 2 mm high is 120 mm from a +40 mm focal length lens. A second lens of +10 mm focal length is 100 mm from the first lens. Find the position and size of the final image.

Ans.: Real erect image 0.333 mm high, 13.33 mm from the second lens.

15. The sun (subtense=0.5°) is imaged, in turn, by a +10 D and -5 D lens. The lenses are separated by 25 mm. Find the position and size of the image.

Ans.: (a) $u_2' = +120$ mm, (b) $y_2' = -1.396$ mm.

16. A telephoto lens consists of a positive 85 mm focal length lens and a negative 50 mm lens separated by 70 mm. An object located 5 meters from the first lens is to be photographed. How far is the film plane from the second lens when the image is in focus?

Ans.: 24.56 mm.

17. A picture on a wall is to be projected onto an opposite wall 12 feet away. The image is to be enlarged five times. Find the focal length of the necessary lens.

Ans.: 20 inches.

18. A plano-cx lens of index 1.5, has a focal length of 64 mm. What is the *reflecting* power of the convex surface?

Ans.: -62.5 D

19. A photograph of a statue located 20 meters away is taken with a camera with a 1000 mm focal length telephoto lens. The size of the image on the film is measured and found to be 26.316 mm long. How large is the statue?

Ans.: 50 cm.

20. A thin equiconvex lens has radii of 100 mm. The lens index is 1.5, and it is surrounded by air. A real object, 10 mm high, is 200 mm from the lens. Find: a) the lens power *F* in diopters; b) f and f' in mm; c) image distance AM' in mm; d) the lateral magnification Y; e) image height M'Q' in mm.

Ans.: a) +10 D; b) +100 and -100 mm; c) +200 mm; d) -1; e) -10mm

21. Repeat Prob. 20 with the lens in water of index 4/3.

Ans.: a) +3.33 D; b) +400 and -400 mm; c) -400 mm; d) +2; e) +20 mm.

22. Repeat Prob. 20 with air in front of the lens and water behind it.

Ans.: a) +6.67 D; b) +150 and -200 mm; c) +800 mm; d) -3; e) -30 mm.

23. Repeat Prob. 20 with water in front and air behind the lens.
Ans.: a)+6.66 D; b)+200 and -150 mm; c)∞; d)∞; e)∞.

24. Repeat Prob. 20, with the surrounding medium glass, and the lens made of air.
Ans.: a) -10 D; b) -150 and +150 mm; c) -85.7142 mm; d) 0.4286; e) +4.286mm.

25. Find the prismatic effects of decentering a +4 D lens: (a) up 3 mm, (b) out 5 mm, (c) down 2.5 mm, and in 5 mm. (Answers are in prism diopters). Repeat for a -4 D. lens.
Ans.: (a) 1.2 base-up. (b) 2 base-out. (c) 1 base-down and 2 base-in.

26. Find the decentration of a +8 D lens that produces 2.4 prism-diopters base-up and 6.8 prism-diopters base-out.
Ans.: 3 mm up and 8.5 mm out.

27. A 50-mm focal length lens with a 20-mm aperture diameter is used to view a page of print, 80 mm away. The eye is 20 mm behind the lens and has a 5-mm pupil. Find: a) the aperture stop, (b) entrance and (c) exit pupils, (d) the field stop, (e) the real and (f) apparent angular fields of view, (g) the linear FOV (the diameter of the page of print that can be seen).
Ans.: (a) the eye is the A.S., (b) 33.33 mm to right of the lens, (c) 20 mm to right of the lens, (d) the lens is the F.S., (e) 33.40°, (f) 53.13°, (g) 68 mm.

28. A 20-mm focal length lens with a 10-mm aperture is used to view a page of print, 60 mm away. The eye is 40 mm behind the lens and has a 5-mm aperture. Find: a) the aperture stop, (b) entrance pupil, (c) exit pupil, (d) the field stop, (e) the real and (f) apparent angular fields of view, (g) the linear FOV (the diameter of the page of print that can be seen).
Ans.: (a) the lens is the A.S., (b) and (c) coincident with the lens, (d) the eye is the F.S., (e) and (f) 7.15°, (g) 7.5 mm.

CHAPTER 8

ROTATIONALLY SYMMETRICAL SYSTEMS

8.1 INTRODUCTION

A rotationally symmetrical optical system consists of a series of refractive and/or reflective surfaces whose centers of curvature lie on a single optical axis. All meridional sections containing the optical axis are identical. Examples of rotationally symmetrical systems range from a single lens, for example a simple magnifier, or a series of multi-element lenses such as a modern 16 element zoom lens. See Figure 8.1. The surfaces may be spherical or aspherical, i.e., paraboloidal, ellipsoidal or hyperboloidal. Astigmatic lenses, discussed in Chapter 9, are not rotationally symmetrical.

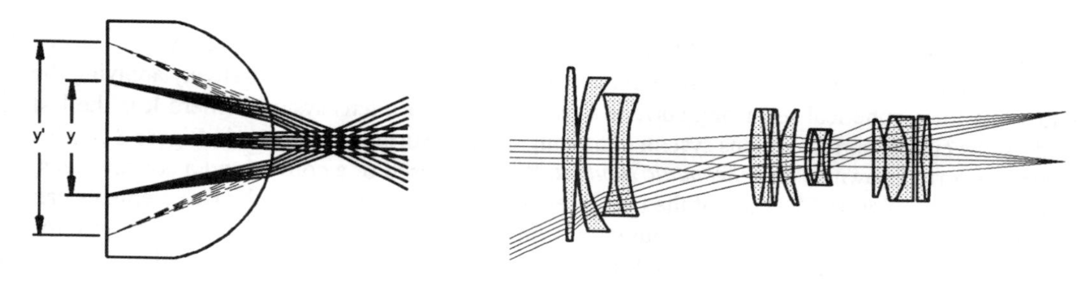

Figure 8.1 a) Thick single lens used as a magnifier. b) Sixteen element zoom camera lens.

Whatever the number of lens elements, a system is *thick* when it has a sufficiently large thickness-to-focal-length ratio (d/f) to result in significant inaccuracies in focal length, were **f** and **f'** measured from the focal points to the outer vertices of the lens.

Image formation by thick systems may be solved by the iterative application of the image equations and graphical constructions for reflecting and refracting surfaces and thin lenses.

8.2 PARAXIAL CONSTRUCTION OF THE IMAGE

Figure 8.2 illustrates the construction of successive images of object $\mathbf{Q}_1$, produced by two spherical refracting surfaces, or a thick lens. Object rays 1 and 2, with predictable paths through F_1' and F_1, after refraction at the first surface, become rays 1' and 2', and appear to meet at $\mathbf{Q}_1'$. This, in turn, is virtual object $\mathbf{Q}_2$ for the second surface. We can predict that ray 2' will be refracted through F_2'. Its path is shown by ray 2". However, it is not obvious how ray 1' will be refracted. To determine its path we construct ray 4', parallel to ray 1', but intersecting F_2. This ray will emerge parallel to the axis and intersect the second focal plane of the second surface at $\mathbf{S}_2'$. Since ray 1' also must intersect $\mathbf{S}_2'$, we have established that 1" is the path of the refracted ray. Rays 1" and 2" intersect at $\mathbf{Q}_2'$.

$\mathbf{Q_2'}$ is also found by drawing the undeviated ray 3', through $\mathbf{C_2}$. This ray becomes 3", and intersects ray 1" at $\mathbf{Q_2'}$.

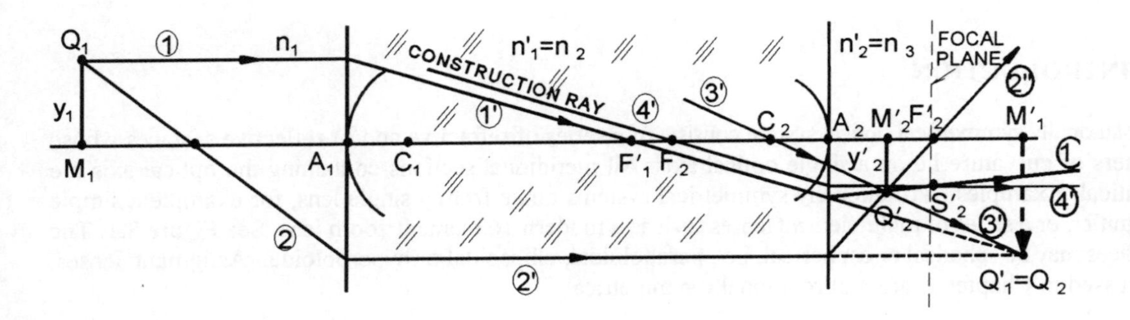

Figure 8.2 Construction of the image of an off-axis point produced by two refracting surfaces.

8.3 PARAXIAL CALCULATION OF IMAGES THROUGH AN OPTICAL SYSTEM

The position and size of an image formed by a thick lens system can be found by iteratively applying the below equations for a spherical refracting surface at each surface of a lens system, where **k** is the last surface and **j** is the current surface. A lens system consisting of three surfaces is shown in Figure 8.3. This is a common lens form known as an *achromatic doublet*. It is composed of a convex and a concave lens, each of a different kind of glass. The radii of the two inner surfaces are equal so that the two elements can be cemented together. Consequently, this doublet has three surfaces, i.e., **k=3.** In particular, **j=1** for computations at the first surface, then **j=2** and, finally, **j=3=k** at the last surface of this doublet.

Refraction equation:

Spherical refracting surface or mirror.
$$\frac{n_j'}{u_j'} = \frac{n_j}{u_j} + \frac{n_j' - n_j}{r_j} \qquad 8.1a$$

Series of thin lenses.
$$\frac{1}{u_j'} = \frac{1}{u_j} + \frac{1}{f_j} \qquad 8.1b$$

Transfer equation:
$$u_{j+1} = u_j' - d_j \qquad 8.2$$

Lateral magnification equation:
$$Y_j = \frac{y_j'}{y_j} = \frac{n_j}{n_j'} \cdot \frac{u_j'}{u_j} \qquad 8.3$$

The total lateral magnification of the system is equal to the product of the magnifications at each surface.
$$Y = Y_j \cdot Y_{j+1} \cdots Y_k \qquad 8.4$$

The Lagrange Invariant or Smith-Helmholtz equation is also applicable to a series of surfaces.
$$n_j y_j \theta_j = n_{j+1} y_{j+1} \theta_{j+1} \qquad 8.5$$

where, **j** takes all values from **1** to **k**. The object-space product equals the image-space product, i.e., for the doublet lens shown in Figure 8.3.

$$n_1 y_1 \theta_1 = n_4 y_4 \theta_4$$

or, in short

$$n y \theta = n' y' \theta' \qquad 8.6$$

NOTE: When there are one or more mirrors in an optical system, multiply the index of refraction before reflection by -1 to obtain the index after reflection. The index after a single reflection remains negative *until* the light is reflected by a second mirror. The index is multiplied again by -1, and it becomes positive. If light in negative index space encounters a lens, interchange the positions of the first and second focal points of the lens. Always set up problems so that light initially travels from left to right. All object, image and focal distances, etc. measured from left-to-right are positive, otherwise they are negative, despite the actual direction light is traveling.

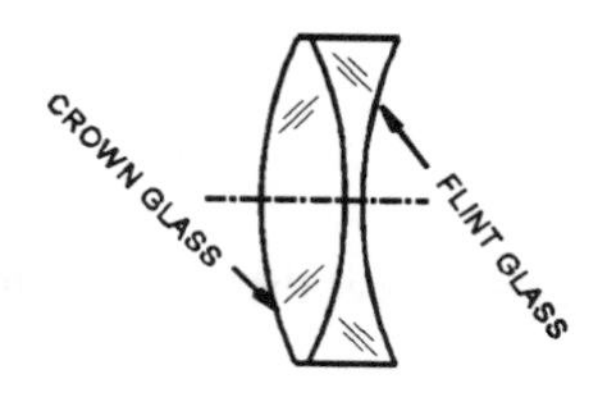

Figure 8.3 A cemented achromatic doublet has three refracting surfaces.

EXAMPLE 1: A 20 mm tall object is located 100 mm in front of a +50 mm focal length thin lens. A concave spherical mirror, with a radius of curvature of 25 mm is located 75 mm from the lens. Find the position and size of the image produced by a single refraction and reflection. See Figure 8.4.

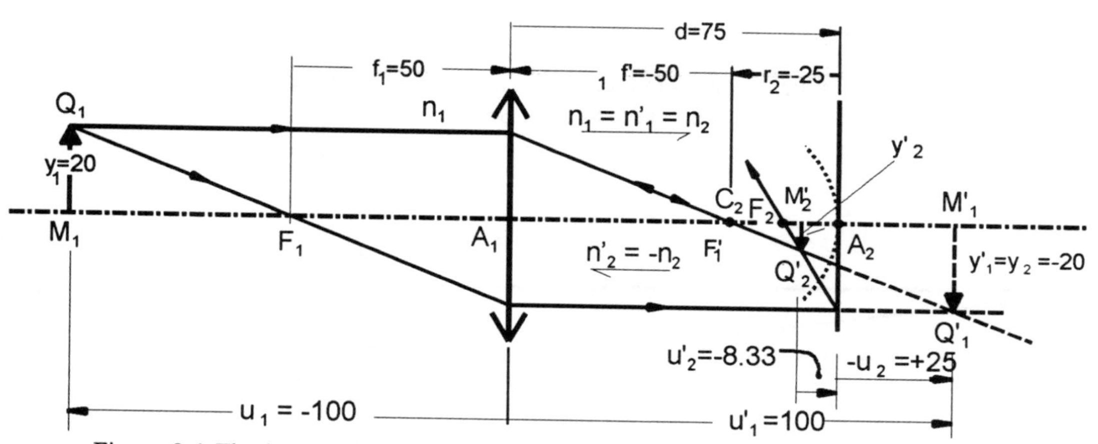

Figure 8.4 The image produced by refraction through a thin lens and reflection by a spherical mirror.

1. Find the image produced by the lens: Given: $u_1 = -100$ mm, $f_1 = +50$ mm, $y_1 = +20$ mm, $A_1A_2 = d_1 = +75$ mm, $n_1 = 1 = n_1'$.

Apply Eq. 8.1. $$\frac{1}{u_1'}=\frac{1}{u_1}+\frac{1}{f_1}=\frac{1}{-100}+\frac{1}{50}$$

and so, $u_1' = +100$ mm

2. Find Y_1; the lateral magnification produced by the lens:

$$Y_1=\frac{1}{1}\cdot\frac{+100}{-100}=-1$$

thus, $y_1' = -20\text{ mm} = y_2$

3. Find the new object distance: Apply Eq. 8.2. $u_2 = u_1' - d_1 = +100 - 75 = +25$ mm. $= A_2M_2$.

4. The data for the reflected image are: $u_2 = +25$ mm, $n_2 = 1 = -n_2'$, $r_2 = -25$ mm, $y_1' = y_2 = -20$ mm.

Apply Eq. 8.1a. $$-\frac{1}{u_2'}=\frac{1}{+25}+\frac{-2}{-25}$$

the image distance is $u_2' = -8.33$ mm

5. Find Y_2; the lateral magnification produced by the mirror:

$$Y_2 = +1/3$$

thus, $y_2' = -6.67\text{ mm} = y_3$

EXAMPLE 2: Given the same system as in Example 1, above, find the position and size of the image produced after refracting through the lens a second time. See Figure 8.5.

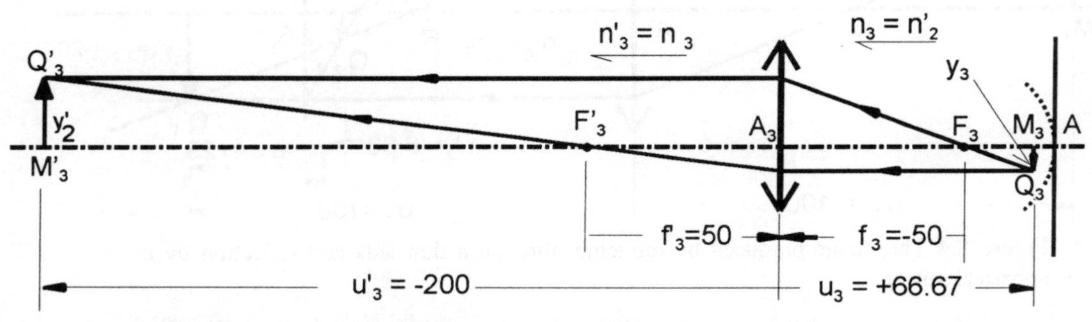

Figure 8.5 Same system as in previous figure. Reflected image is refracted through lens a second time.

6. Continue from step 5. Refract through the lens. Image $\mathbf{y_2'}$ becomes object $\mathbf{y_3}$. The data are:

$n_3 = n_2' = -1 = n_3'$, $d_2 = A_2A_3 = -75$ mm, $f_3 = F_3A_3 = -50$ mm $= -f_3'$, $y_3 = y_2' = -6.67$ mm

7. Find object distance with Eq. 8.2. $u_3 = u_2' - d_2 = -8.33 + 75 = +66.67$ mm $= A_3M_3$

Apply Eq. 8.1a $$-\frac{1}{u_3'} = -\frac{1}{66.67} + \frac{-1}{-50}$$

the image distance is $u_3' = -200$ mm $= A_3M'_3$

8. Find the size of the image. $Y_3 = u_3'/u_3 = -200/\ 66.67 = -3 = y_3'/\ y_3$

The final image height is $y_3' = -3(-6.67) = +20$ mm.

Alternatively, the system magnification is: $Y = Y_1Y_2Y_3 = (-1)(1/3)(-3) = +1$

Thus, the final image height is: $y' = +20$ mm

8.4 THE VERGENCE FORM OF THE REFRACTION AND TRANSFER EQUATIONS

The vergence form of Equations 8.1a and b is $U'_j = U_j + F_j$ 8.7

Where, $$U_j = \frac{n_j}{u_j}\ ;\ U_j' = \frac{n_j'}{u_j'}\ ;\ F_j = \frac{n_j}{f_j}\ .$$

This equation is applicable to reflection and refraction at single surfaces and lenses, provided the appropriate indices are used in finding the values of the reciprocal reduced vergences.

The **Transfer Equation** in vergence form is

$$U_{j+1} = \frac{U_j'}{1 - \frac{d_j}{n_j'}U_j'} = \frac{U_j'}{1 - c_jU_j'} \qquad 8.8$$

Lateral magnification is $$Y = \frac{U_j}{U_j'} \cdot \frac{U_{j+1}}{U_{j+1}'} \cdots \frac{U_k}{U_k'} = Y_j \cdot Y_{j+1} \cdots Y_k \qquad 8.9$$

8.4.1 Iterative Use of the Vergence Equations.

We will use the refraction and transfer vergence equations to solve the previous example. This is sometimes referred to as the step-along method. The specific equations are summarized as follows:

First refraction through the lens. $U'_1 = U_1 + F_1$

Transfer to the mirror. $U_2 = \dfrac{U_1'}{1 - c_1 U_1'}$

Reflection at mirror. $U'_2 = U_2 + F_2$

Transfer to lens. $U_3 = \dfrac{U_2'}{1 - c_2 U_2'}$

Second refraction through lens. $U'_3 = U_3 + F_3$

EXAMPLE 3: Repeat Examples 1 and 2 using the vergence equations.

1. First pass through the lens: Given: $u_1 = -100$ mm, $f_1 = +50$ mm, $y_1 = +20$ mm, $A_1A_2 = d_1 = +75$ mm, $n_1 = 1 = n_1'$.

Based on the above data, $U_1 = 1/-0.1 = -10$ D, and $F_1 = 1/0.05 = +20$ D.

Thus, $U_1' = -10 + 20 = +10$ D

2. The reflection at the mirror:

According to the Transfer Equation, $U_2 = \dfrac{+10}{1 - \dfrac{-0.075}{1} \cdot 10} = +40D$

This is the object vergence at the mirror, where: $n_2 = 1 = -n_2'$; and $r_2 = -25$ mm.

The reflective power of the mirror is: $F_2 = -2/-0.025 = +80$ D

Therefore, the image vergence is: $U_2' = +40 + 80 = +120$ D.

3. Second pass through the lens: To transfer to the lens, we note that $d_2 = A_2A_3 = -0.075$ and $n_2' = n_3 = -1$.

$$U_3 = \frac{+120}{1 - \dfrac{-0.075}{-1} \cdot 120} = + \frac{120}{-8} = -15D$$

This is the object vergence at the lens. The power of the lens F_3 must equal F_1.

$$F_3 = -1/-0.075 = +20 \text{ D}$$

The image vergence is $U'_3 = -15 + 20 = +5$ D.

and the image distance is $u'_3 = -1/5 = -0.2$ m $= -200$ mm $= A_3M'_3$

Finally, the lateral magnification is $Y = \frac{-10}{+10} \cdot \frac{+40}{+120} \cdot \frac{-15}{+5} = +1$

Compare these results with the previous solutions.

PROBLEMS

1. A 10 mm high object is located 100 mm in front of a 50 mm thick equiconvex lens with 75 mm radii and 1.5 index of refraction. Find (a) the position, (b) the lateral magnification and (c) the size the image formed by the lens. Construct the image point $\mathbf{Q}_2'$.

Ans.: (a) $A_2M_2' = 272.727$ mm; (b) Y = -2.455; (c) $y_2' = -24.545$ mm.

2. Repeat Problem 1, except make the lens equiconcave.

Ans.: (a) $A_2M_2' = -57.534$ mm, (b) Y = +0.37; (c) $y_2' = +3.699$ mm.

3. A 5mm high object, in air, is located 300 mm in front of a 25 mm thick lens made of index 1.5.The radii $r_1 = +100$ mm and $r_2 = -80$ mm. The rear surface of the lens is in water of index 1.3. Find a) the position, b) the lateral magnification and c) the size of the image. Construct the point Q_2'.

Ans.: (a) $A_2M_2' = 308.475$ mm; (b) Y = -0.814; (c) $y_2' = -4.068$ mm.

4. Find the positions of the focal points of a 25 mm thick positive meniscus lens, with radii of +50 and +100mm, and 1.5 index. Draw ray paths. Ans.: $A_1F = -185.714$ mm; $A_2F' = +142.857$ mm

5. Find the positions of the focal points of a 50 mm thick positive lens, with radii of +10 and -15mm, and 1.5 index. Draw ray paths.

Ans.: $A_1F = +4$ mm; $A_2F' = -24$ mm

6. Two thin lenses of +30 and -60 mm focal lengths, respectively, are separated by 15 mm. A 3 mm high object is located 120 mm from the first lens. Find (a) the position and (b) the size of the image formed by the lens. Construct the image point $\mathbf{Q}_2'$.

Ans.: (a) $A_2M_2' = +42.857$ mm, (b) $y_2' = -1.714$ mm.

7. Find the positions of the focal points of the lens system in Problem 6. Draw ray paths.

Ans.: $A_1F = -50$ mm; $A_2F' = +20$ mm

8. A 5 mm high object is located 25 mm from a concave mirror of 20 mm radius. The reflected rays then strike a convex mirror of 30 mm radius. The two mirrors are separated by 50 mm. Find (a) the position and (b) the size of the image formed after reflection at each mirror. Construct the image point $\mathbf{Q}_2'$.

Ans.: (a) $A_2M_2' = -10.345$ mm, (b) $y_2' = -1.035$ mm. The image is virtual and inverted.

9. Two concave mirrors face each other and are 75 mm apart. Their radii are 30 and 60 mm, respectively.

An object 2 mm high is placed 20 mm from the first mirror. Find (a) the position and (b) the size the image formed after reflection at each mirror. Construct the image point $\mathbf{Q}'_2$.

Ans.: (a) $A_2M'_2 = -30$ mm, (b) $y'_2 = -12$ mm. The image is virtual and inverted.

10. Repeat Problem 9 for two convex mirrors. Let the object be 10 mm high.

Ans.: (a) $A_2M'_2 = -22.0755$ mm, (b) $y'_2 = 1.132$ mm. Virtual and erect image.

11. A 5 mm high object is 30 mm from a convex mirror of 25 mm radius. A concave mirror of equal radius is 50 mm away. Find (a) the position and (b) the size the image formed after reflection at each mirror. Construct the image point $\mathbf{Q}'_2$.

Ans.: (a) $A_2M'_2 = 15.873$ mm, (b) $y'_2 = -0.397$ mm. The image is real and inverted.

12. An object 10 mm high is 200 mm in front of a thin positive lens of 100 mm focal length. A convex mirror of 75 mm focal length is 50 mm behind the lens.
Find (a) the position and (b) the size the image formed after refraction by the lens, reflection by the mirror and refraction by the lens.

Ans.: A real inverted image of the same size as the object, in the object plane.

13. An object is placed 80 mm in front of a +50 mm focal length lens. A convex mirror follows the lens and is 60 mm from it. A real inverted image of the same size as the object is formed in the object plane. What is the focal length of the mirror? Draw rays.

Ans.: -36.67 mm

14. An object is placed 100 mm in front of a +50 mm focal length lens. A spherical mirror is located 200 mm from the lens. A real inverted image of the same size as the object is formed in the object plane. What is the focal length of the mirror? Draw the rays.

Ans.: concave mirror of 50 mm focal length.

CHAPTER 9

ASTIGMATIC LENSES

9.1 ASTIGMATIC IMAGES

The refracting power of a spherical surface is related to the radius of curvature,

$$F = \frac{n' - n}{r} \qquad 9.1$$

Because the curvature of the surface is the reciprocal of the radius, i.e., **R = 1/r,** refracting power is related to surface curvature,

$$F = (n'-n)R \qquad 9.2$$

Consequently, investigating the curvature of certain surfaces that are of optical and, in particular, optometric interest will be useful. They are spherical, cylindrical and toroidal surfaces of revolution. A sphere is generated by the revolution of a circle about an axis in the plane of the circle through the circle's center of curvature. Paraxial theory shows that spherical refracting surfaces form *point* images because these surfaces have constant curvatures in all meridians. The spherical surface is rotationally symmetric about the optical or X-axis. See Figure 9.1.

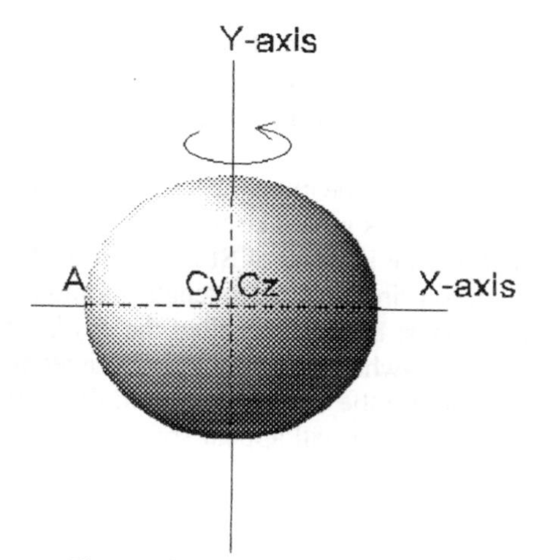

Figure 9.1 A sphere generated about the Y-axis, ($r_y = r_z$).

A cylinder is generated by a line parallel to the cylinder axis. The surface of a cylinder generated about the **y**-axis has zero curvature in the **xy** or *tangential* section and maximum curvature in the **xz** or *sagittal* section. See Figure 9.2. *The section of zero curvature is orthogonal or 90⁰ to the section of maximum curvature.* These sections of maximum and minimum curvature are the *principal* sections of the cylinder. A point object will be imaged by a cylindrical refracting surface as a *line parallel to the cylinder* axis. The cylinder has bilateral symmetry about the **xy** or tangential section.

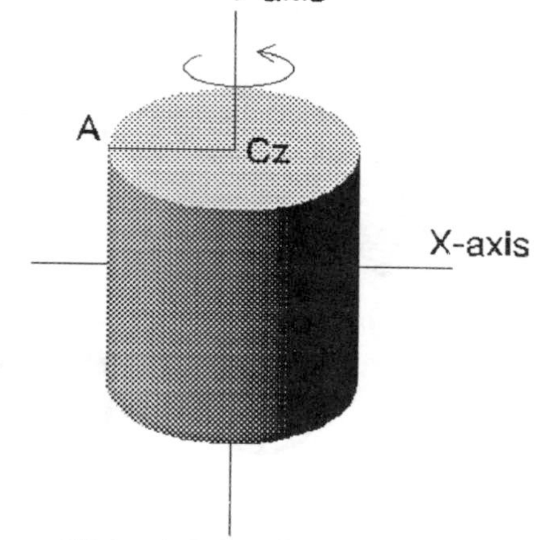

Figure 9.2 A cylinder generated about the Y-axis, ($r_y = \infty$).

A donut shaped *toric* surface will be generated by a circular disk attached to an axis of revolution, as seen in Figure 9.3a. The donut shape is produced when the radius of the disk is shorter than the distance **AO** from the vertex to the generating axis, i.e., **AC<AO**. When the radius of the disk is greater than the distance to the generating axis, that is, **AC>AO**, the toric surface is barrel or football shaped. See Fig. 9.3b. If **AC** is negative, the toric surface becomes a pulley or capstan.

Of course, any conic curve, such as an ellipse or parabola, may be substituted for the circle to produce aspheric torics.

A toric surface, generated about the **y**-axis, also has bilateral symmetry. The two principal sections of the toric surface intersect at the vertex **A**. In the donut shaped toric shown in Figure 9.3a the tangential radius of curvature $\mathbf{r}_y$ is shorter than the sagittal radius of curvature $\mathbf{r}_z$. This surface forms two line images of a point at right angles to each other and at different distances along the axis. The horizontal line focus is produced by the tangential curvature and lies closer to the surface. Refraction by the sagittal curvature produces the vertical line focus; it is farther from the surface. The distance between these line foci is called the *Interval of Sturm*. Line foci are *astigmatic* images, where the prefix *a* means not, and *stigma* means point.

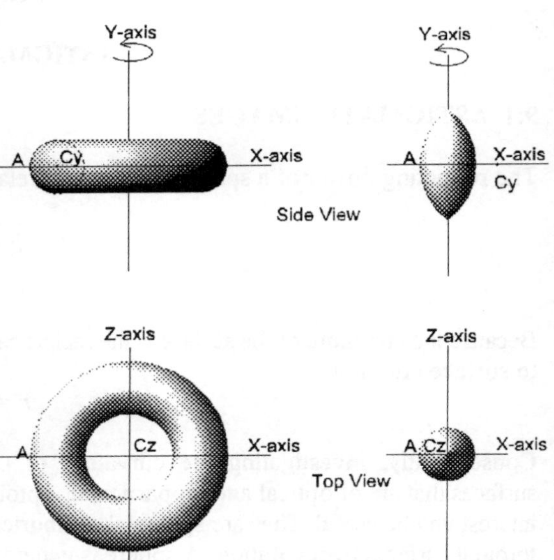

Figure 9.3 Torics surfaces generated about the Y-axis: (a) $r_y < r_z$; (b) $r_y > r_z$.

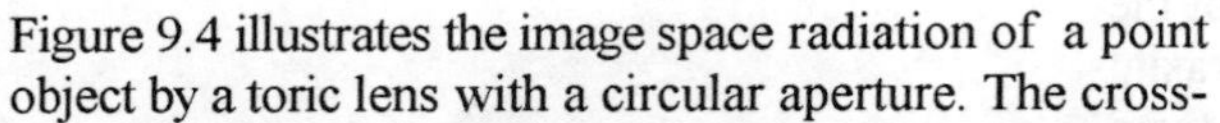

Figure 9.4 illustrates the image space radiation of a point object by a toric lens with a circular aperture. The cross-sections along the interval of Sturm are transformed from a vertical line at V that widens into a vertical oval, becomes a circle at C, then lengthens into a horizontal oval and line. The circular cross-section, where the image is least blurred, is called the *circle of least confusion (clc)*. All points on an extended object, such as a spoked-wheel, will be imaged as intervals of Sturm by an astigmatic lens. On a surface at one end of the interval are the vertical line foci; the horizontal line foci will be formed at the other end. Vertical elements in the object will appear sharp in the former plane because the focal lines are aligned with these elements, horizontal elements will appear sharp at the latter plane.

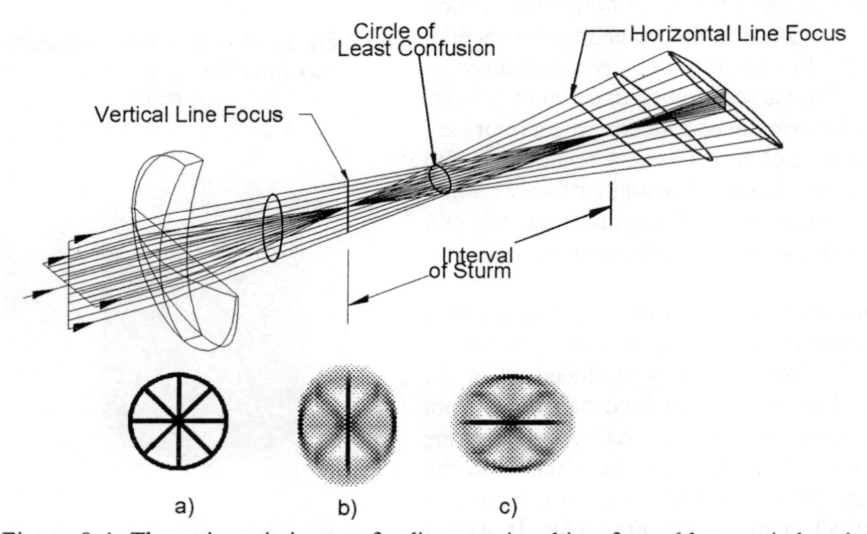

Figure 9.4 The astigmatic image of a distant point object formed by a toric lens is shown at the top. Below: (a) The object is a spoked wheel. (b) and (c) are astigmatic images at the vertical line and horizontal line foci, respectively.

9.2 CURVATURE AND POWER IN PRINCIPAL AND NORMAL SECTIONS

Let the tangential and sagittal sections of a toric surface be the *principal sections*. As shown in Figure 9.5, they correspond to *minimum* and *maximum* curvature, respectively. Their centers of curvature, c_z and c_y lie on the optical axis, and their curvatures are R_y and R_Z. Any other normal section, i.e., a section containing the optical axis, at an inclination θ will have a center of curvature c_θ that lies on the axis between c_z and c_y and a curvature R_θ, where

$$R_\theta = R_y \cos^2\theta + R_z \sin^2\theta \qquad 9.3$$

Angle θ is measured from the *y-axis*. The term *meridian* is used to denote a section. In a meridian 90° from the above section, i.e., at (θ+90°) the curvature is

$$R_{(\theta+90)} = R_y \sin^2\theta + R_z \cos^2\theta \qquad 9.4$$

By adding equations 9.3 and 9.4 we will obtain the sum of the refracting powers in the two normal sections:

$$R_\theta + R_{(\theta+90)} = R_y + R_z \qquad 9.5$$

Thus, the sum of the curvatures of any two normal sections at 90° to each other (orthogonal) is constant and equal to the sum of the principal curvatures.

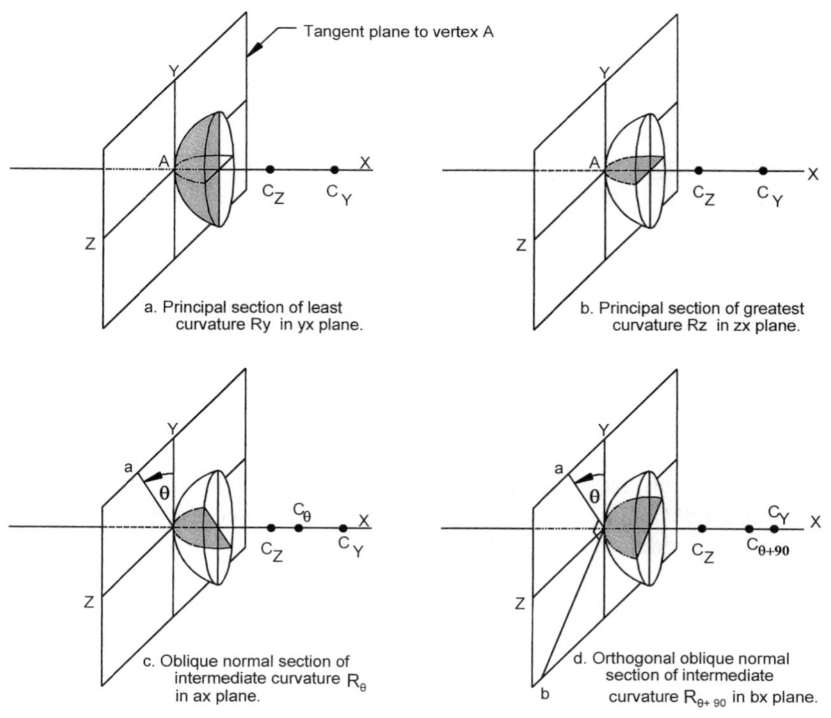

Figure 9.5 Normal sections of a toric surface.

Example 1: The principal curvatures of a toric surface that separates air from glass of index 1.5 are R_y = +4.00D and R_z = +10.00 D. The curvature R_{30} = +5.50 D. Find the curvature at the 120° meridian.

According to Eq. 9.5, $$R_{\theta+90} = R_y + R_z - R_\theta$$

$$R_{120} = +4.00 + 10.00 - 5.50 = +8.50 \text{ D}$$

If *curvature* varies from a maximum to a minimum, so too will *power* vary because refracting power is proportional to curvature. In other words, $F_y = (n'-n)R_y$, $F_z = (n'-n)R_z$, and $F_\theta = (n'-n)R_\theta$

Thus, we can rewrite Eqs. 9.3, 9.4, and 9.5 in refracting power:

$$F_\theta = F_y \cos^2\theta + F_z \sin^2\theta \tag{9.3a}$$

$$F_{(\theta+90)} = F_y \sin^2\theta + F_z \cos^2\theta \tag{9.4a}$$

$$F_\theta + F_{(\theta+90)} = F_y + F_z \tag{9.5a}$$

This last equation says that the sum of the refracting powers in any two orthogonal normal sections is a constant and equal to the sum of the principal refracting powers.

A point object **M** will be imaged at $\mathbf{M'_y}$ and at $\mathbf{M'_z}$ as horizontal and vertical lines by the tangential and sagittal sections of the toric surface. The applicable vergence equations for meridional refraction by the surface are:

$$U'_y = U_y + F_y \tag{9.6a}$$

$$U'_z = U_z + F_z \tag{9.6b}$$

and $$U'_\theta = U_\theta + F_\theta \tag{9.6c}$$

The object distance for all meridians is the same for a *point* object, therefore, the vergences $U_y = U_z = U_\theta = U$. An astigmatic image that serves as an object for successive optical elements will have vergences that vary with meridian, i.e., $U_y \neq U_z \neq U_\theta$.

9.3 POWER IN OBLIQUE MERIDIANS OF A CYLINDER

How does power vary with meridian for the cylindrical surface? As shown in Figure 9.6, the y-axis corresponds to the axis of the cylindrical surface. The curvature and power are zero along the y-axis, i.e., the tangential meridian. Curvature and power are maximal along the z-axis, i.e., the sagittal meridian. The power of a cylinder is designated ***F***. It is the maximal power, and in this case lies along the sagittal section of the surface. The sag of this section is AD = s, and DE = h is the semi-chord. At an oblique meridian θ, where *θ is measured from the axis of the cylinder*, i.e., the **y-axis**, the sag is also equal to **AD = s**, but the semi-chord **DP = $\mathbf{h_\theta}$**. According to the approximate equation for sags, the curvatures of the sagittal and oblique sections are

$$R = \frac{2s}{h^2}, \qquad R_\theta = \frac{2s}{h_\theta^2}$$

From which we find, $\dfrac{R_\theta}{R} = \dfrac{h^2}{h_\theta^2} = \sin^2\theta$

thus, $R_\theta = R\sin^2\theta$, or $\boldsymbol{F_\theta = F\sin^2\theta}$ 9.7a

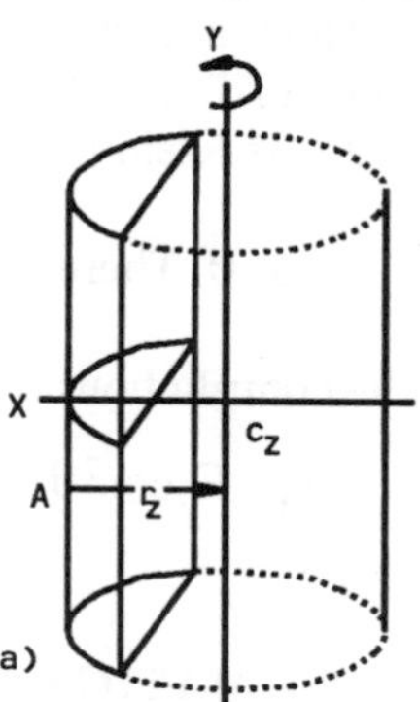

The power in meridian θ+90 may be found in a similar way. According to Figure 9.7,

$$\frac{R_{(\theta+90)}}{R} = \frac{h^2}{h_{(\theta+90)}^2} = \cos^2\theta$$

and $\boldsymbol{F_{\theta+90} = F\cos^2\theta}$ 9.8

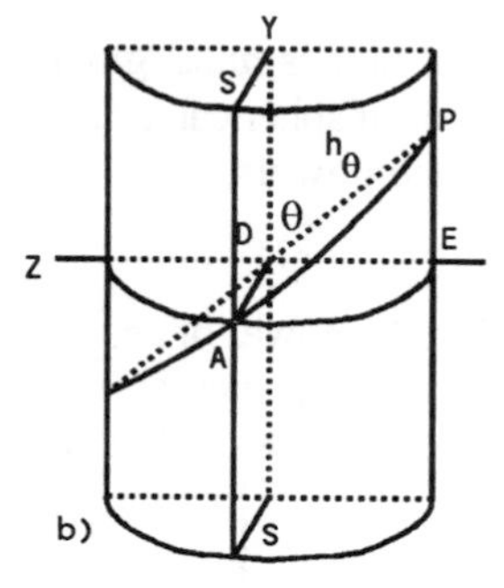

Add Eqs. 9.7 and 9.8, $F_\theta + F_{(\theta+90)} = F\sin^2\theta + F\cos^2\theta$

Consequently,

$$\boxed{F_\theta + F_{(\theta+90)} = F} \qquad 9.9$$

*This says that the sum of the powers in any two perpendicular meridians of a cylindrical refracting surface is a constant that is equal to the maximum or cylinder power **F**.*

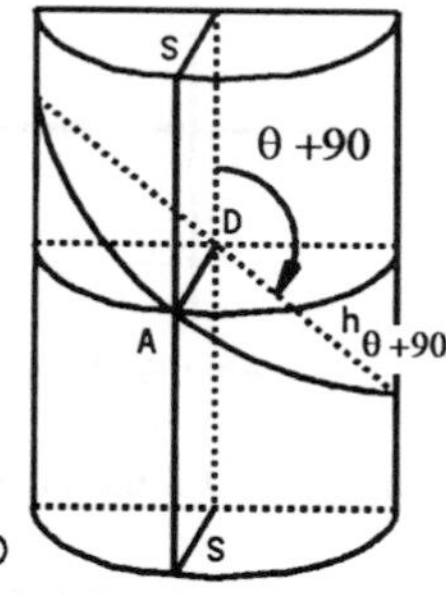

Figure 9.6 Sections of a cylinder: (a) principal section with radius r_z; (b) normal oblique section at angle θ; (c) normal oblique section at angle (θ+90°).

9.3.1 THE OBLIQUE POWER EQUATION

The equation for power along an oblique meridian of a toric surface, Eq. 9.3a, can be derived by considering the toric surface as equivalent to a cross-cylinder. That is, two cylinders with their axes 90° apart. Let the first cylinder be oriented with its axis along the Y-axis, as shown in Fig. 9.6. The power of this cylinder is along the z meridian.

$$F_{cyl} = F_z$$

According to Eq. 9.7a the power at an oblique meridian is $F_\theta = F_z \sin^2\theta$, where the oblique meridian AP ≡ DP at angle θ is measured from the cylinder axis (Y-axis). See Fig 9.7.

The axis of a second cylinder is along the Z-axis. Its power is $F_{cyl} = F_y$. Nevertheless, the oblique power to be found is still along AP. This meridian is at an angle of 90-θ from the axis of *this* cylinder. Thus, according to Eq. 9.7a, the power at meridian AP is

$$F_{(\theta)} = F_y \sin^2(90-\theta)$$

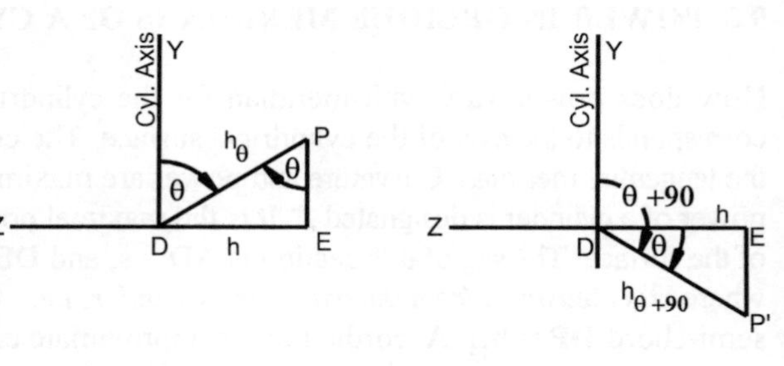

Figure 9.7 Diagram for solving for power along normal oblique sections at angles θ and θ+90°.

But sin(90-θ) = cos θ. Therefore, $F_{(\theta)} = F_y \cos^2\theta$ 9.7b

The sum of the contributions of both cylinders to the oblique power along meridian AP is

$$F_\theta = F_y\cos^2\theta + F_z\sin^2\theta \qquad (cf.\ Eq.\ 9.3a)$$

9.4 THIN ASTIGMATIC LENSES

A spherical lens is *an*astigmatic, i.e., it forms point images. The two surfaces of astigmatic lenses are combinations of spherical, cylindrical and toric surfaces. Since the power of a thin lens is equal to the sum of the surface powers, i.e.,

$$\boldsymbol{F} = \boldsymbol{F_1} + \boldsymbol{F_2},$$

we may find the power of an astigmatic lens by adding the surface powers along any given meridian. The powers in the two principal meridians for each surface are shown on the *optical cross diagrams* for each surface. See Fig. 9.8. Note that *the power meridian is 90° to the cylinder axis.*

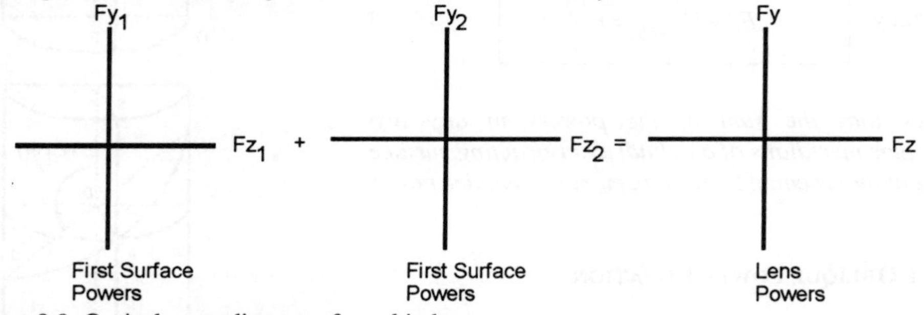

Figure 9.8 Optical cross diagrams for a thin lens.

The tangential power is $F_y = F_y1 + F_y2$ 9.10a

The sagittal power is $F_z = F_z1 + F_z2$ 9.10b

The oblique power, per Eq. 9.3, is $F_\theta = F_y\cos^2\theta + F_z\sin^2\theta$ 9.3a

The vergence equations are applicable to a toroidal lens by simply substituting the tangential, sagittal and oblique powers (Eqs. 9.10) as appropriate. That is, given a *point* object **M**, the vergence equations are,

$$U'_y = U + F_y, \qquad U'_z = U + F_z \qquad U'_\theta = U + F_\theta \qquad 9.11$$

9.4.1 Forms of Astigmatic Lenses

Astigmatic lenses can be manufactured in four general forms: plano-cylindrical, cross-cylindrical, sphero-cylindrical, and toric. Spectacle lenses commonly are toric. We will apply Eqs 9.10 and 9.11 to these forms.

1. A plano-cylinder has a cylindrical and a flat surface. The cylinder may be convex or concave, as shown in Figure 9.9. According to the diagram, for the convex cylinder,

the tangential power is, $\quad F_y = F_y1 + F_y2 = 0 + 0 = 0$

the sagittal power is, $\quad F_z = F_z1 + F_z2 = F_z1 + 0 = F$ (the cylinder power)

and the oblique power (Eq. 9.3b), is $\quad F_\theta = F_y\cos^2\theta + F_z\sin^2\theta = F_{cyl}\sin^2\theta$

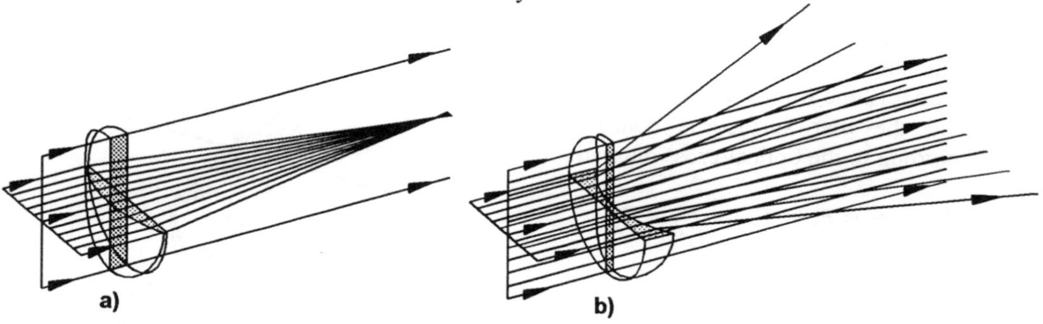

Figure 9.9 (a) Plano-convex and (b) plano-concave cylindrical lenses. The tangential rays encounter zero power; the sagittal pencils are converged or diverged as shown.

2. The **cross-cylinder** is a lens that is ground with cylindrical surfaces on front and back. In Figure 9.10, the front surface cylinder axis is at 180° and the rear surface cylinder axis is at 90°.

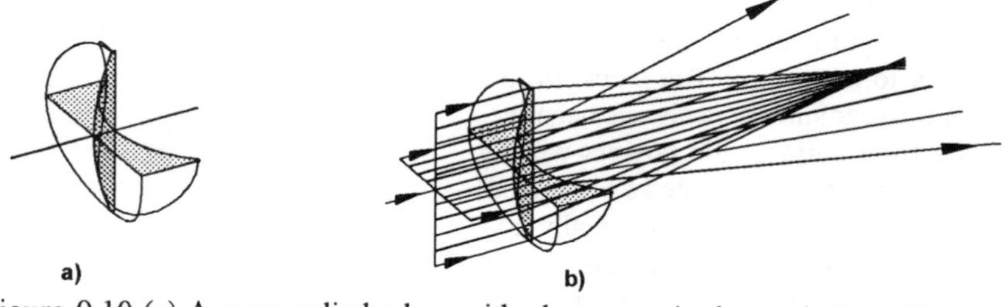

Figure 9.10 (a) A cross-cylinder lens with plus power in the vertical meridian, and minus power in the horizontal meridian. (B) Ray pencils.

Figure 9.11 illustrates a lens with a convex front-surface cylinder, generated about a vertical axis, and a concave rear-surface cylinder with a horizontal axis. See the optical crosses for the meridional powers.

The tangential power is, $\boldsymbol{F_y} = \boldsymbol{F_y}1 + \boldsymbol{F_y}2 = 0 + \boldsymbol{F_y}2 = \boldsymbol{F_{y2}}$

The sagittal power is, $\boldsymbol{F_z} = \boldsymbol{F_z}1 + \boldsymbol{F_z}2 = \boldsymbol{F_z}1 + 0 = \boldsymbol{F_z}1$

The oblique power is, $\boldsymbol{F_\theta} = \boldsymbol{F_y}\cos^2\theta + \boldsymbol{F_z}\sin^2\theta = \boldsymbol{F_y}2\cos^2\theta + \boldsymbol{F_z}1\sin^2\theta$

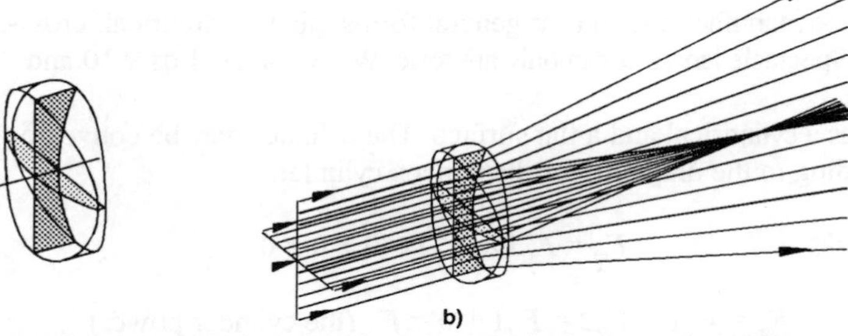

Figure 9.11 (a) A cross cylinder with plus power in the horizontal meridian and minus power in the vertical meridian. (b) Ray pencils.

3. A **sphero-cylinder**, as its name implies, is a lens form comprising a cylindrical and a spherical surface. In Figure 9.12 the front surface is convex-cylindrical and the rear surface is concave-spherical. According to the optical-cross diagram,

The tangential power is $\boldsymbol{F_y} = \boldsymbol{F_y}1 + \boldsymbol{F}_2 = \boldsymbol{F}_2$

The sagittal power is $\boldsymbol{F_z} = \boldsymbol{F_z}1 + \boldsymbol{F}_2$

The oblique power is $\boldsymbol{F_\theta} = \boldsymbol{F}_2\cos^2\theta + (\boldsymbol{F}_{z1}+\boldsymbol{F}_2)\sin^2\theta$

and, $\boldsymbol{F_\theta} = \boldsymbol{F_z}1\sin^2\theta + \boldsymbol{F}_2$

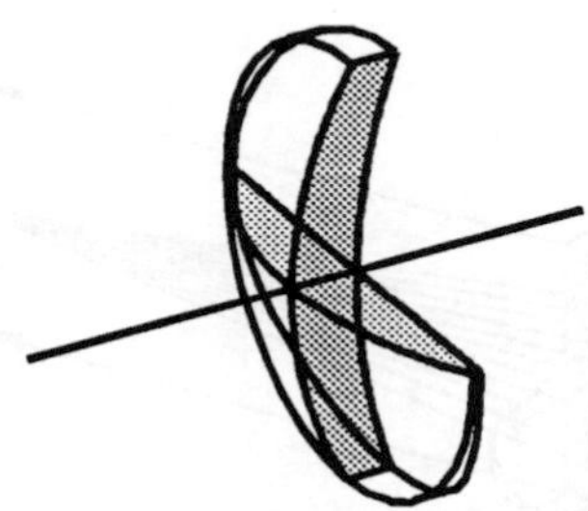

Figure 9.12 A sphero-cylindrical lens. The front surface is spherical, the rear surface is a cylinder with axis at 180°.

4. **Toric lenses** are the most common form used for ophthalmic lenses. They are generally meniscus lenses with a spherical and a toric surface. The front-surface power is positive and the rear surface is always negative. If the toric is on the front surface, the powers will be maximally and minimally positive along the two principal meridians. A rear toric surface will be maximally and minimally negative at the two principal meridians. Figure 9.13 illustrates a very unlikely spectacle form. The front surface is a saddle-shaped toric and the rear surface is concave-spherical. The student should draw the optical-cross diagram and find the principal and oblique powers for a toric lens.

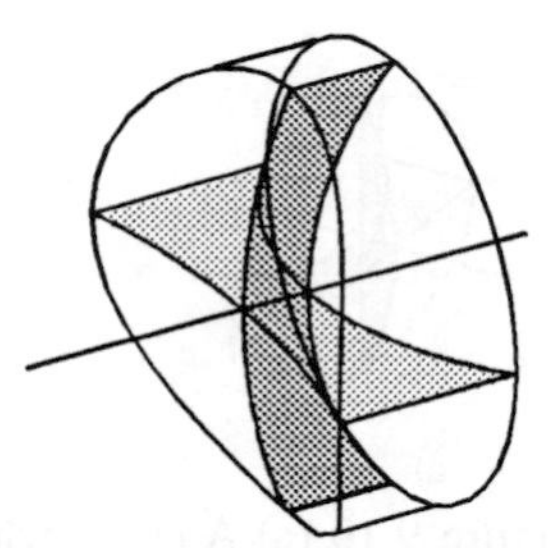

Figure 9.13 A toric lens with a saddle shaped front surface and spherical rear surface.

9.5 TRANSPOSITION of FLAT PRESCRIPTIONS

Lens prescriptions generally are written as sphero-cylinders, in either minus or plus cylinder form. For example, the prescription of a +2 diopter sphere, combined with a -1 diopter cylinder with its axis at 90° is written as follows:

$$+2.00 \text{ DS} \circ -1.00 \text{ DC x } 90$$

where the symbol $\circ$ is read "combined with". This Rx is in *minus-cylinder* form. Because the first number represents the spherical component, and the second number, next to the axis symbol is the cylindrical component, the Rx may be written simply as:

$$+2.00\ -1.00\text{x}90.$$

This notation describes a *flat prescription*. It does not suggest the actual form (toric, sphero-cylinder, cross-cylinder) much less the curvatures to be used in fabricating the lens. Identical prescriptions may be written in minus-cylinder, plus-cylinder and cross-cylinder forms. For example, the above prescription in plus cylinder form is

$$+1.00 \text{ DS} \circ +1.00 \text{ DC x } 180$$

and in cross-cylinder form it is,

$$+2.00 \text{ DC x } 180 \circ +1.00 \text{ DC x } 90$$

The procedure for writing the same prescription in these three forms is called *flat transposition*. The prototype Rx is

$$Q_{sph} \circ P_{cyl} \text{ x } \phi \qquad 9.12$$

1. Rules to transpose one sphero-cylinder to another.

(a) the new sphere equals Q + P; (b) the new cylinder equals -P; (c) the new axis equals $\phi \pm 90°$.

or

$$(Q+P)_{sph} \circ -P_{cyl} \text{ x } (\phi \pm 90°) \qquad 9.13$$

Example 2: Transpose -4.00 DS $\circ$ -2.00 DC x 135° to plus cylinder form.

Result: -6.00 DS + 2.00 DC x 45°

Note: The decision to add or subtract 90° must result in an axis between 0 and 180°.

2. Rules to transpose a sphero-cylinder to a cross-cylinder.

(a) Q_{sph} becomes the Q_{cyl}, (b) its axis equals $\phi \pm 90°$, (c) Q+P becomes the second cyl. (d) axis equals ϕ.

or

$$Q_{cyl} \text{ x } (\phi \pm 90°) \circ (Q+P)_{cyl} \text{ x } \phi$$

Example 3: Transpose the minus cylinder Rx in the previous example to cross-cylinder form.

Result: -4.00 DC x 45 $\circ$ -6.00 DC x 135

3. Rules to transpose a cross-cylinder to sphero-cylinder form. The prototype flat Rx of a cross-cylinder is

$$P_{cyl} \times \phi ⊂⊃ R_{cyl} \times (\phi \pm 90°) \quad 9.14$$

(a) P cyl becomes the P sph., (b) the new cylinder equals R-P, (c) the new axis is the same as the R cylinder axis, i.e., $\phi \pm 90°$.

or $\quad P_{sph} ⊂⊃ (R-P)_{cyl} \times (\phi \pm 90°)$

Example 4: Transpose the previous cross-cylinder back to sphero-cylinder form.

Result: -4.00 DS⊂⊃-2.00 DC x 135

9.6 THE CIRCLE OF LEAST CONFUSION AND THE SPHERICAL EQUIVALENT LENS

The circle of least confusion (**clc**) is that part of the interval of Sturm where there is a minimum blur. In Figure 9.14, the rays in the two principal sections of the lens have been superposed. The **clc** is at the narrowest (waist) diameter of the light distribution. According to this diagram

$$\frac{d}{D} = \frac{a}{f_y} = \frac{b}{f_z}, \text{ where, } a = f_c - f_y, \text{ and } b = f_z - f_c.$$

Thus,

$$\frac{f_c - f_y}{f_y} = \frac{f_z - f_c}{f_z}$$

Substituting power for focal length results in

$$F_c = \frac{F_y + F_z}{2} \quad 9.15$$

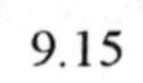

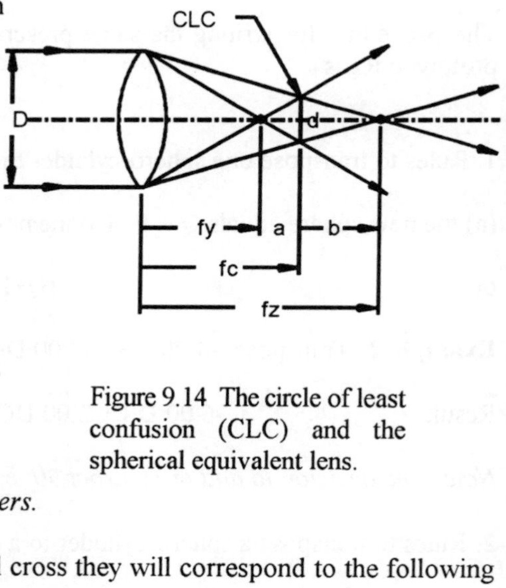

Figure 9.14 The circle of least confusion (CLC) and the spherical equivalent lens.

*The **clc** is dioptrically at the mean of the two principal powers.*

Now if we put the principal powers F_y and F_z on an optical cross they will correspond to the following sphero-cylinder:

$$F_{y\,sph} ⊂⊃ (F_z - F_y)_{cyl} \times \phi$$

It is sometimes useful to approximate an astigmatic prescription or to compare various sphero-cylinder lens corrections by using a *spherical equivalent (SE)* lens. The spherical lens most nearly equal to a sphero-cylinder, is found by adding one-half the cylinder power to the sphere,

$$SE = Q + \frac{P}{2} \quad 9.16$$

where, $Q = F_y$, and $P = F_z - F_y$. Therefore, the spherical equivalent can be rewritten as

$$SE = F_y + \frac{F_z - F_y}{2}$$

However, this is identical to Eq.9.15, i.e., *the spherical equivalent Rx places the* ***circle of least confusion*** *on the retina.*

$$SE = \frac{F_y + F_z}{2} = F_c \qquad 9.17$$

9.7 THE CIRCLE OF LEAST CONFUSION FOR ANY OBJECT DISTANCE

The ray paths, for a near object, in the two principal sections of a toric lens are superposed in Figure 9.15. According to this diagram

$u_c' = u_y' + a$, therefore, $a = u_c' - u_y'$

and $u_z' = u_c' + b$, therefore, $b = u_z' - u_c'$

From similar triangles and the above equations, we find

$$\frac{d}{D} = \frac{a}{u_y'} = \frac{u_c' - u_y'}{u_y'}$$

and

$$\frac{d}{D} = \frac{b}{u_z'} = \frac{u_z' - u_c'}{u_z'}$$

thus,

$$\frac{u_c' - u_y'}{u_y'} = \frac{u_z' - u_c'}{u_z'}$$

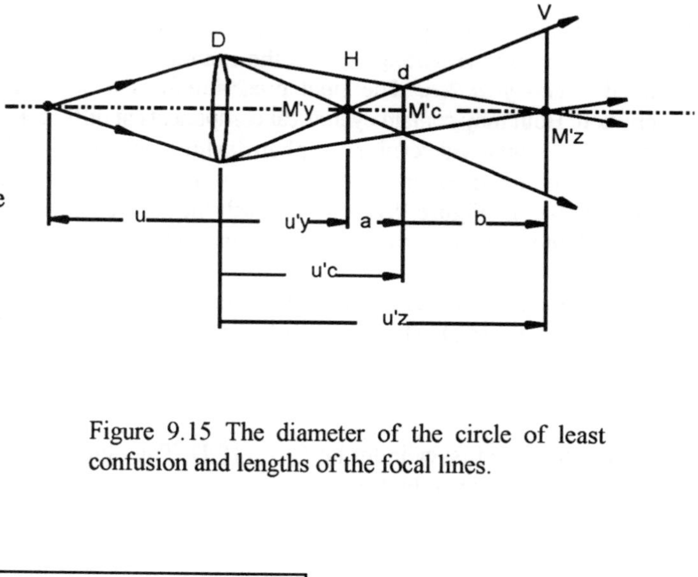

Figure 9.15 The diameter of the circle of least confusion and lengths of the focal lines.

From which we find

$$U_c' = \frac{U_y' + U_z'}{2} \qquad 9.18$$

The dioptric position of the **clc** *is equal to the mean of the dioptric positions of the two line foci.* Do not confuse this with the geometrical midpoint of the Interval of Sturm.

9.8 THE LENGTHS OF THE FOCAL LINES AND THE DIAMETER OF THE CLC

In Figure 9.15 let us assume that the lens has a circular shape, and that the focal line H is formed by the tangential section of the lens. It will be a horizontal line image. Focal line V will be formed by the sagittal section and will be a vertical line image. Evidently, from the similar triangles

$$\frac{H}{D}=\frac{u_z^{\prime}-u_y^{\prime}}{u_z^{\prime}};\qquad \frac{V}{D}=\frac{u_z^{\prime}-u_y^{\prime}}{u_y^{\prime}};\qquad \frac{d}{D}=\frac{u_z^{\prime}-u_c^{\prime}}{u_z^{\prime}}$$

From which we will find the lengths of the line foci and the diameter of the **clc** to be

$$H=D\frac{U_y^{\prime}-U_z^{\prime}}{U_y^{\prime}};\qquad V=D\frac{U_y^{\prime}-U_z^{\prime}}{U_z^{\prime}}\qquad d=D\frac{U_y^{\prime}-U_z^{\prime}}{U_y^{\prime}+U_z^{\prime}} \tag{9.19}$$

9.9 OBLIQUELY COMBINED CYLINDERS

Patients with very strong sphero-cylindrical lens corrections can be refracted advantageously while wearing their lenses because the starting point of the examination is very near their best correction. This procedure is called *over-refraction.* The final prescription is a combination of the new findings and the old prescription. Both may contain cylindrical components at different axes. It can be shown that two obliquely combined cylinders will result in a sphero-cylindrical Rx. It will suffice to show graphically that, just as the sum of the powers of any two orthogonal sections of two crossed-cylinders equal the sum of the two principal powers, (Figure 9.16), so too will the sum of any two orthogonal sections of two *obliquely-crossed* cylinders (Figure 9.17). The combination is equivalent to a sphero-cylindrical Rx.

In Figure 9.16c, the cross-cylinder comprises a +10 DC axis 180, (Figure 9.16a) combined with a +5 DC axis 90°, (Figure 9.16b). This Rx is written as follows:

+10 DC x 180 ⌔ +5 DC x 90

Let the power of the first cylinder be $F_1 = +10$ D. The oblique powers at angle θ (measured from the axis of the cylinder) are given by Eq. 9.7a.

$$F_\theta = F_1 \sin^2\theta$$

The power contributed by the second cylinder, $F_2 = +5$ D, at the same meridians is given by Eq. 9.7b.

$$F_\theta = F_2 \cos^2\theta$$

		Meridian									
Thus, at	$\theta =$	0°	15°	30°	45°	60°	75°	90°	105°	120°	
Power due to	$F_1 =$	0.0	0.67	2.5	5.0	7.50	9.33	10.0	9.33	7.5	Diopters
Power due to	$F_2 =$	5.0	4.67	3.75	2.5	1.25	0.33	0.0	0.33	1.25	Diopters
Total power	$F_\theta =$	5.0	5.33	6.25	7.5	8.75	9.66	10.0	9.66	8.75	Diopters

As noted above, the sum of the powers in any two orthogonal meridians equals the sum of the powers of the two cylinders. For example

$$F_{\theta}+F_{(\theta+90)}=F_{y}+F_{z}$$

$$F_{30}+F_{120}=F_{1}+F_{2}$$

$$+6.25+8.75=+10+5=+15$$

A similar graphical analysis, for *obliquely* combined cylinders is shown in Figure 9.17. Such a prescription is written as follows:

$$F_{1\,cyl} \text{ x } \phi \circ F_{2\,cyl} \text{ x } (\phi+\gamma)$$

where, F_1 is the power of the cylinder with the smaller axis slope.
F_2 is the power of the other cylinder
ϕ is the meridian along which the axis of cylinder F_1 lies.
γ is the angle between the axes of the two cylinders

The object is to convert these cylinders into **$Q_{sph} \circ P_{cyl}$ x ($\phi+\alpha$),** where, α is measured from the axis of cylinder F_1.

We will graphically analyze the following prescription: +10 DC x 60 $\circ$ +10 DC x 90. The power contributed by each cylinder at oblique meridians with respect to their axes, as found with the $F \sin^2\theta$ equation, is shown in Figures 9.17a and b. When the powers are summed at each meridian, as shown in Figure 9.17c and tabulated below, where 0° corresponds to the 0-180° meridian, we obtain:

	Meridian							
θ* =	0°	15°	30°	45°	60°	75°	90°	
Power due to F_1 =	7.5	5.0	2.5	0.67	0.0	0.67	2.5	Diopters
Power due to F_2 =	10.0	9.33	7.5	5.0	2.5	0.67	0.0	Diopters
Total power F_θ =	17.5	14.33	10.0	5.67	2.5	1.33	2.5	Diopters

* Angle measured from the x-axis.

Power is zero along the axes of each cylinder. The angle between the axes is acute, therefore, we should expect the minimum power of the obliquely combined cylinders to lie between these axes. Because we chose $F_1 = F_2 = +10$ D, the meridian of minimum power will be midway between these axes, i.e., at 75°. Minimum power is +1.33 D.

Similarly, the power of each cylinder is maximum at 90° to the axes, i.e., at 150 and 180°. Therefore, the meridian of maximum power of the obliquely combined cylinders will be at 165°. According to Figure 9.17c maximum power is +18.67 D. The sum of these powers and, in fact, the sum of the powers of any two orthogonal meridians equals $F_1 + F_2$.

$$F_{\theta}+F_{(\theta+90)}=F_{y}+F_{z}$$

$$F_{min}+F_{max}=F_{1}+F_{2}$$

$$+1.33+18.67=+10+10=+20$$

This means that even obliquely combined cylinders may be replaced with a sphero-cylinder lens.

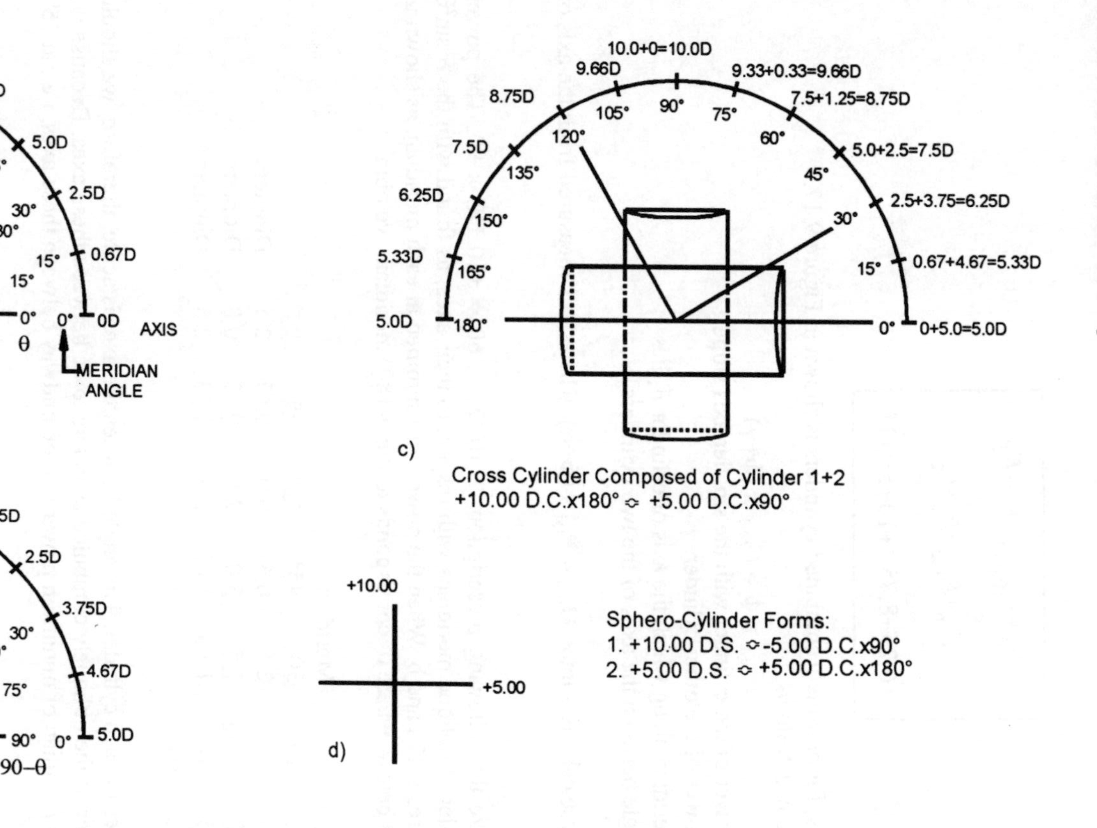

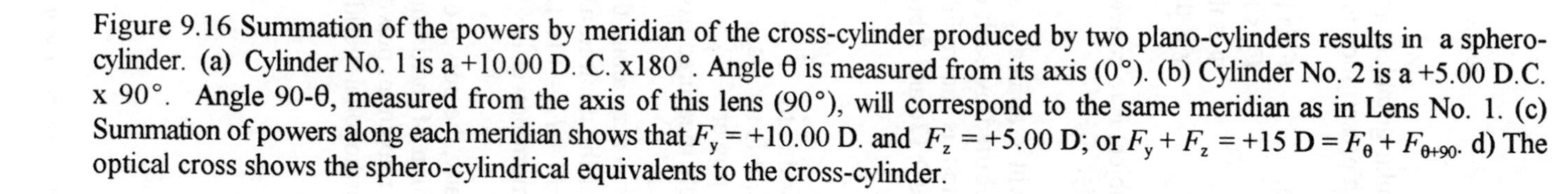

Figure 9.16 Summation of the powers by meridian of the cross-cylinder produced by two plano-cylinders results in a sphero-cylinder. (a) Cylinder No. 1 is a +10.00 D. C. x180°. Angle θ is measured from its axis (0°). (b) Cylinder No. 2 is a +5.00 D.C. x 90°. Angle 90-θ, measured from the axis of this lens (90°), will correspond to the same meridian as in Lens No. 1. (c) Summation of powers along each meridian shows that $F_y = +10.00$ D. and $F_z = +5.00$ D; or $F_y + F_z = +15$ D $= F_\theta + F_{\theta+90}$. d) The optical cross shows the sphero-cylindrical equivalents to the cross-cylinder.

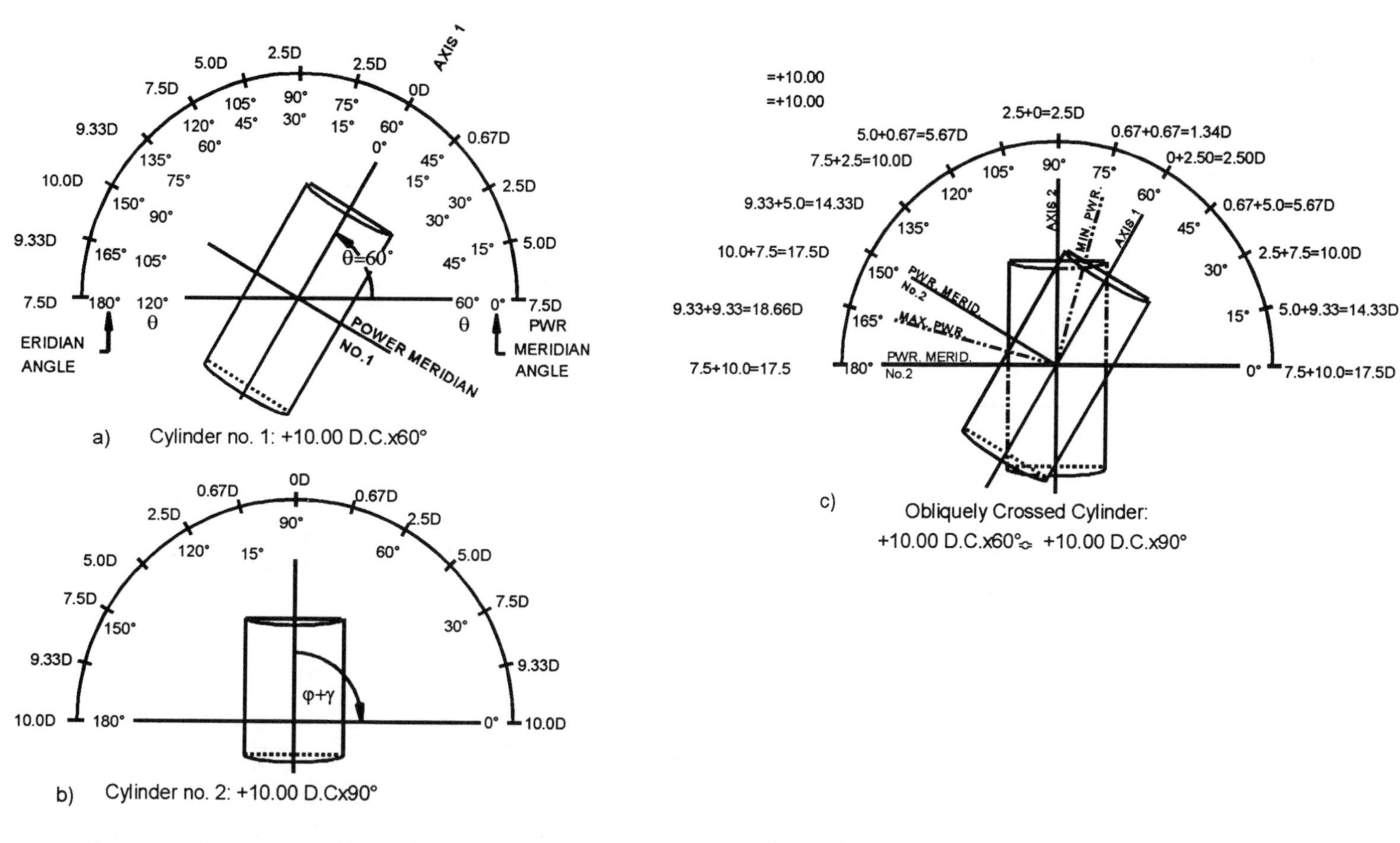

Figure 9.17 Summation of the powers by meridian of two obliquely crossed cylinders results in a sphero-cylinder. (a) Cylinder No. 1 is a +10.00 D. C. x60°, (b) Cylinder No. 2 is a +10.00 D.C. x 90°, (c) Summation of powers along each meridian shows that the sum of the powers in any two orthogonal meridians $F_\theta + F_{\theta+90} = +20$ D and, in particular, $F_{MAX} + F_{MIN} = +18.67$ D + 1.33 D = 20D. Note the meridian of minimum power lies between the axis 1 and axis 2; the meridian of maximum power lies between the power meridians of the cylinders.

Figure 9.18a summarizes the conditions for the obliquely crossed cylinders. The equivalent lens, shown on the optical cross in Figure 9.18b is

$$Q_{sph} \circ P_{cyl} \times \varphi \qquad 9.12$$

or

$$+1.33 \text{ DS} \circ +17.33 \text{ DC} \times 75^\circ$$

From the optical cross it is clear that F_{max} = Q + P = +18.67 D, and F_{min} = Q = +1.33 D. Therefore, the spherical power Q is directly obtained from

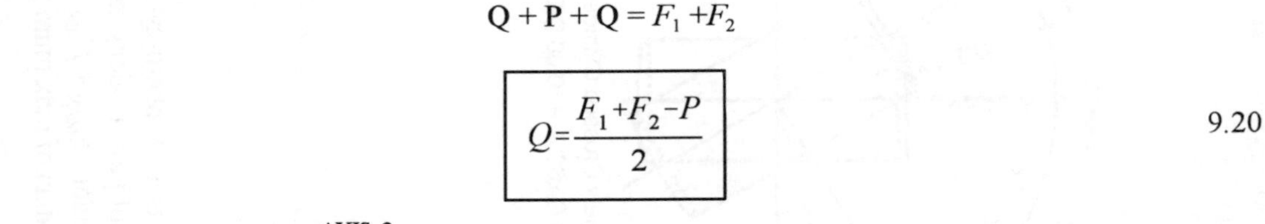

$$Q + P + Q = F_1 + F_2$$

or

$$\boxed{Q = \frac{F_1 + F_2 - P}{2}} \qquad 9.20$$

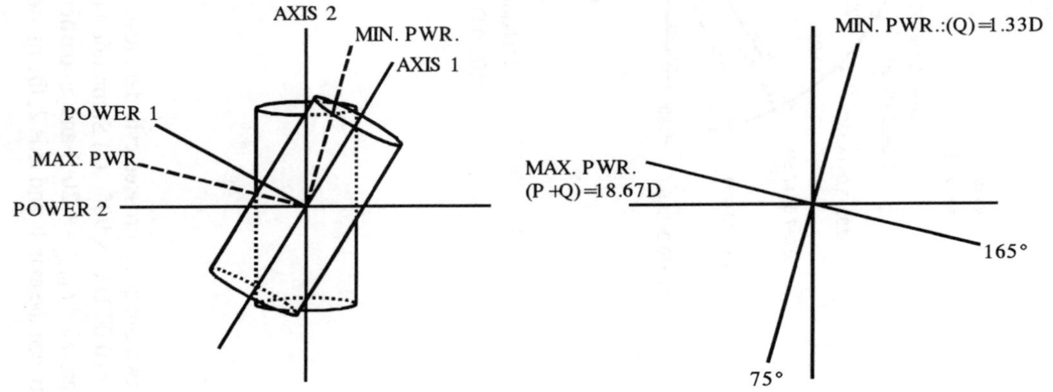

Figure 9.18 (a) The meridian of minimum power lies between the cylinder axes; the meridian of maximum power is orthogonal and between the power meridians. (b) Resultant sphero-cylinder.

Recalling that the oblique powers of a cylinder are referred to angles measured from their axes, Figure 9.18 shows that the minimum power is

$$F_{min} = F_\alpha = Q = F_1 \sin^2\alpha + F_2 \sin^2(\gamma-\alpha) \qquad 9.21a$$

The maximum power is $$F_{max} = F_{(\alpha+90)} = P + Q = F_1 \cos^2\alpha + F_2 \cos^2(\gamma-\alpha) \qquad 9.21b$$

Differentiating Eq. 9.19a and setting it equal to zero will yield,

$$\frac{F_1}{sin2(\gamma-\alpha)} = \frac{F_2}{sin2\alpha} \qquad 9.21c$$

The cylinder power **P** is equal to (P+Q)-Q, that is, Eq. 9.21b-9.21a.

$$P = F_1 \cos^2\alpha + F_2 \cos^2(\gamma-\alpha) - F_1 \sin^2\alpha - F_2 \sin^2(\gamma-\alpha)$$

Using the identity, $\cos^2 x - \sin^2 x = \cos 2x$, and eliminating the F_2 term by means of Eq.9.21c, we obtain

$$\frac{F_1}{sin2(\gamma-\alpha)} = \frac{P}{sin2\gamma}$$

But, this may be combined with Eq. 9.20 to obtain the following

$$\frac{P}{sin2\gamma} = \frac{F_1}{sin2(\gamma-\alpha)} = \frac{F_2}{sin2\alpha}$$

Inspection of the above equation shows that it is the Law of Sines with respect to twice the angles. Consequently, Figure 9.19 can be used to find **P** and angle **α**.

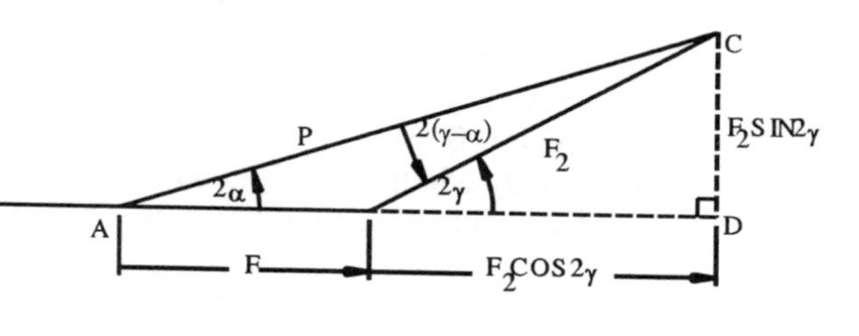

Figure 9.19 A vector diagram for constructing the cylindrical power P and angle α, based on doubling the angles α and γ.

As indicated in this figure, vectors AB $= F_1$ and BC $= F_2$ are drawn with an angle of 2γ between them. The cylindrical power P = vector AC. The Pythagorean theorem, applied to the right triangle ADC, yields

$$P^2 = AD^2 + DC^2$$

From which we obtain

$$P = +\sqrt{F_1^2 + F_2^2 + 2F_1F_2\cos 2\gamma} \qquad 9.22$$

The angle α is given by the triangle BDC,

$$\tan 2\alpha = \frac{F_2 sin2\gamma}{F_1 + F_2 cos2\gamma} \qquad 9.23$$

Because this equation solves for 2α it is important to set up problems in conformity with the conditions used in the above derivation. That is, obliquely combined cylinders should both be of the same sign and the angle between their axes should be acute. Otherwise, introducing an error of 90° in the axis of the new prescription is possible. Bennett's alternative equation for angle α, Eq. 9.24 given below, avoids this problem. Therefore, the following equations for finding Q, P, and α are recommended:

$$\boxed{P = +\sqrt{F_1^2 + F_2^2 + 2F_1F_2 cos2\gamma}} \qquad 9.22$$

$$\boxed{Q = \frac{F_1 + F_2 - P}{2}} \qquad 9.20$$

$$\boxed{\tan\alpha = \frac{-F_1 + F_2 + P}{F_1 + F_2 + P}\tan\gamma} \qquad 9.24$$

Procedure for combining oblique cylinders:

1. Transpose each Rx to cylinder forms of the same sign.
2. Chose the cylinder with the smaller axis for $F1$.
3. Compute γ = axis F_2 - axis F_1.
4. Solve for P: (Eq.9.20).
5. Solve for Q: (Eq. 9.21). Note: If the original cylinders also had spherical components, Q_1 and Q_2, add these to Q to obtain the total new spherical component.
6. Add α to ϕ to obtain the new axis. The new axis must lie between 0 and 180°.

Example 5: Combine -1.50 -1.50x30 and +4.00 - 4.00x150

Compare the magnitudes of the axes to choose F_1: 30° <150°, therefore, F_1 = -1.50,

$F_2 = -4.00$, $\gamma = 120°$, $\phi = 30°$, $Q_1 = -1.50$, and $Q_2 = +4.00$.

Solve for P (Eq. 9.20):	P = +3.50 D
Solve for Q (Eq. 9.21):	Q = -4.50 D
Solve for new sphere (add Q, Q_1 and Q_2):	Q sph = -2.00 D
Solve for α (Eq. 9.22):	$\alpha = 40.89°$
Add ϕ to find the new axis:	$\alpha + \phi = 70.89°$

The sphero-cylindrical Rx is: -2.00 +3.50x70.89°

REVIEW QUESTIONS

1. In which section is the maximum curvature of a cylinder generated about the **z** axis?
2. Distinguish between donut, football and capstan toric surfaces. Which have principal curvatures of the same sign?
3. Describe the Interval of Sturm. What is the circle of least confusion?
4. How are the powers in any two normal orthogonal sections of a toric surface related to the powers of the two principal sections? Does this relationship change if the surface is cylindrical?
5. List four astigmatic lens forms. How do they differ?
6. Does flat transposition necessarily indicate the lens form?
7. What is the value of the spherical equivalent lens?
8. Is the clc geometrically midway between the line foci of an astigmatic lens?
9. What method of eye refraction requires obliquely combining cylinders to determine the final correction lenses?

PROBLEMS

1. A cylindrical surface, generated about the **y**-axis, separates air from glass (n=1.5). The radius of the cylinder is 166.667 mm. (a) What is the refracting power at a normal oblique section at 60° to the axis? (b) At what meridian is this power? (c) What is the power in an orthogonal section?

Ans.: (a) 2.25 D, (b) 30°, (c) 0.75 D.

2. The principal curvatures of a toric surface separating air from glass (n=1.5) are +13.333 and +26.667 M^{-1}. Find the image distances corresponding to (a) the tangential power, (b) the sagittal power, (c) the clc and (d) the interval between the two line foci?

Ans.: (a) 225 mm, (b) 112.5 mm, (c) 150 mm, (d) 112.5 mm.

3. A toric surface between air and glass has radii r_y = +10 cm and r_z = +25 cm. Find (a) the principal powers, (b) the power in a section at 30° to the **y**-axis, and (c) the power at the orthogonal meridian.

Ans.: (a) F_y = +5 D, F_z = +2 D, (b) F_{30} = +4.25 D, (c) F_{120} = +2.75 D.

4. A toric surface between air and glass has radii r_y = +10 cm and r_z= -25 cm. Find (a) the principal powers, (b) the power in a section at 30° to the **y**-axis, and (c) the power at the orthogonal meridian.

Ans.: (a) F_y = +5 D, F_z = -2 D, (b) F_{30} = +3.25 D, (c) F_{120} = -0.25 D.

5. The principal powers of a toric surface separating air from glass (n=1.5) are, F_y = +10 D and F_z = + 5 D. Find the normal oblique powers midway between the principal sections.

Ans.: +7.5 D.

6. The principal powers of a refracting surface are F_y = -2.5 D and F_z = +2 D. The power at oblique meridian θ is +1.5 D. Find θ.

Ans.: 70.53°

7. The principal powers of a refracting surface are F_y = +1.5 D and F_z = +0.75 D. The power at oblique meridian θ is +0.8 D. Find θ.

Ans.: 75°

8. Transpose the Rx: -4.00⌔+2.00x135° to minus cylinder and cross-cylinder forms.

Ans.: -2.00⌔-2.00x45°, -4.00x45°⌔-2.00x135°

9. Transpose the Rx: +3.50⌔+1.75x22° to minus and cross-cylinder form.

Ans.: +5.25⌔-1.75x112, +3.5x112°⌔+5.25x22°

10. Transpose the Rx: -1.62⌔-0.75x145° to plus and cross-cylinder form.

Ans.: -2.37⌔+0.75x55°, -1.62x55°⌔-2.37x145°

11. Transpose the Rx. -1.75x30°⌔-2.25x120° to sphero-cylindrical forms.

Ans.: -1.75⌔-0.50x120°, -2.25⌔+0.50x30°

12. Transpose the Rx. -1.75x130°⌔+4.00x40° to sphero-cylindrical forms.

Ans.: -1.75⌔+5.75x40°, +4.00⌔-5.75x130°

13. A point object is at a very great distance from a thin astigmatic lens having principal powers: F_y = +2.00 D, and F_z = +5.00 D. The diameter of the lens is 50 mm. Find (a) the positions of the line foci, (b) the lengths of the horizontal and vertical lines, and (c) the position and diameter of the clc.

Ans.: (a) Distances: H line =500 mm, V line=200 mm, (b)Lengths: H line=75 mm, V line=30 mm, (c)Distance to clc=285.714 mm, Diameter of clc=21.429 mm.

14. A point object is 1 meter from a thin astigmatic lens having principal powers: $F_y = +5.00$ D, and $F_z = +2.00$ D. The diameter of the lens is 50 mm. Find (a) the positions of the line foci, (b) the lengths of the horizontal and vertical lines, and (c) the position and diameter of the clc.

Ans.: (a) Distances: H line =250 mm, V line=1000 mm, (b)Lengths: H line=37.5 mm, V line=150 mm, (c)Distance to clc=400 mm, Diameter of clc=30 mm.

15. Find the sphero-cylindrical equivalent of the following lenses:

(a) +2.25x50° and -3.25x130°. Ans.: +2.21-5.42x134°
(b) -1.50x40° and -2.50x85°. Ans.: -0.54-2.92x69.5°
(c) +2.00-1.50x140° and -3.00+1.5x25° Ans.: +0.36-2.72x127.5°
(d) +5.00-1.50x15° and -1.00-1.00x170° Ans.: +1.61+2.28x95.2°

16. A thin toric lens has powers: $F_y = +5$ D, and $F_z = +9$ D. The interval between the real image line foci is 125 mm. How far is the point source? Ans.: u = -1000 mm

17. A second toric lens is placed at a distance of 500 mm from the lens in Prob. 16. This lens forms a point image at a distance of 1000 mm. Find the principal powers of this lens.
Ans.: $F_y = +5$ D, $F_z = +3.67$ D.

18. A +10.00⌓+5.00x90 lens is placed 15 mm in front of a reduced eye. The cornea of this eye is a toric single spherical refracting surface separating air from index 4/3. Its power in the tangential meridian is +50 D. The lens corrects the refractive error of the eye, i.e., a sharp image of a distant point is formed on the retina. What is the power of the cornea in the sagittal meridian? +42.41 D.

19. Find the spherical equivalent prescriptions of the following Rx's:

(a) +4.00⌓+2.00x45 Ans.: +5.00 D.
(b) -4.00⌓-2.00x135 Ans.: -5.00 D.
(c) -2.00⌓+4.00x90 Ans.: plano
(d) +3.00+4.00x180 Ans.: +5.00 D.

CHAPTER 10

THICK LENS SYSTEMS: PART I

10.1 INTRODUCTION

A thin lens has a thickness that is negligibly small compared to its focal length. When the thickness of a single lens, or the separations between a series of thin lenses cannot be neglected without introducing significant errors, we must treat it as a thick lens system. Although we may calculate object/image conjugates and magnification at each surface or element of a thick lens system to find the final image, such a procedure is cumbersome. It must be repeated each time the object position is varied. Furthermore, it does not describe the lens system as a whole. In this chapter we will see how a complex optical system can be represented by six *cardinal points*. This paraxial treatment of thick lenses is known as Gaussian optics.

10.2 THE CARDINAL POINTS

The cardinal points comprise the first and second: *focal points* (**F** and **F'**), *principal points* (**H** and **H'**) and *nodal points* (**N** and **N'**). The transverse planes containing the *focal points* are the focal surfaces. They are defined as before, i.e., an object at infinity is imaged in the second focal plane, an object in the first focal plane is imaged at infinity. The focal planes are ***not*** conjugate to each other.

The *principal points* locate a pair of principal planes. These transverse planes are *conjugate* to each other. An object at the first principal plane will be imaged at the second principal plane with a *lateral magnification* **Y = +1**. Therefore, the object and image are *congruent.* The particular value of the principal points is that they serve as the reference points for specifying object/image distances and focal lengths.

The *nodal points* are two axial points having an *angular magnification* $\mathbf{M_a} = \theta'/\theta = +1$. That is, the slope of the object-space ray at **N** is the same as the slope of the image-space ray at **N'**. The ray is undeviated.

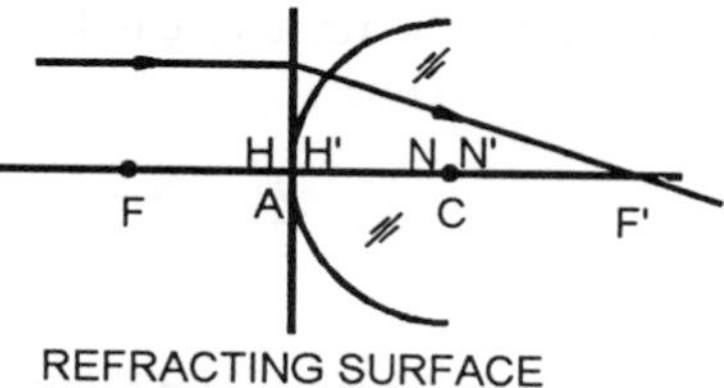

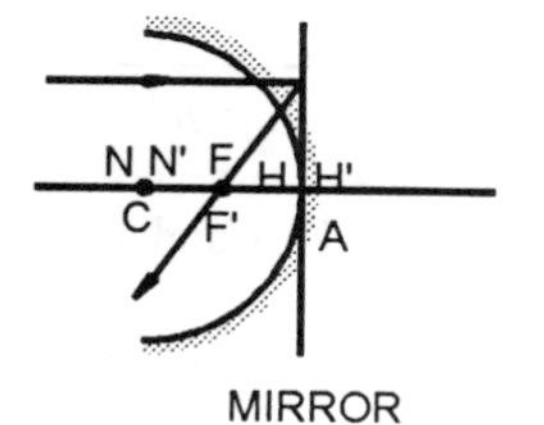

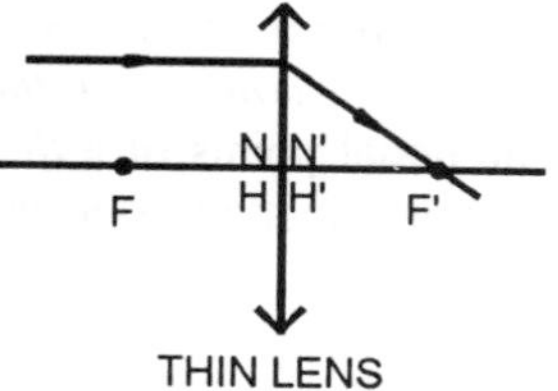

Figure 10.1 Cardinal points of thin elements.

10.2.1 THE PRINCIPAL PLANES AND POINTS

The definition that the principal planes are characterized by a lateral magnification of +1 applies not only to thick lenses but to spherical refracting surfaces, spherical mirrors and thin lenses as well. The Newtonian equation for lateral magnification shows where the principal planes for these components are.

$$Y = \frac{f}{x} = \frac{x'}{f'} = +1$$

Therefore, **f = x**, or **FA = FM**, and **A ≡ M**. Also, **f' = x'**, or **F'A = F'M'**, and **A ≡ M'**. In other words, the lateral magnification equals **+1** when the object **M**, and the image **M'** coincide at the vertex **A**. Therefore, *the principal planes H and H' coincide at the vertex of these components.* Figure 10.1.

The position of the second principal plane of a thick optical system is at the intersection of the axial object ray with the image ray through **F'**. The first principal plane is at the intersection of the object ray from **F** with the image ray to infinity. See Figure 10.2. Principal planes represent thin lens equivalents to the thick lens. Object and image distances are now **u = HM** and **u' = H'M'**, not AM and AM'. The equivalent focal lengths (efl) are **f = FH** and **f' = F'H'**. *A ray incident at some height at the first principal plane is transferred across the hiatus to emerge with the same height at the second principal plane.* Thus, no one plane can replace a thick lens.

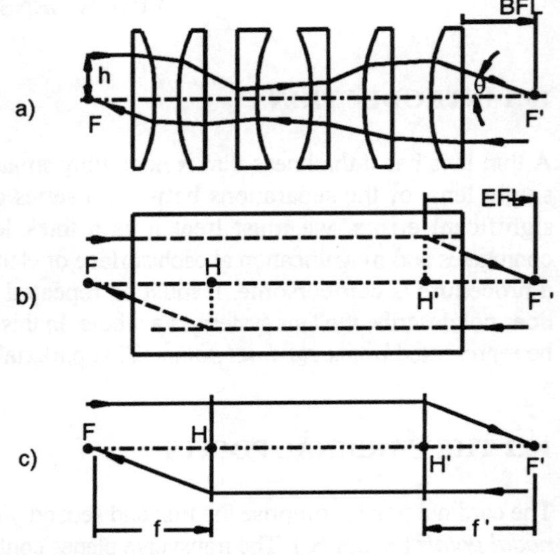

Figure 10.2 Principal planes of a complex lens system. (a) Ray paths from infinity to F and F', (b) The lens system is replaced by a black box in which two planes, found by the intersections of the incident and emergent rays, are located, (c) The lens system is replaced by the two planes.

10.2.2 The Nodal Points

The positions of the nodal points of spherical refracting surfaces, spherical mirrors and thin lenses, can be found with the Lagrangian Invariant: **n'y'θ' = nyθ**, which, for the ray through the nodal points, becomes **n'y' = n y**. Thus,

$$\frac{y'}{y} = \frac{n}{n'} \qquad 10.1$$

That is, *the lateral magnification at the nodal points is n/n'*. But, in general, the lateral magnification is

$$\frac{y'}{y} = \frac{nu'}{n'u} \qquad 10.2$$

Combining Eqs. 10.1 and 10.2 results in **u' = u**. Substituting this into the refraction equation

$$\frac{n'}{u'} = \frac{n}{u} + \frac{n'-n}{r}$$, will result in $$\frac{n'-n}{u} = \frac{n'-n}{r}.$$

Consequently, **u'= u = r**, or **AM' = AM = AC = AN = AN'**. In other words, *the nodal points of a spherical refracting surface coincide at the center of curvature*. This also is true for a spherical mirror. See Figure 10.1. The nodal points of a *thin lens in a uniform medium* can be found directly with the Newtonian magnification equation. Because **n' = n**, Eq. 10.1 equals **+1**.

$$Y = \frac{n}{n'} = \frac{f}{x} = \frac{x'}{f'} = +1$$

Consequently, *the nodal points, like the principal points (see Fig. 10.1), coincide at the vertex of a thin lens in a uniform medium.*

These conclusions about the nodal points were graphically demonstrated earlier when we noted that rays through the centers of curvature of spherical refracting surfaces and spherical mirrors, and the optical centers of lenses are undeviated.

When object and image space indices differ for thin systems such as single spherical refracting surfaces and spherical mirrors, the principal points and the nodal points do *not* coincide with each other. However, **H** coincides with **H'**, and **N** coincides with **N'** at the center of curvature.

10.2.3 Construction of the Nodal Points

Assume that we have located the principal and focal points of a thick lens system, **(n' > n)**, as shown in Figure 10.3. Collimated off-axis rays, with a slope of θ, are incident. Ray 1, through **F**, will be transferred across the hiatus between **H** and **H'** and emerge parallel to the axis. Somewhere in the collimated pencil of rays there is a ray (ray 2) that will emerge undeviated from the lens. By definition, this ray must head toward the first nodal point **N**. It appears to strike the first principal plane at some height **h**. Ray 2 will appear to emerge from the second principal plane with the same height **h** and slope θ . It will intersect the axis at the second nodal point **N'**. Note that the principal and nodal points do *not* coincide.

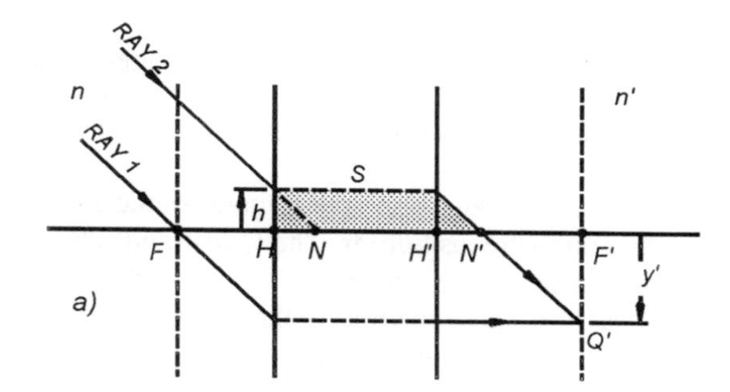

Several important relationships may be found with Figure 10.3. The shaded rectangle and parallelogram in Figure 10.3a share a common side **s**, therefore, the principal and nodal points have the same separation, i.e.,

$$\mathbf{HH' = NN'} \tag{10.3}$$

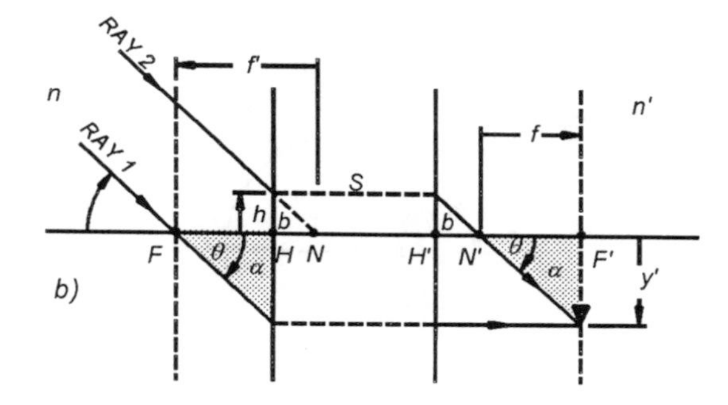

The first focal length is **FH**. It is found with the congruent triangles marked **a**, in Figure 10.3b. They show that

$$\mathbf{f = FH = N'F'} \tag{10.4}$$

The **b** triangles also are congruent, therefore,

$$\mathbf{HN = H'N'} \tag{10.5}$$

or *the nodal points are equidistant from the corresponding principal points.*

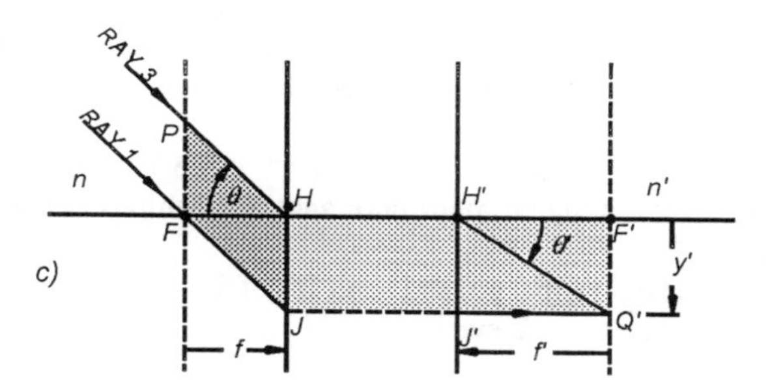

The second focal length **F'H'** is obtained by combining Eqs. 10.4 and 10.5, i.e.,

$$\mathbf{f' = F'H' = NF} \tag{10.6}$$

Figure 10.3 Cardinal points of a thick system.

The relationship between **f** and **f'** is indicated in Figure 10.3c. A third ray in the collimated pencil, one that heads toward the axial point **H**, will emerge from **H'** with a new slope θ', and proceed to **Q'** in the second focal plane. The shaded parallelogram and rectangles have *vertical sides of equal length*. Therefore, PF = HJ = H'J' =F'Q' = y'. Now, **θ = y'/f**, and **θ' = y'/-f'**. Consequently,

$$\frac{\theta'}{\theta} = -\frac{f}{f'} \qquad 10.7$$

Since **y = y'**, the invariant, **nyθ = n'y'θ'**, at the principal planes becomes **nθ = n'θ'**, or **θ'/θ = n/n'**.

Substitute this into Eq. 10.7 to obtain $\frac{f}{f'} = -\frac{n}{n'}$ In other words, the focal lengths of a thick lens system are proportional to the object and image-space indices and opposite in sign, exactly as for thin systems.

EXAMPLE OF CARDINAL POINTS

A *schematic eye* is an optical model of the eye. It is a thick lens system comprising cornea, aqueous humor, lens and vitreous humor. The Gullstrand schematic eye, for example, contains six refracting surfaces and six indices. See Figure 10.4. We will study this eye model, in detail, later. For now, a *reduced eye* will serve as an instructive example of the cardinal points of an optical system. The reduced eye is a simplified model of the eye. It is an aqueous filled globe with a spherical refracting surface serving as the cornea.

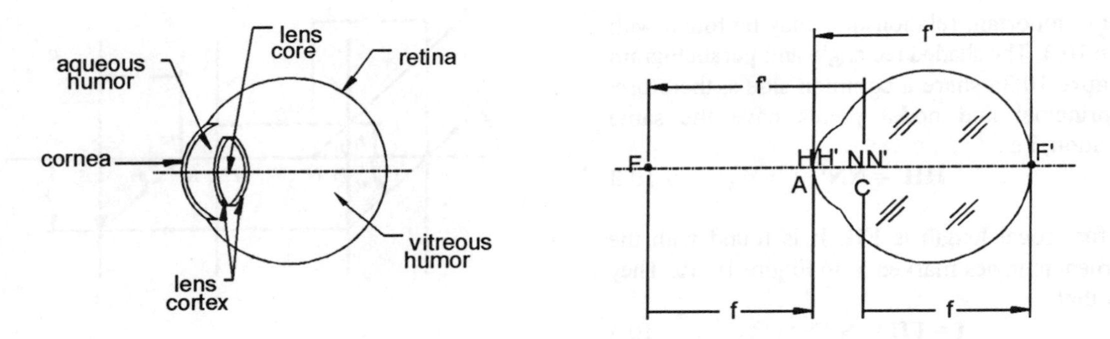

Figure 10.4 Schematic eye-optical system shown on the left. The cornea and lens are shown centered on the optical axis. The fovea is about 5° off-axis. The reduced eye is shown on the right. Note the relationships for the first and second focal lengths.

Let the radius of the "cornea" **r = 5.7 mm**, **n = 1.0** and **n'= 4/3**. Find the power of the eye, its length (assuming emmetropia), and the positions of the cardinal points.

The principal points **H** and **H'**, of the reduced eye, coincide with the vertex of the "cornea". The refracting power of the eye is

$$F = \frac{n'-n}{r} = \frac{333.33}{5.7} = +54.48D.$$

Therefore, the focal lengths are $f = n/F = 17.1$ mm = FH

and $$f' = -n'/F = -22.8 \text{ mm} = F'H'$$

To find the nodal points we utilize the identities, **FH = N'F'**, and **N'F' = N'H' + H'F'.**

or, $$17.1 = N'H' + 22.8$$

Therefore, **H'N' = 5.7 mm**, and **N'** coincides with **c**. Also, **F'H' = NF**, and **NF = NH + HF.** and so, -22.8 = NH - 17.1. **HN = 5.7 mm**, and **N** coincides with **N'** at the center of curvature **C**. It should be obvious that the nodal points of a **spherical refracting surface** coincide with **C**, because rays through **C** are undeviated.

10.2.4 Other "Cardinal" Points

There are other points with notable properties. They are of secondary importance. The conjugate points where the lateral magnification is equal to -1 are sometimes called the *anti-principal* points. They are located as far from the focal points as the principal points but on opposite sides. Similarly, we can define a pair of *anti-nodal* points at which the angular magnification is -1. They coincide with the anti-principal points.

10.3 THE NEWTONIAN EQUATIONS FOR A THICK SYSTEM

As with the thin system, the image of an object point Q, formed by a thick system, may be constructed by finding the intersection of the rays through **F** and **F'**, as shown in Figure 10.5, where

FM = x,	F'M' = x'	HM = u,	H'M' = u'
MQ = y,	M'Q' = y'	FH = f	F'H' = f'

According to the **(a)** triangles, $\dfrac{y'}{y} = \dfrac{f}{x}$. The **b** triangles show that $\dfrac{y'}{y} = \dfrac{x'}{f'}$.

Therefore, **xx' = ff'**. Thus, the Newtonian equations are identical for thin and thick systems.

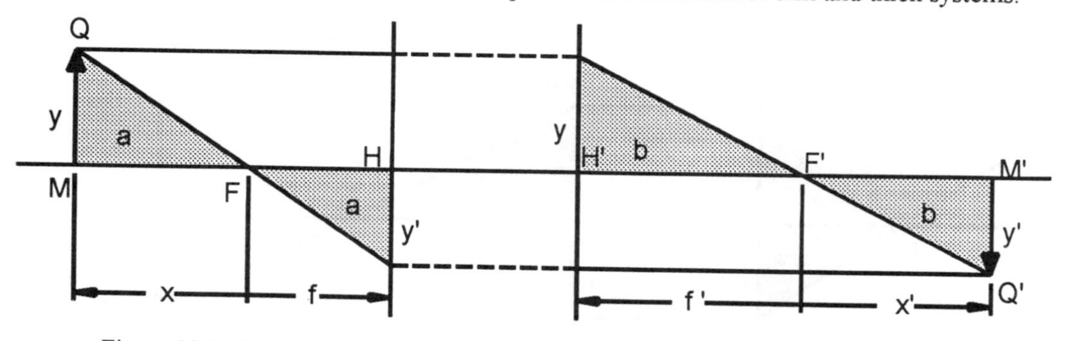

Figure 10.5 Construction for Newtonian equations.

10.4 THE REFRACTION EQUATIONS FOR A THICK SYSTEM

The refraction equation for a thick system may be derived from the Newtonian equation xx' = ff'. The distances x and x' are replaced, as follows:

$$\mathbf{x = FM = FH + HM = f + u}$$

and

$$\mathbf{x' = F'M' = F'H' + H'M' = f' + u'}$$

therefore,

$$\mathbf{(f+u)(f'+u') = ff'}$$

from which we find

$$fu' + f'u + uu' = 0$$

By substituting f' = -n'f/n; and multiplying by n/(uu'f) we will obtain the familiar refraction equation

$$\frac{n'}{u'} = \frac{n}{u} + \frac{n}{f} \qquad 10.8$$

where u= HM, u' = H'M' and f = FH.

10.5 THE LATERAL MAGNIFICATION OF A THICK LENS

Lateral magnification can be derived with Figure 10.6. It provides two sets of similar triangles from which the following proportions may be obtained:

$$\textit{The } \mathbf{a} \textit{ triangles give}, \; y - y' = \frac{uy'}{f}; \; \textit{the } \mathbf{b} \textit{ triangles give } y - y' = -\frac{u'y}{f'}. \qquad 10.9\text{-}10$$

Set Eqs. 10.9 and 10.10 equal to each other and solve for y'/y.

$$\frac{y'}{y} = -\frac{u'f}{uf'}$$

but, -f/f' = n/n' ,therefore,

$$Y = \frac{y'}{y} = \frac{nu'}{n'u}$$

and the equations for the lateral magnification are identical for thin and thick lens systems.

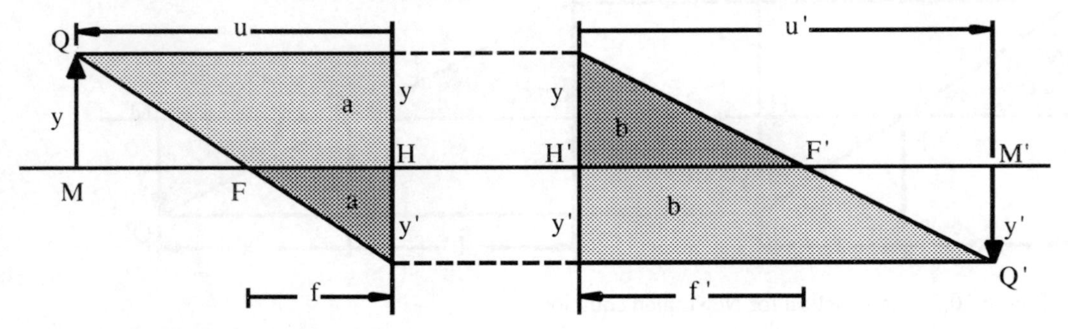

Figure 10.6 Construction for lateral magnification equations for a thick lens.

10.6 SUMMARY OF VERGENCE EQUATIONS FOR A THICK LENS SYSTEM

Power:	**F = n/f = -n'/f'**
Object vergence:	**U = n/u**
Image vergence:	**U' = n'/u'**
Vergence equation:	**U'= U +** F
Lateral magnification:	**Y = U/U'**

10.7 LATERAL, AXIAL AND ANGULAR MAGNIFICATION

Optical systems provide three types of magnification. *Lateral magnification* is exemplified by the action of a slide projector that throws a much magnified image of a slide onto a screen. *Axial or longitudinal magnification* alters the axial thickness of an image of an object that has depth. In a camera, it blurs images of objects that are nearer or further than the subject being photographed. Its effect in microscopy is obvious. *Angular magnification* is used to describe telescopes and microscopes, i.e., instruments that are *afocal* in object and/or image space.

10.7.1 LATERAL MAGNIFICATION: Y

We have dealt in depth with this idea. It will suffice to summarize the most common applicable equations.

$$Y = \frac{y'}{y} = \frac{nu'}{n'u} = \frac{f}{x} = \frac{x'}{f'} = \frac{U}{U'}$$

10.7.2 AXIAL MAGNIFICATION: X

Axial magnification is the ratio of a small image displacement **dx'**, to the corresponding object displacement **dx**. See Figure 10.7. Equivalently, **dx** may represent an axial object, i.e., an object that extends in depth along the optical axis. Its image, **dx'**, extends along the axis and its extent is related to the extent of the object by

$$\mathbf{X = dx'/dx} \qquad 10.11$$

X may be found by differentiating **xx' = ff'.** First, solve for **x'**,

$$x' = \frac{ff'}{x} = ff'x^{-1}$$

$$dx' = -ff'x^{-2}dx$$

$$\frac{dx'}{dx} = \frac{-ff'}{x^2}$$

$$but, \quad ff' = xx'$$

$$thus, \quad \frac{dx'}{dx} = \frac{-xx'}{x^2} = \frac{x'}{x}$$

and the axial magnification is

$$X = -\frac{x'}{x} \qquad 10.12$$

Axial and lateral magnifications are related. According to the Newtonian equations, x = f/Y, and x'= f'Y. Substitute these into Eq. 10.12,

$$X = -\frac{f'Y}{f/Y} = -\frac{f'}{f}Y^2$$

thus

$$X = \frac{n'}{n}Y^2 \qquad 10.13$$

When n and n' are the same, the axial magnification is equal to the square of the lateral magnification.

Figure 10.7 illustrates the effect of lateral and axial magnification on three small cubes; a, b and c. Their square meridional sections become trapezoidal when imaged by the lens. The cubes are imaged as truncated pyramids; a', b' and c'. Image a' is laterally and axially minified; image c' is magnified in both dimensions.

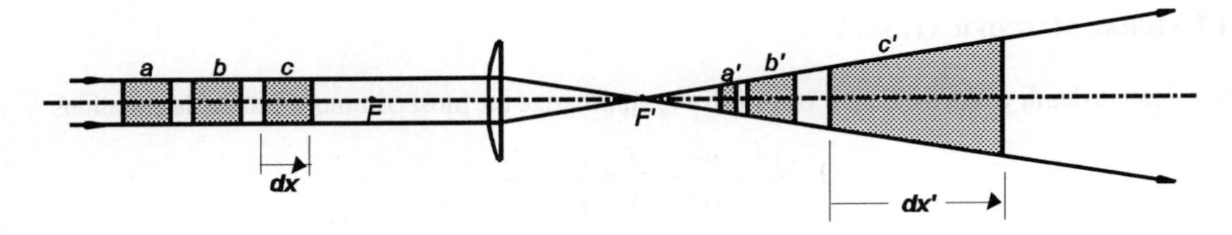

Figure 10.7 Lateral and axial magnification transforms cubical objects into truncated pyramids.

10.7.3 ANGULAR MAGNIFICATION: M_a

The ratio of the slopes of the image and object rays is the angular magnification, $\mathbf{M_a = \theta'/\theta}$. See Figure 10.8. According to the invariant,

$$\frac{\theta'}{\theta} = \frac{ny}{n'y'} = \frac{n}{n'Y}$$

and according to Eq. 10.13,

$$\frac{n}{n'} = \frac{Y^2}{X}$$

Consequently, all three magnifications are related as follows:

$$M_a = \frac{Y}{X} \qquad 10.14$$

Angular magnification is used to specify the magnifying powers of telescopes and microscopes.

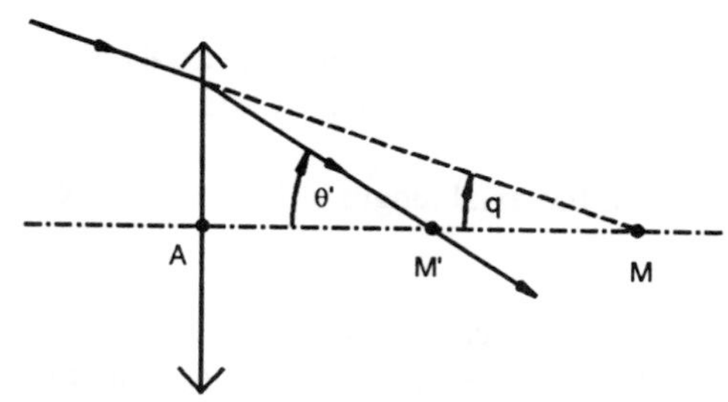

Figure 10.8 Angular magnification.

EXAMPLE 1: A 2 mm long pin lies on the optical axis, and its midpoint is 30 mm from a thin lens with a 10 mm focal length. Find the length of the image of the pin.

Given: dx = 2, u = -30 mm, f = 10 mm. Therefore, x = FM = -20 mm. Find x'.

$$x' = -f^2/x = -100/-20 = 5 \text{ mm}$$

According to Eq. 10.12

$$X = -x'/x = -5/-20 = 0.25$$

According to Eq. 10.1

$$dx' = dx(X) = 2(0.25) = 0.50 \text{ mm}$$

Alternate solution: Find Y and solve Eq. 10.13.

$$Y = f/x = 10/-20 = -0.50$$

$$X = n'Y^2/n = 0.25$$

Therefore,

$$dx' = 0.50 \text{ mm}$$

Find the angular magnification of the ray from M to M'.

$$M_a = Y/X = -0.50/0.25 = -2$$

REVIEW QUESTIONS

1. What are the advantages of reducing a thick lens system to its cardinal points?
2. What are the definitions of the principal and nodal points?
3. Where are the principal and nodal points of thin lenses, mirrors and spherical refracting surfaces?
4. Define the object and image distances, **u** and **u'**, for a thick system.
5. What is axial magnification, how is it related to lateral magnification, and can it be negative?
6. Define angular magnification.

PROBLEMS

1. A reduced eye consists of a spherical refracting surface of 7 mm radius, separating air from aqueous humor of index 1.336. Find: (a) **FH**, (b) **F'H'**, (c) *F*, (d) **NF**, (e) **N'F'**, (f) **HN**, (g) **H'N'**, (h) **AH'**, (i) **AN'**.
Ans.: (a)+20.833 mm, (b) -27.833 mm, (c) +48 D, (d) -27.833 mm, (e) +20.8337 mm, (f) 7 mm, (g) 7 mm, (h) 0 mm, (i) 7mm.

2. The midpoint of a 9 mm long axial object (for example, a pin) is located 100 mm from a thin lens of 25 mm focal length. Find Y, X and dx'. Ans.: (a) Y =-1/3, (b) X = 1/9, (c) 1 mm.

3. Find the axial length of the image by computing the image positions of the ends of the object in Prob. 2. Ans.: 1.004 mm

4. Find the object distance (from the midpoint of the object) which results in an axial magnification of +1, for the lens in Prob. 2. How long is the image if the end points of the object are used?
Ans.: u =-50 mm or u = 0 mm, 9.30 mm.

5. Replace the lens in Problem 2 with a 25 mm focal length convex spherical refracting surface which separates air from index 1.5. What is (a) the lateral magnification, (b) the axial magnification, and (c) the axial length of the image? Ans.: (a) Y =-1/3, (b) X = 0.167, (c) 1.5 mm.

6. Find the axial length of the image by computing the image positions of the ends of the object in Prob. 5. Ans.: 1.505 mm

7. Replace the axial object with a hollow transparent cube, 9 mm on a side, centered 100 mm from the lens in Problem 2. Find the area of the transverse image faces (a) nearer the lens and (b) further from the lens.
Ans.: (a) 8.01 sq. mm, (b) 10.19 sq. mm.

8. Consider only the midpoint of the object in Problem 2. Let the diameter of the lens be 20 mm. Find the angular magnification of a ray from the midpoint that goes through the edge of the lens? Ans.: -3

9. A glass sphere of index 1.5 has a diameter of 200 mm. A 10 mm high object is located 500 mm in front of the first surface of the sphere. Find the position and size of the image produced by the sphere using a surface by surface calculation. Ans.: $A_2M_2' = u_2' = +100$ mm, $y_2' = -3.33$ mm.

10. Move the object to -250 mm in front of the sphere and repeat Problem 9.
Ans.: $A_2M_2' = u_2' = +162.5$ mm, $y_2' = -7.5$ mm.

11. Find the positions of the principal and nodal points of the above sphere as a thick lens. Hint: find the nodal points. Next, determine where the principal points are for a system in a uniform medium. Assume that the focal length of the sphere f = +150 mm. Find: (a) FA_1 and (b) $F'A_2$ (See Eqs.10.4 and 10.5). (c) Find the Newtonian object and image distances for the object whose position was given in Prob. 9.
Ans.: (a) +50 mm, (b)-50 mm, (c) x = FM = -450 mm, x'= F'M' = +50 mm, $A_2M_2' = +100$ mm.

12. Repeat Problem 11 for the object position in Problem 10.
Ans.: x = FM = -200 mm, x'= F'M' = +112.50 mm, $A_2M_2' = +162.5$ mm.

CHAPTER 11

THICK LENS SYSTEMS: PART II

11.1 THE GULLSTRAND EQUATIONS

The Gullstrand equations reduce a thick lens system to an equivalent thin lens. The equations provide the equivalent power F, and the positions of the cardinal points. Principal planes represent equivalent thin lenses. When treating with object space, the thin lens occupies the first principal plane. In image space it occupies the second principal plane. Focal lengths and object and image distances are measured with respect to the principal planes.

First, as an overview of the terminology, consider a simple thick lens, as shown in Figure 11.1. The object space index is $\mathbf{n_1}$, the lens index is $\mathbf{n_2}$, and $\mathbf{n_3}$ is the image space index. In Figures 11.1**a** and **b**, an axial ray from infinity strikes the first surface at a height $\mathbf{h_1}$, and is refracted toward $\mathbf{F_1'}$ (the second focal point of that *surface*). $\mathbf{F_1'A_1}$ is the second focal length of the *first surface*. The refracted ray next strikes the second surface at height $\mathbf{h_2}$ and is refracted to $\mathbf{F'}$ (the second focal point of the *lens*). $\mathbf{A_2F'} = \mathbf{f_v'}$ is the *back focal length* (BFL) of the lens. The intersection of the object and image rays, obtained by their projections, is the position of the *second principal plane* $\mathbf{H'}$. The second focal length of the lens $\mathbf{f'} = \mathbf{F'\,H'}$. Note: **F** and $\mathbf{F'}$ denote *focal points*. Italicized F denotes refracting *power* in diopters.

Figures 11.1**c** and **d**, similarly illustrate the path of the ray imaged at ∞. These diagrams will supply the position of the *first principal point* **H** and the *first focal length* $\mathbf{f} = \mathbf{FH}$.

11.1.1 EQUIVALENT POWER

We will use the vergence equations at each surface to find the second focal point of the *lens*, $\mathbf{F'}$. This will lead to the back focal length, and then to equivalent power. At the first surface,

$$F_1 = \frac{n_1}{f_1} = -\frac{n_1'}{f_1'}\,; \qquad U_1 = \frac{n_1}{u_1}\,; \qquad U_1' = \frac{n_1'}{u_1'} \qquad 11.1\text{-}3$$

For an object at infinity, $U_1' = U_1 + F_1 = F_1;\ \ also\ \ u_1' = \dfrac{n_1'}{F_1}$

The transfer equation, $\mathbf{u_2 = u_1' - d}$, gives the object distance respect to the second surface.

$$u_2 = \frac{n_2 - dF_1}{F_1} \qquad 11.4$$

$$\therefore\ U_2 = \frac{n_2}{u_2} = \frac{n_2F_1}{n_2 - dF_1} \qquad 11.5$$

and substitution into the vergence equation, $U_2' = U_2 + F_2$ will result in

$$U_2' = \frac{F_1 + F_2 - \dfrac{d}{n_2}F_1F_2}{1 - \dfrac{d}{n_2}F_1} \qquad 11.6$$

$$but; \quad u_2' = \frac{n_2'}{U_2'} = \frac{n_3}{U_2'} \qquad 11.7$$

Substitute Eq. 11.6 into 11.7 to find the image distance. In this case u_2' is the back focal length f_v' or the distance A_2F'.

$$A_2F' = f_v' = \frac{n_3(1 - \frac{d}{n_2}F_1)}{F_1 + F_2 - \frac{d}{n_2}F_1F_2} \qquad 11.8$$

The *reduced thickness* $\mathbf{d/n_2}$ is called **c**. We will rewrite Eq. 11.8 as a reduced back focal length by dividing both sides by n_3.

$$\frac{A_2F'}{n_3} = \frac{1 - cF_1}{F_1 + F_2 - cF_1F_2} \qquad 11.9$$

According to the similar triangles in Figure 11.1a:

$$\frac{h_2}{h_1} = \frac{A_2F_1'}{A_1F_1'} = \frac{-(f_1'+d)}{-f_1'} = 1 + \frac{d}{f_1'}$$

But, $f_1' = -n_2/F_1$, and $d/n_2 = c$, thus

$$\frac{h_2}{h_1} = 1 - cF_1 \qquad 11.10a$$

According to the similar triangles in Figure 11.1b:

$$\frac{h_2}{h_1} = \frac{A_2F'}{-f'} = \frac{A_2F' \cdot F}{n_3} \qquad 11.10b$$

where F is the equivalent power of the thick lens.

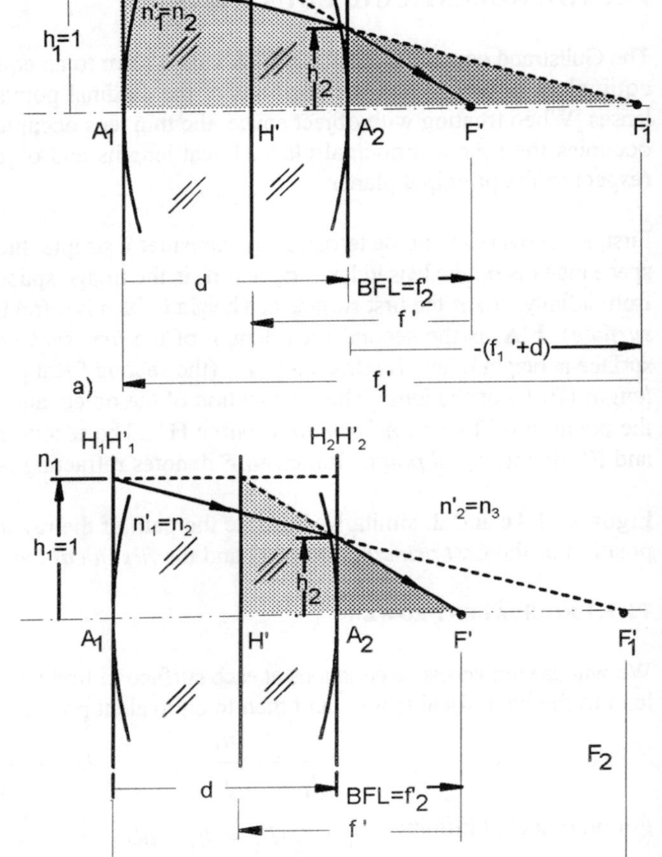

Figure 11.1 a,b. Refraction by Thick Lenses.

Set Eq. 11.10a equal to 11.10b.

$$\frac{A_2F'}{n_3} = \frac{1 - cF_1}{F} \qquad 11.11$$

Equations 11.9 and 11.11 are equal, therefore, the denominators on the right sides of these equations must be equal. Consequently, the *equivalent power* of a thick lens is

$$F = F_1 + F_2 - cF_1F_2 \qquad 11.12$$

The equivalent power F = n/f = -n'/f' ; where f = FH, f' = F' H' , n = n_1 and n' = n_3.

11.1.2 THE POSITION OF THE SECOND FOCAL POINT F'

Since the denominator in Eq. 11.9 is equal to the power of the lens ***F***, the reduced *back focal length* is equal to

$$\frac{A_2F'}{n_3} = \frac{1 - cF_1}{F} \qquad 11.11$$

11.1.3 THE POSITION OF THE SECOND PRINCIPAL POINT H'

The position of the *second principal point* from the second vertex is **A₂H'** .

$$A_2H' = A_2F' + F'H' = \frac{n_3(1-cF_1)}{F} - \frac{n_3}{F}$$

Solve for the reduced distance A_2H'/n_3

$$\frac{A_2H'}{n_3} = -\frac{cF_1}{F} \qquad 11.13$$

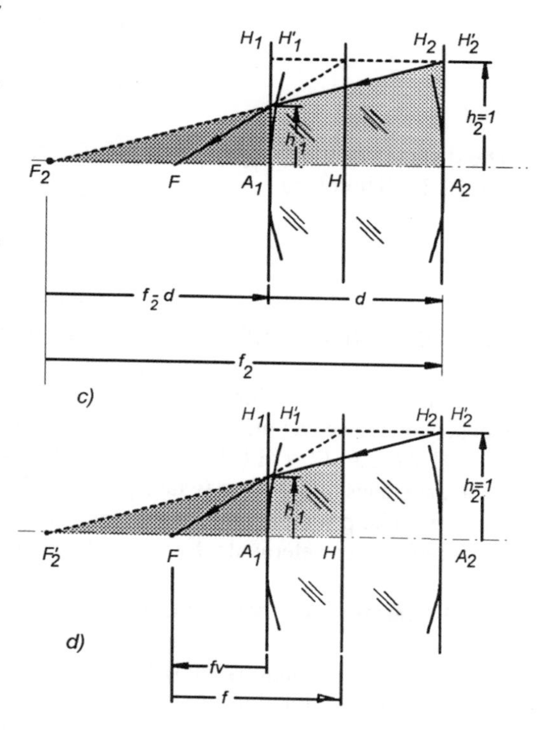

Figure 11.1 c,d. Refraction by Thick Lenses.

11.1.4 THE POSITIONS OF THE FIRST FOCAL AND PRINCIPAL POINTS, F AND H

The path of a ray from **F** to infinity, shown in Figures 11.1c and d, will enable us to find the positions of the first focal and principal points. The shaded similar triangles in Figure 11.1c will help find the distance from the front vertex A_1 to the first focal point of the lens F, i.e., the *front focal length*, $\mathbf{A_1F = f_v}$.

$$\frac{h_1}{h_2} = \frac{F_2A_1}{F_2A_2} = \frac{f_2 - d}{f_2} = 1 - \frac{d}{f_2} = 1 - cF_2 \qquad 11.14a$$

where $f_2 = n_2/F_2$. According to the shaded triangles in Figure 11.1d

$$\frac{h_1}{h_2} = \frac{FA_1}{FH} = -\frac{A_1F}{f} = \frac{A_1F \cdot F}{-n_1} \qquad 11.14b$$

Equations 11.14a and b are equal, therefore, the reduced *front focal length* is,

$$\frac{A_1F}{n_1} = -\frac{1-cF_2}{F} \qquad 11.15$$

The distance of the *first principal point* from the front vertex, $A_1H = A_1F + FH$, where A_1F is given by Eq. 11.15. Therefore,

$$\frac{A_1H}{n_1} = \frac{cF_2}{F} \qquad 11.16$$

11.2 DISTANCES REFERRED TO PRINCIPAL PLANES

The principal planes of each surface of the lens coincide at the surface, as shown in Figures 11.1a and 2. Consequently, we can make the following substitutions in the above equations: $A_1H = H_1H$; $A_1F = H_1F$; $A_2H' = H_2'H'$; $A_2F' = H_2'F'$, and $d = H_1'H_2$. When two or more thick lenses form an optical system the principal planes of each lens (which probably do not coincide with the lens vertices) become the reference surfaces. The principal and focal planes of the system are located with respect to the principal and focal planes of the lens elements, for example, H_1H, H_1F, $H_2'H'$, $H_2'F'$ and $d = H_1'H_2$.

11.2.1 SUMMARY OF THE GULLSTRAND EQUATIONS

The Gullstrand equations apply to a thick lens, which may be in a non-uniform medium or to two thin lenses, surrounded and separated by an airspace, as shown in Figure 11.2b. In the latter case the object and image space indexes $\mathbf{n_1}$ and $\mathbf{n_3}$ = one, and $\mathbf{c} = \mathbf{d}$, the separation between the lenses.

	THICK LENS IN NON-UNIFORM MEDIUM:	THIN AIR-SPACED LENSES OR THICK LENS IN AIR
Equivalent Power:	$F = F_1 + F_2 - cF_1F_2$	$F = F_1 + F_2 - cF_1F_2$
First principal plane.	$\frac{A_1H}{n_1} = \frac{H_1H}{n_1} = \frac{cF_2}{F}$	$A_1H = \frac{cF_2}{F}$
First focal plane.	$\frac{A_1F}{n_1} = \frac{H_1F}{n_1} = -\frac{1-cF_2}{F}$	$A_1F = -\frac{1-cF_2}{F}$
Second principal plane.	$\frac{A_2H'}{n_3} = \frac{H_2'H'}{n_3} = -\frac{cF_1}{F}$	$A_2H' = -\frac{cF_1}{F}$
Second focal plane.	$\frac{A_2F'}{n_3} = \frac{H_2'F'}{n_3} = \frac{1-cF_1}{F}$	$A_2F' = \frac{1-cF_1}{F}$

11.2.1.1 The Positions of the Nodal Points N and N'

The equivalent power and the positions of the focal and principal points (F, F′ , H and H′) are found with the above Gullstrand Equations. *When a uniform medium surrounds the lens, the nodal points N and N′ coincide, respectively, with H and H′.* If $n_1 \neq n_3$, the nodal points are found with the following identities:

$$\mathbf{N'F'} = \mathbf{FH} \qquad 11.17a$$

and

$$\mathbf{NF} = \mathbf{F'H'} \qquad 11.17b$$

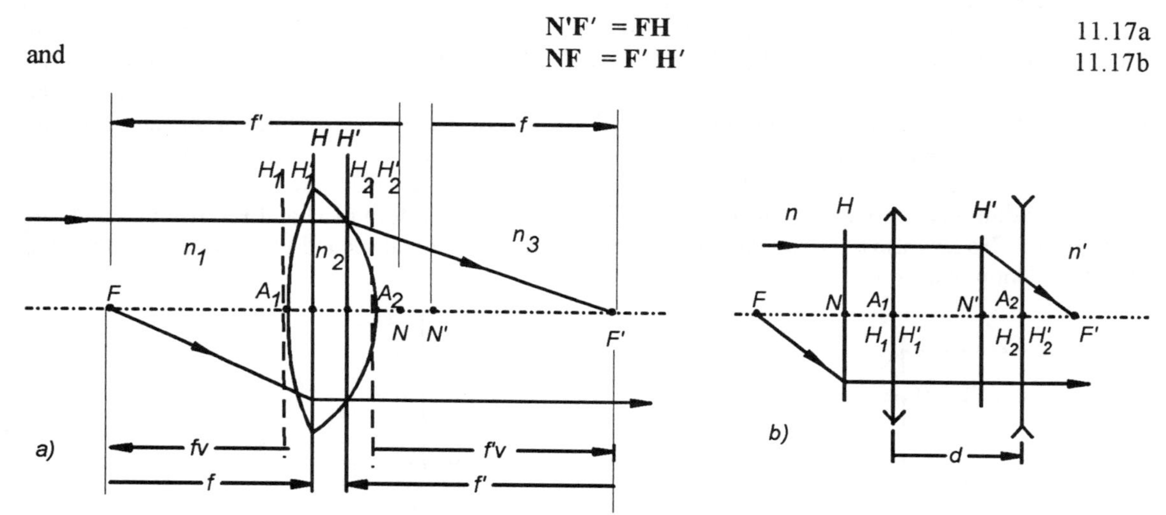

Figure 11.2 a) Cardinal points of a thick lens system with different object and image space media, i.e. $n_1 \neq n_3$. b) Cardinal points of air spaced thin lenses n=n'.

EXAMPLE 1: Find the equivalent power and the positions of the cardinal points of a 25-mm thick biconvex lens, made of glass of index=2.0, with radii of 100 and 50 mm. The object and image space indices are 1.0 and 1.5. See Figure 11.2.

Given: $r_1 = 0.10$ m $\quad r_2 = -0.05$ m
$n_1 = 1.00 = n \quad n_2 = 2.00 \quad n_3 = 1.50 = n'$
$d = 0.025$m $\quad \therefore \quad c = 0.0125$

Calculate surface powers: $F_1 = 10.00$ D and $F_2 = 10.00$ D

Solve Eq. 11.12 for equivalent power. $F = 18.75$ D

Find the focal lengths. $FH = f = n/F = 1/18.75 = 0.053$ m $= 53.3$ mm
$F'H' = f' = -n'/F = -1.5/18.75 = -0.080$ m $= -80.0$ mm

Find A_1F, A_1H, A_2F' and A_2H' with Eqs. 11.16, 11.17, 11.13, and 11.14.

$A_1F = -0.0467$ m $= -46.67$ mm = FFL $\qquad A_1H = 0.0067$ m $= 6.67$ mm
$A_2F' = 0.0700$ m $= 70.00$ mm = BFL $\qquad A_2H' = -0.0100$ m $= -10.00$ mm

Find the nodal points: A_1N and A_2N' with Eqs. 10.7 and 10.9.

$$A_1N = A_1F + FN = A_1F\text{-}f' = -46.67 + 80 = 33.33 \text{ mm.}$$
$$A_2N' = A_2F' + F'N' = A_2F' - f = 70.0 - 53.33 = 16.67 \text{ mm.}$$

11.2.2 THE EFFECT OF LENS THICKNESS ON THE POSITIONS OF THE PRINCIPAL PLANES

Varying lens thickness will shift the positions of the principal planes. A series of equiconvex lenses with surface powers of $F_1 = F_2 = +10\text{D}$ ($r_1 = -r_2 = 50$ mm, n = 1.5) and of increasing thickness are shown in Figure 11.3. When d = 0 (thin lens), H, H′, A_1 and A_2 all coincide at the optical center of the lens. See Fig. *11.3a.* When d = 50mm, H moves to the right of A_1, and H′ moves to the left of A_2. See Fig. *11.3b.* The lens becomes spherical when d = 100 mm. H and H′ have now moved to the center of the lens. See Fig. *11.3c.* With increased thickness they cross each other. See Fig. *11.3d.* In Fig. *11.3e*, the thickness is 300 mm, and the lens is a unit-power (x1) telescope. The lens is *afocal*; H′ and F are at -∞, and H and F′ are at +∞. An additional increase in thickness, as in Fig *f*, results in a return of H and H′ from -∞ and +∞, respectively. This lens also has a real intermediate focal point F_1' and a real second focal point F′. Note that the equivalent power is *negative.*

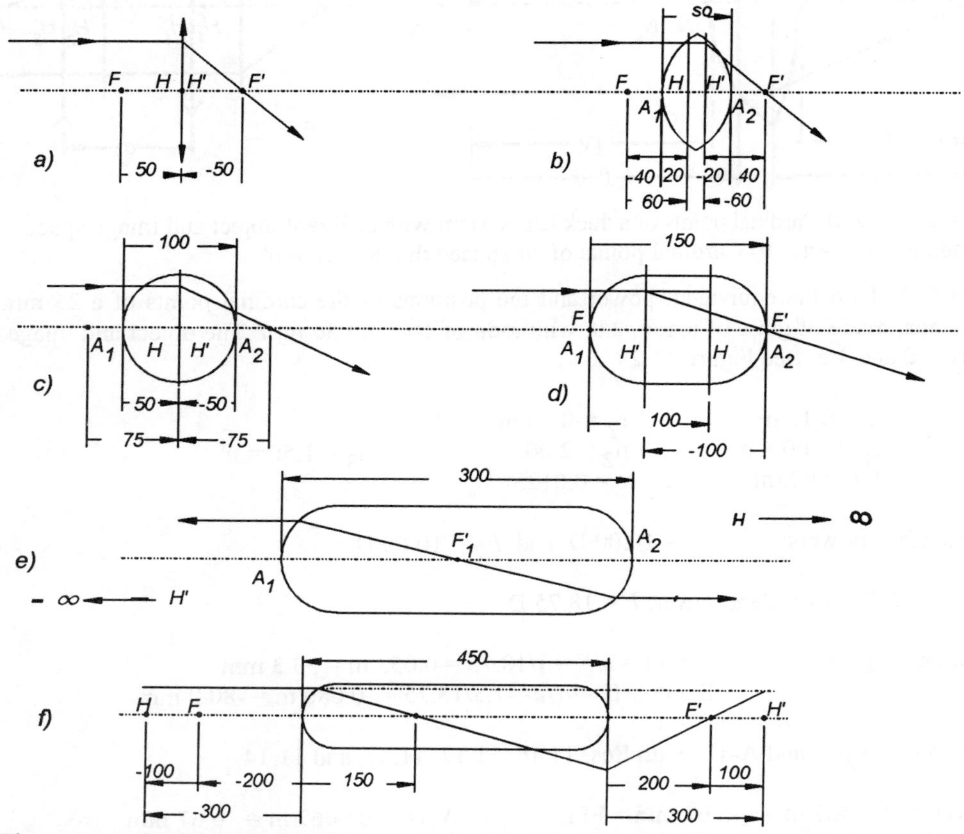

Figure 11.3 Shift in Principal Planes of a Positive Lens as Thickness is increased

11.2.3 THE EFFECT OF LENS SHAPE ON THE POSITION OF THE PRINCIPAL PLANES

The principal planes of an equiconvex lens lie within the lens about 1/3rd of the thickness from the vertices. As the lens is bent to convex and steepening meniscus forms, H and H′ migrate toward the convex surface, even passing through it. See Figure 11.4, and Table 1. This shift in the principal planes is used in designing wide-angle and telephoto lenses.

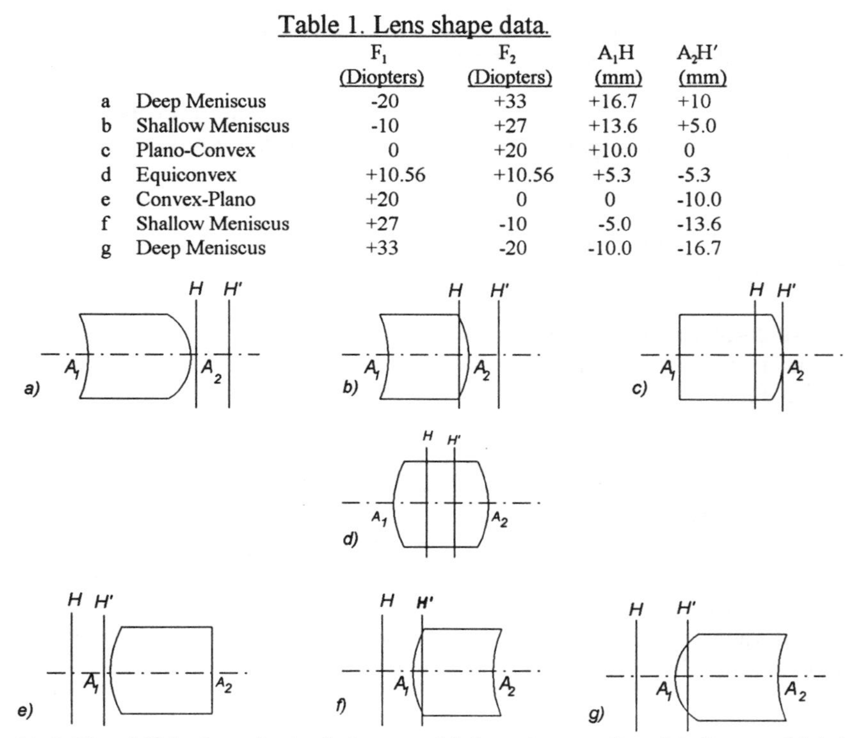

Table 1. Lens shape data.

		F_1 (Diopters)	F_2 (Diopters)	A_1H (mm)	A_2H' (mm)
a	Deep Meniscus	-20	+33	+16.7	+10
b	Shallow Meniscus	-10	+27	+13.6	+5.0
c	Plano-Convex	0	+20	+10.0	0
d	Equiconvex	+10.56	+10.56	+5.3	-5.3
e	Convex-Plano	+20	0	0	-10.0
f	Shallow Meniscus	+27	-10	-5.0	-13.6
g	Deep Meniscus	+33	-20	-10.0	-16.7

Figure 11.4 The shift in the principal planes, with lens shape, of a +20 diopter thick lens, n=1.5, d=15mm.

11.2.4 TELEPHOTO AND WIDE-ANGLE LENSES

A single lens reflex (SLR) camera is so named because viewing and photographing is done through the same lens. A plane mirror between the lens and the film reflects light to the eye via a viewing system. When the shutter is snapped the mirror swings out of the way to expose the film. The lens must have a minimum back focal length of about 36 mm to provide room for the mirror, yet (wide-angle) lenses with focal lengths as small as 8 mm can be used. It also possible to buy (telephoto) lenses for these cameras with focal lengths of 200 mm and more, yet the lengths of the lens barrels are fractions of the focal lengths.

Telephoto and wide-angle lenses show how we may shift principal planes about to provide desired optical properties. A *basic* telephoto lens consists of a front positive lens separated from a rear negative lens. The reverse, i.e., a front negative and rear positive lens is the basic wide-angle lens design.

EXAMPLE 2: Telephoto lens.
Let f_1 =33.33mm, f_2 = -12.50mm and d = 25mm, see Figure 11.5. Find the EFL, A_1H, A_2H', A_1F and A_2F'.

Results: f = 100mm = EFL
A_1H =-200mm
A_2H' = - 75mm
A_1F = -300mm
A_2F' = 25mm = BFL

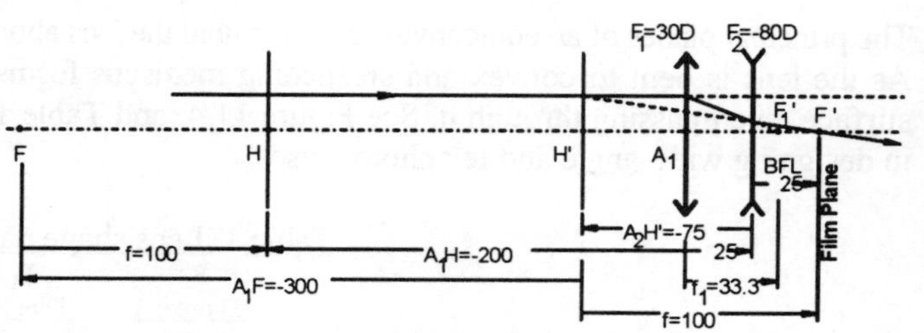

Figure 11.5 The principal planes of a Telephoto lens are shifted away from the film plane.

The EFL is four times the BFL. The overall length of the lens to film (OVL) = A_1F' = 50mm, thus, the telephoto effect = f/OVL = 2x.

EXAMPLE 3: Wide-angle or reverse telephoto lens.
Let f_1 = -12.5mm, f_2 =33.33mm and d = 25mm. Reverse Figure 11.6.

Results: f = 100mm = EFL, A_1H = 75mm, A_2H' = 200mm, A_1F = -25mm, A_2F' = 300mm = BFL. Thus, the BFL is 3 times the EFL.

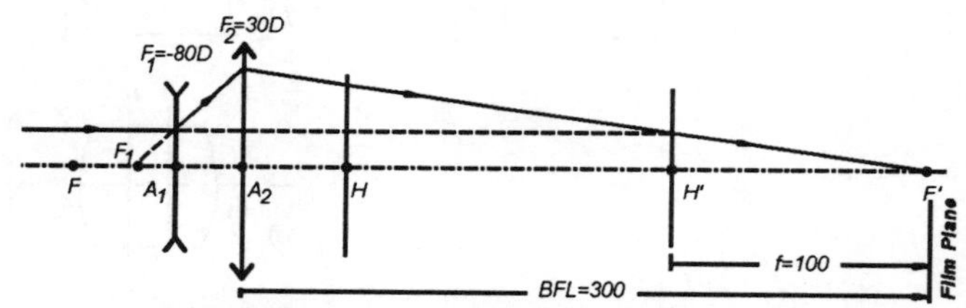

Figure 11.6 The principal planes of a Wide-Angle (reverse Telephoto) lens are shifted toward the film plane.

The shift in the points H and H′ for Nikon lenses that range from wide-angle to telephoto, is illustrated in Figure 11.7

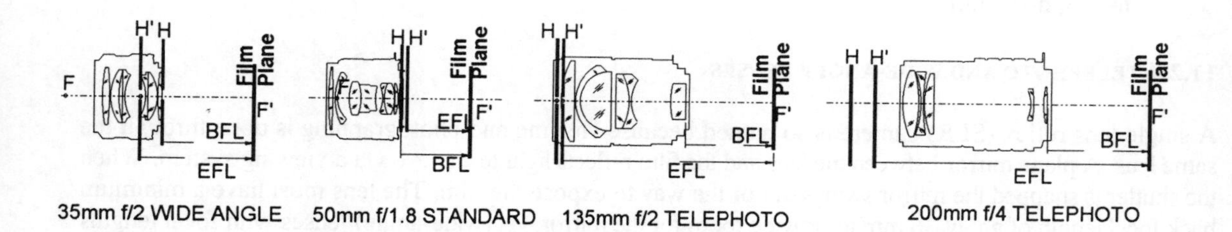

Figure 11.7 Positions of principal planes of photographic lenses. With increasing focal length, the principal planes move away from the film plane but the flange focal length remains constant.

11.3 VERTEX POWER

We describe thick lens systems, ordinarily, in terms of *equivalent power* F and *equivalent focal length* **EFL** = **f**. Ophthalmic lenses, however, are specified by their *back vertex powers* F'_v

$$F'_v = \frac{n'}{f'_v} \qquad 11.18$$

where, n' is the index in image-space and f'_v is the back focal length, A_2F'. According to Eq. 11.11, the back vertex power of a lens in air is,

$$\boxed{F'_v = \frac{F}{1 - cF_1}} \qquad 11.19$$

The back vertex power of a thick lens is equivalent to a thin lens of that power, placed at the second vertex A_2. The second focal points of these two lenses will coincide.

The front focal length $\mathbf{f_v}$ and the *front vertex power* F_v may be obtained from Eq. 11.15. Because our sign convention will cause a *positive* lens to have a *negative* front focal length, if the front vertex power is to be positive it must be defined as follows, where, $f_v = A_1F$:

$$F_v = -\frac{n}{f_v} \qquad 11.20$$

The front vertex power is

$$\boxed{F_v = \frac{F}{1 - cF_2}} \qquad 11.21$$

The front and back vertex powers will be equal if the lens is symmetrical and in a uniform medium.

EXAMPLE 4: The front and back surface powers of a lens are +10 and -2D, respectively. The lens is 9 mm thick and made of index 1.5. Find F, F_v, f_v, F'_v, and f'_v. Given: $F_1 = +10$, $F_2 = -2$, c = 0.006.

Solve Eq. 11.12, for equivalent power, and Eqs. 11.18 - 21 for vertex powers and focal lengths.

$F = +8.12$ D.

$F_v = +8.024$ D $\qquad$ $f_v = -124.63$ mm

$F'_v = +8.638$ D $\qquad$ $f'_v = +115.764$ mm

11.4 VERTEX POWER AND OPHTHALMIC LENSES

Although equivalent power is generally applicable to thick lenses, it is not a suitable measure for ophthalmic lenses because the equivalent power is referenced to principal planes that shift as a function of bending. Therefore, lenses of equal equivalent power but different bendings will have focal points that vary in distance from the lens surface. This is incompatible with the basic condition for correcting myopia and hyperopia, namely, to place the second focal point F′ of the correcting lens at the far point F.P. of the eye, as shown for a thin + 10 D lens in Figure 11.8. The lens is in the *spectacle plane*, at a *vertex distance* of 15 mm in front of the cornea. Also shown is a thick equiconvex lens of the same equivalent power positioned with its back vertex A_2 at a distance of 15 mm from the cornea. The second principal point H′ of this lens is shifted 5.13 mm to the left of A_2, consequently, F′ is shifted 5.13 mm from the far point, and the condition for correction is not satisfied. Still worse, the +10 D meniscus lens shifts F′ through a distance of 17.1 mm away from the far point. Clearly, the condition for correction is not satisfied. Equivalent power is not a satisfactory unit for prescribing lenses.

Referring to the thin lens diagram, we see that the +10 D power must be exerted at the spectacle plane. That is, the 100-mm focal length must be measured from the spectacle plane. The correction lens must have a *back vertex power* of +10 D, whatever its form and thickness.

The equivalent (F), back vertex (F'_v), and effective (F_e) powers of a lens depend on lens shape. A spectacle lens corrects ametropia when its second focal point (F′) coincides with the far point (F.P.) of the eye. The three lenses shown here have an equivalent power F = +10 D and a vertex distance of 15 mm. (a) The thin lens properly corrects the eye; $F'_v = F = +10D$; $F_e = +11.77$ D. (b) The thick equiconvex lens has $F'_v = +10.54$ D; $F_e=+12.52$D; and overcorrects by 0.76 D. (c) The meniscus lens has $F'_v = +12.06$D, $F_e = +14.72$ D, and overcorrects by 2.97 D. (Note: <u>Equiconvex</u>: $F_1=F_2=5.13$ D; $A_1H=5.13$mm; $A_1F= -94.87$mm; $A_2H' = -5.13$mm; $A_2F' = 94.87$mm. <u>Meniscus</u>: $F_1= 17.08$D; $F_2=-8.54$D; $A_1H=-9.54$mm; $A_1F=-108.54$mm; $A_2H' =-17.08$ mm; $A_2F' = 82.92$mm.)

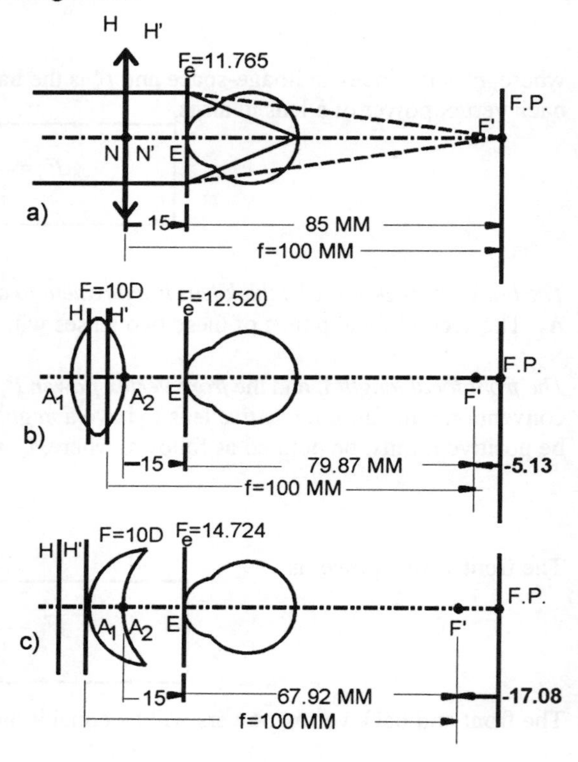

Figure 11.8 The effect of lens shape on power.

11.5 EFFECTIVE POWER

A lens or spherical refracting surface focuses collimated light at F′. The *effective power* F_e of that lens corresponds to the power of a thin lens, placed a distance **d** behind or in front of the former lens, that focuses collimated light to that same F′. In Figure 11.9, a single spherical refracting surface of power *F* separates air from index n. Its second focal length **f′** = **F′ A**. The effective power at a distance **d** = **AB**

is $F_e = -\frac{n}{f_e'}$. Substituting $f_e' = F'B = f' + d = d - n/F$ into this equation results in the equation for

Effective power.

$$F_e = \frac{F}{1-cF} \tag{11.22}$$

The distance **d** replaces **c** in Eq. 11.22 for finding the effective power of a thin lens in air.

As indicated in Figure 11.8 the effective power of the thin lens at the cornea is 11.77 D. This is a more accurate measure of the refractive error of the eye. It is equivalent to the power of a contact lens that would correct this eye.

EXAMPLE 5: Find the effective power at the cornea if the thin +10 D spectacle lens is worn at a vertex distance of 15mm. Given: $F = +10$ D; $d = 0.015$ m.

Solve Eq. 11.22. $F_e = +11.765$ D

The equiconvex lens, however, has an effective power at the cornea of 12.52 D. Thus, it overcorrects the hyperopia by 0.75 D. The meniscus lens overcorrects by nearly 3 D!

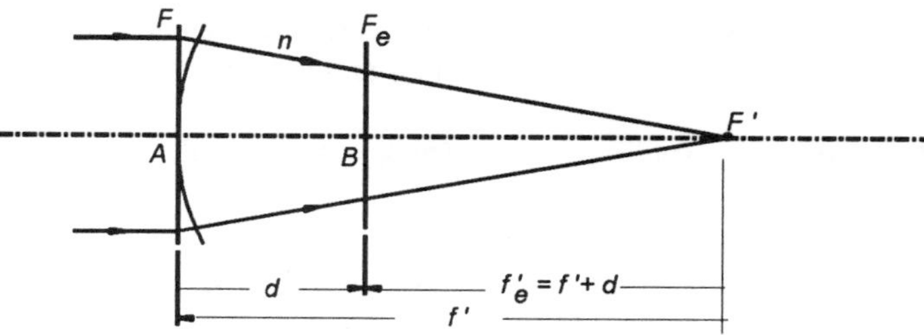

Figure 11.9 Effective power. Power F is impressed on the incident wavefront at A. The vergence of the light brings it to a focus at F'. The effective power F_e at B is simply the vergence of the light at B or $1/f_e'$.

11.5.1 THE RELATIONSHIP OF EFFECTIVE POWER TO VERTEX POWER

The back vertex power of a thick lens is given by Eq. 11.19. $F_v' = \dfrac{F}{1 - cF_1}$.

Another way to view vertex power is that it is equal to the effective power of the first lens surface at the second surface, plus the power of the second surface.

$$F_v' = F_{e1} + F_2 = \frac{F_1}{1 - cF_1} + F_2$$

From which we will obtain the back vertex power equation. $F_v' = \dfrac{F}{1 - cF_1}$.

11.6 COMBINATION OF TWO THICK LENS

The thick lens equations enable us to calculate the cardinal points of two air-spaced thin lenses or one thick lens bounded by two surfaces. We may separately find the cardinal points of additional elements in
an optical system by the iterative use of these equations for each additional thick lens. As this process proceeds, the principal planes of the components must be used to determine the reduced distances between components. The distance **d** is the step from the second principal plane of the first component to the first principal plane of the second component. The medium between components is used to calculate the

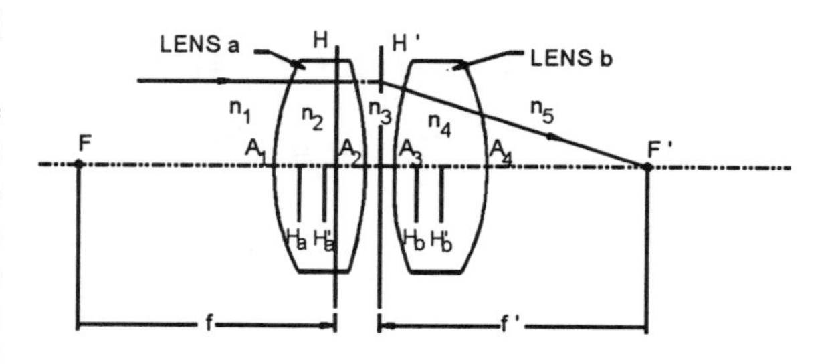

Figure 11.10 Combination of two thick lenses.

reduced thickness. A final set of calculations will then yield the cardinal points of the system. The following example shows this procedure for two thick lenses separated by air.

EXAMPLE 6. See Figure 11.10 The lenses are in air and labeled a and b. They are separated by distance A_2A_3 = 5 mm. The data for each lens and the results of solving the Gullstrand Equations are:

Lens a:	Radii	Thickness	Index	Power (D)
			$n_1 = 1.0$	
	$r_1 = +50$ mm			$F_1 = +10$
		A_1A_2=15mm	$n_2 = 1.5$	
	$r_2 = -100$ mm			$F_2 = +5$
			$n_3 = 1.0$	

Results for Lens a: $F_a = +14.50$ D;
$A_1H_a = +3.45$ mm; $A_1F_a = -65.52$ mm.
$A_2H_a' = -6.90$ mm; $A_2F_a' = +62.07$ mm.

Lens b:	Radii	Thickness	Index	Power (D)
			$n_3 = 1.0$	
	$r_3 = +75$ mm			$F_3 = +6.67$
		A_3A_4=12mm	$n_4 = 1.5$	
	$r_2 = -150$ mm			$F_4 = +3.33$
			$n_5 = 1.0$	

Results for Lens b: $F_b = +9.82$ D;
$A_3H_b = +2.71$ mm; $A_3F_b = -99.10$ mm.
$A_4H_b' = -5.43$ mm; $A_4F_b' = +96.38$ mm.

The final step is to combine these data to obtain the power and find the principal points of the system. Given: $F_a = +14.50$ D; $F_b = +9.82$ D.

$$d = H_a'H_b = H_a'A_2 + A_2A_3 + A_3H_b = +6.90 + 5.00 + 2.71 = +14.61 \text{ mm}$$

$$c = d/n_3 = 0.01461/1.0 = 0.01461$$

Solve the thick lens equations:

$$F = +14.50 + 9.82 - 0.01461(14.50)(9.82) = +22.24D.$$

$$FH = f = +44.96mm; \quad F'H' = f' = -44.96mm$$

$$H_aH = \frac{.01461(9.82)}{22.24} = 0.00645m = 6.45mm$$

$$H_b'H' = -\frac{0.01461(14.50)}{22.24} = -.00953m = -9.53mm$$

Also, $H_aF = -38.51$ mm; $H_b'F' = +35.44$ mm
$A_1F = -35.06$ mm; $A_4F' = 30.01$ mm.

11.7 THE GULLSTRAND SCHEMATIC EYE

The schematic eye is an optical model of the human eye. Gullstrand's schematic eye is one of the most widely accepted models. The eye consists of a thin cornea, followed by aqueous humor, a crystalline lens, vitreous humor, and retina. The actual crystalline lens of the eye has a gradient index, that is, the index of refraction gradually increases from the outer surface to the center of the lens. Gullstrand's lens model comprises an outer cortex of 1.386 index surrounding a central core of 1.406 index. See Figure 11.11. The data for the unaccommodated 24 mm long eyeball with 1 diopter of axial hyperopia are:

THE GULLSTRAND SCHEMATIC EYE

SURFACE	RADIUS (mm)	MEDIUM	THICKNESS (mm)	INDEX	CUMULATIVE THICKNESS
		Air		$n_1 = 1.000$	
1. Anterior cornea	+7.700				
		Cornea	$A_1A_2 = 0.500$	$n_2 = 1.376$	$A_1A_2 = 0.500$
2. Posterior cornea	+6.800				
		Aqueous	$A_2A_3 = 3.100$	$n_3 = 1.336$	$A_1A_3 = 3.600$
3. Anterior lens cortex	+10.000				
		Cortex	$A_3A_4 = 0.546$	$n_4 = 1.386$	$A_1A_4 = 4.146$
4. Anterior lens core	+7.911				
		Core	$A_4A_5 = 2.419$	$n_5 = 1.406$	$A_1A_5 = 6.565$
5. Posterior lens core	-5.760				
		Cortex	$A_5A_6 = 0.635$	$n_6 = 1.386$	$A_1A_6 = 7.200$
6. Posterior lens cortex	-6.000				
		Vitreous	$A_6A_7 = 16.800$	$n_7 = 1.336$	$A_1A_7 = 24.000$
7. Retina					

11.7.1 PROCEDURE FOR CALCULATING CARDINAL POINTS OF THE SCHEMATIC EYE

The Gullstrand equations handle only two surfaces at a time. Once we obtain the power and cardinal points of a thick lens it, in effect, is reduced to a thin lens. Therefore, we must find the cardinal points of the eye in stages, as follows:

1. Calculate the power and cardinal points of the following, as thick lenses:
   a. the cornea
   b. the anterior cortex
   c. the posterior cortex

2. Calculate the power and cardinal points of the crystalline lens by combining the anterior and the posterior cortex, which have been reduced to "thin" lenses separated by lens core. The separation is equal to the distance between the second principal point of the anterior cortex and the first principal point of the posterior cortex.

3. Calculate the power and cardinal points of the entire eye by combining cornea and lens, which are separated by aqueous. The separation is equal to the distance between the second principal point of the cornea and the first principal point of the lens.

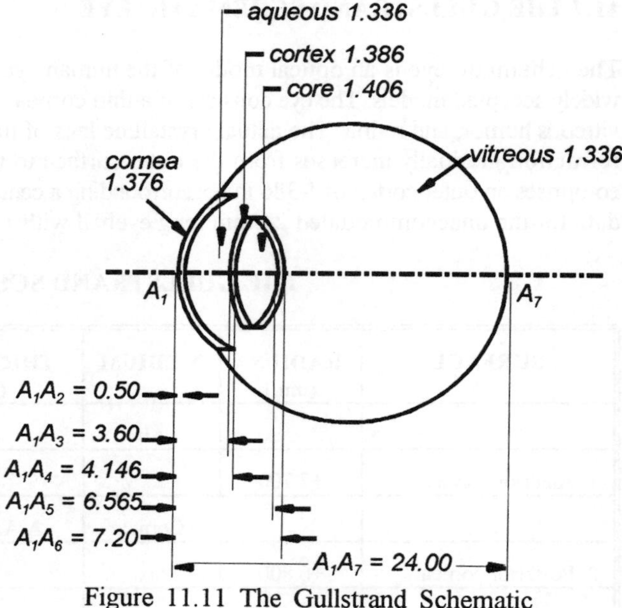

Figure 11.11 The Gullstrand Schematic Eye.

Results

1a. Cornea

Power of anterior surface of cornea	$F_1 = +48.831$ D.
Power of posterior surface of cornea	$F_2 = -5.882$ D.
Reduced thickness of cornea	c = 0.000363 m
Power of cornea	$F_c = +43.053$ D.
Position of first principal point, h_c	$A_1H_c = -0.0496$ mm
Position of second principal point, H'_c	$A_2H'_c = -0.550$ mm
H'_c referred to first vertex of eye	$A_1H'_c = -0.0501$ mm

1b. Anterior lens cortex

Power of first surface of anterior cortex	$F_3 = +5.000$ D.
Power of second surface of anterior cortex	$F_4 = +2.528$ D.
Reduced thickness of anterior cortex	c = 0.000394 m
Power of anterior cortex	$F_{ac} = +7.523$ D.
Position of first principal point, H_{ac}	$A_3H_{ac} = +0.177$ mm
H_{ac} referred to first vertex of eye	$A_1H_{ac} = +3.777$ mm
Position of second principal point, H'_{ac}	$A_4H'_{ac} = -0.368$ mm
H'_{ac} referred to first vertex of eye	$A_1H'_{ac} = +3.778$ mm

1c. Posterior lens cortex

Power of first surface of posterior cortex	$F_5 = +\ 3.472$ D.
Power of second surface of posterior cortex	$F_6 = +\ 8.333$ D.
Reduced thickness of posterior cortex	c = 0.000458 m
Power of posterior cortex	$F_{pc} = +11.792$ D.
Position of first principal point, H_{pc}	$A_5H_{pc} = +0.455$ mm
H_{pc} referred to first vertex of eye	$A_1H_{pc} = +7.020$ mm
Position of second principal point, H'_{pc}	$A_6H'_{pc} = -0.180$ mm
H'_{pc} referred to first vertex of eye	$A_1H'_{pc} = +7.020$ mm

2. The lens

Power of anterior cortex	$F_{ac} = +\ 7.523$ D.
Power of posterior cortex	$F_{pc} = +11.792$ D.
Reduced thickness = (7.020-3.778)/1.406	c = 0.002306 m
Power of lens	$F_L = +19.110$ D.
Position of first principal point, H_L	$H_{ac}H_L = 1.904$ mm
H_L referred to first vertex of eye	$A_1H_L = +5.678$ mm
Position of second principal point, H'_L	$H'_{pc}H'_L = -1.215$ mm
H'_L referred to first vertex of eye	$A_1H'_L = +5.805$ mm

3. The Eye

Power of cornea	$F_c = +43.053$ D.
Power of lens	$F_L = +19.110$ D.
Reduced thickness = (0.050+5.678)/1.336	c = 0.004287 m
Power of eye	$F = +58.64$ D.
Position of first principal point, H	$H_cH = 1.397$ mm
H referred to first vertex of eye	$A_1H = +1.348$ mm
Position of second principal point, H'	$H'_LH' = -4.205$ mm
H' referred to first vertex of eye	$A_1H' = +1.600$ mm
Separation between principal points	$HH' = 0.252$ mm

The power of the eye is 58.64 D. The focal lengths are:

$$f = FH = 1000/58.64 = 17.053 \text{ mm}$$
$$f' = F'H' = -1336/58.63 = -22.783 \text{ mm}$$

The distance of the first focal point from the front vertex of the cornea is

$$FA_1 = FH + HA_1 = 17.053 - 1.348 = 15.705 \text{ mm}$$

The distance of the second focal point from the front vertex of the cornea is

$$A_1F' = A_1H' + H'F' = 1.6 + 22.783 = 24.383 \text{ mm.}$$

The Gullstrand eye is too short by 0.383 mm. As noted, this corresponds to 1 diopter of hyperopia.

The nodal points of the eye may be found from the identities

$$F'H' = NF = NA_1 + A_1F = -22.783 \text{ mm, thus,}$$
$$\mathbf{A_1N = 7.078 \text{ mm}}$$

and $\quad FH = N'F' = N'A_1 + A_1F' = 17.053$ mm,
thus, $\quad \mathbf{A_1N' = 7.330 \text{ mm}}$

The separation between the two nodal points is 0.252 mm. These results are illustrated in Figure 11.11.

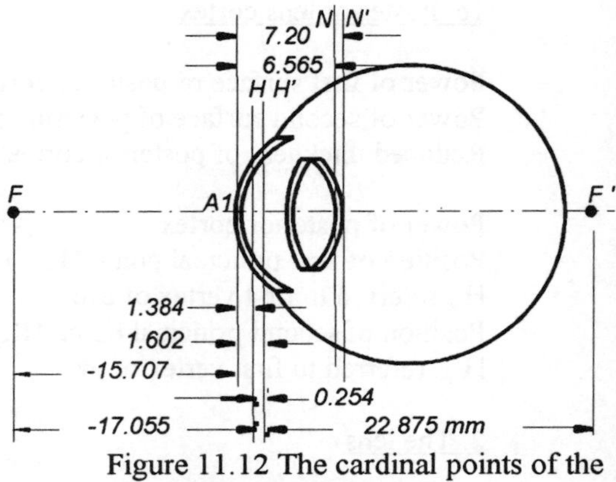

Figure 11.12 The cardinal points of the Gullstrand Schematic eye.

11.8 PURKINJE IMAGES

We have seen that when light strikes a polished glass surface a small portion of the light is reflected as from a mirror. The cornea and crystalline lens of the eye similarly act as spherical mirrors. The images that they form are called the Purkinje images. The virtual image due to reflection at the first surface of the cornea (No. I) is called the corneal reflex. It is the brightest reflection because the difference in indices between cornea and air is the greatest. Illustrations of eyes often depict this image as a small distorted windowpane or a bright spot on the cornea. The position of this image may be found by applying the equation for reflection by a spherical mirror A weak and not easily visible reflection, Purkinje Image II, is produced by the posterior surface of the cornea. The position of this image necessitates refracting the rays twice through the anterior surface of the cornea and reflecting once from the posterior surface, or else computing the cardinal points of such a system.

The third Purkinje image is produced by reflection from the front surface of the lens. Unlike images I and II, Purkinje Image III depends on the state of accommodation of the lens. Its changing position and size, in fact, have been used to show that the lens becomes rounder as it accommodates. An iterative calculation of image III involves five surfaces, refraction twice by the cornea and reflection once at the anterior surface of the lens.

The fourth Purkinje image is a real image produced by reflection at the posterior surface of the lens. It also changes with accommodation. Six refractive surfaces and one reflective surface are encountered in computing this image. Surface by surface calculations are not merely tedious, but must be repeated each time the object position is changed. Although a considerable effort is required to find the cardinal points of such systems, once found the system is reduced to an equivalent thin mirror and the computation of images is simplified.

11.8.1 CALCULATION OF PURKINJE IMAGE I

The front surface of the cornea, treated as a convex mirror of radius +7.7 mm has a reflecting power

$$F = -2/0.0077 = -259.74 \text{ D}.$$

EXAMPLE 7: Find the image of a 6 foot high window, located 2 meters away.

$U = -0.5$ D; $U' = -.5 - 259.74 = -260.24$ D

thus, $u' = 3.843$ mm = image distance.

$Y = U/U' = 0.00192$, $\therefore y' = 72(0.00192) = 0.138$ inch.

11.8.2 CALCULATION OF PURKINJE IMAGE II

The second Purkinje image is formed by the cornea acting as a thick mirror. Light refracts at the anterior surface of the cornea, reflects at its posterior surface and refracts out the anterior surface. Images may be found by iterative calculations at each of these interfaces, but would have to be repeated for each new object position. Instead we will find the image of the posterior corneal surface as formed by refraction by the front surface. This will become an "equivalent mirror" which replaces the cornea. Note: light travels from *right to left!* Also note that vertex A_2 is the object.

Given: $n = 1.376$; $n' = 1.000$;
$u = A_1A_2 = -0.5$ mm $= -0.0005$ m

$\therefore U = 1.376/-.0005 = -2752$ D.
$F = (1.000-1.376)/-0.0077 = 48.831$ D.
and, $U' = -2752 + 48.831 = -2703.169$ D.

thus, $u' = 1000/-2703.169 = -0.37$ mm
$= A_1A_2' = A_1A$,

where the image point A_2' becomes the "vertex" **A** of the "*equivalent thin mirror*". The vertex **A** is 0.37 mm to the right of the front surface of the cornea.

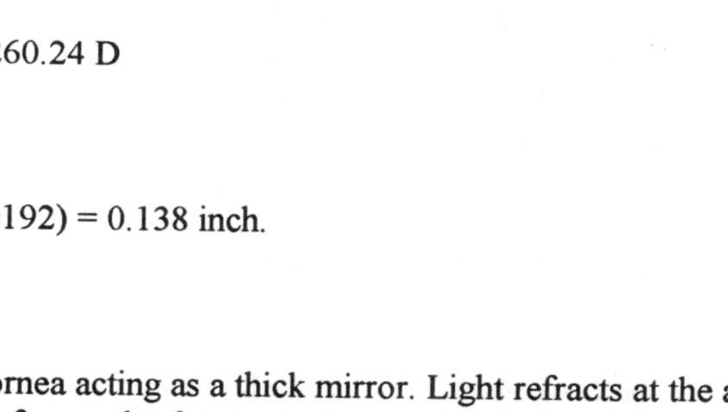

Figure 11.13 Purkinje Image II is produced by refraction at the first surface of the cornea, reflection at the second surface, and refraction at the first surface. The lower figure shows the "equivalent thin mirror".

Next we will find the image, produced by refraction by the anterior surface of the cornea, of the center of curvature C_2 of the posterior surface of the cornea. The radius of curvature of this surface is $A_2C_2 = -6.8$ mm.

$$u = A_1C_2 = A_1A_2 + A_2C_2 = -0.5 - 6.8 = -7.3 \text{ mm}$$

$$U = 1.376/-0.0073 = -188.493 \text{ D.}$$

$$U' = -188.493 + 48.831 = -139.662 \text{ D.}$$

thus, $u' = -7.160$ mm $= A_1C$

Now we restore the conventional light path of *left to right*. The "equivalent thin mirror" is convex to the incident light and the radius of curvature of this "mirror" is measured along the direction that light travels, thus, AC = 7.160 - 0.37 = 6.790 mm. The reflecting power F = -2/0.00679 = -294.545 D. This equivalent thin mirror replaces the corneal thick mirror. Its principal points coincide with A, and its nodal points with C. Object/image calculations are greatly simplified. See Figure 11.13.

EXAMPLE 8: Given a windowpane at -2000 mm from the cornea (U = -0.5 D if we neglect the 0.37 mm distance from A_1 to A), the second Purkinje Image of the windowpane will be located at

U' = -0.5 - 294.545= -295.045 D

thus, u' = - 1000/-295.045 = 3.389 mm = AM'

Measured from the anterior vertex of the cornea, it is 3.389 + 0.37 = 3.759 mm.

The lateral magnification is Y = U/U' = -0.5/-295.045 = 0.00170

Thus the size of the Purkinje image is y' = 72(0.00170) = 0.1222 inch

11.8.3 CALCULATION OF PURKINJE IMAGE III

The third Purkinje image is produced by reflection from the anterior surface of the crystalline lens. Rays are refracted four times (twice refracting through both surfaces of the cornea) and reflected once by the thick mirror comprising the cornea and lens surface. We will again find an "equivalent thin mirror." To simplify the calculation we will replace the cornea with its principal planes H_c and H'_c. The power of the cornea, 43.053 D, was calculated in Section 11.8. The images of the front surface of the lens (A_3) and its center of curvature (C_3) will define the equivalent thin mirror. The procedure is similar to the Purkinje II calculation. Note: 1) Light travels from *right to left*, i.e., from the lens to the cornea. 2) the distances of the principal planes of the cornea from the front of the cornea: A_1H_c and $A_1H'_c$ = +0.05 mm; 3) Vertex A_3 is the object point M.

Given: n = 1.336; n' = 1.000; u = $H_cA_3 = H_cA_1 + A_1A_3$ = -0.05 - 3.6 mm = -.00365 m

U = 1.336/-.00365 = -366.027 D. F = 43.053 D. U' = -366.027 + 43.053 = -322.974 D.

thus, u' = 1000/-322.974 = -3.096 mm = $H'_cA'_3 = H'_cA$

where the image point A'_3 is the vertex **A** of the "equivalent mirror." The distance of the "vertex" from the cornea is $A_1A = A_1H'_c + H'_cA$ = +0.05-3.096 = -3.046.

The vertex **A** of the "equivalent mirror" is 3.046 mm to the right of the front surface of the cornea. See Figure 11.14.

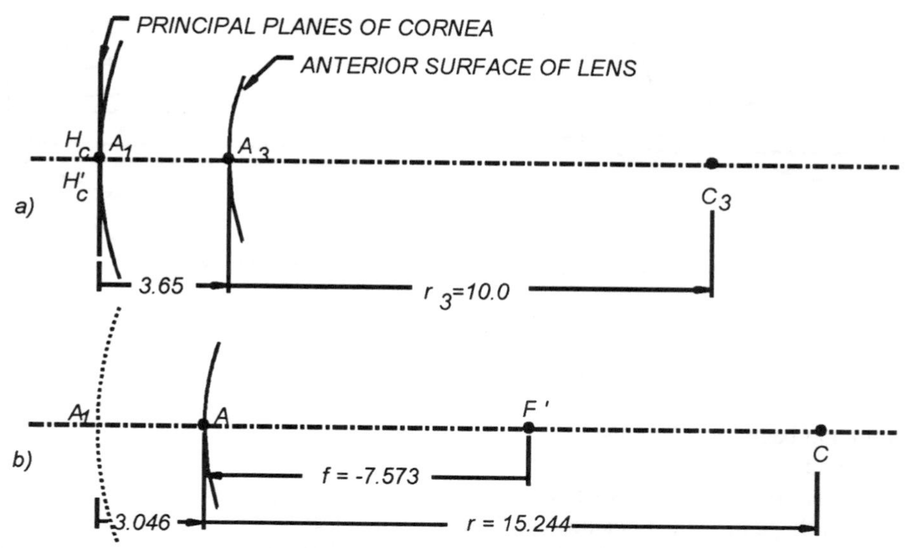

Figure 11.14 Purkinje Image III a) Cornea is represented by H_c and H'_c. b) Equivalent thin mirror is at A, r = AC.

Next we will find the image of the center of curvature C_3 of the posterior surface of the cornea. It has a radius A_3C_3 = 10.0 mm. The object distance u = H_cC_3 = -13.65 mm.

U = 1.336/ -0.01365 = -97.875 D. $\quad F$ = 43.053 $\quad$ U' = -97.875 + 43.053 = -54.822 D.

thus, $\quad$ u' = 1000/-54.822 = -18.241 mm = $H'_cC'_3 = H'_cC$

where the image point C'_3 is the center of curvature **C** of the "equivalent mirror". The distance of the center of curvature C from the cornea is $A_1C = A_1H'_c + H'_cC$ = +0.05 -18.241 = -18.191 mm

The radius of curvature of the "mirror" is AC = $AA_1 + A_1C$ = +3.046 - 18.191 = -15.145 mm. The mirror is actually convex to light traveling from *left to right*; therefore, its radius of curvature is +15.145 mm, and f= -7.573 mm. The *reflecting power of the equivalent thin mirror* is F = -2000/15.145 = -132.057 D.

EXAMPLE 9: Find the third Purkinje Image of a 72-inch high windowpane, located 2 meters away

U = -0.5 D $\quad F$ = -132.057 D $\quad$ U' = -.5 - 132.057 = -132.557 D

thus, $\quad$ u' = -1000/-132.557 = 7.544 mm = AM'

Measured from the anterior vertex of the cornea, it is $A_1M' = A_1A$ + AM' = 3.046 + 7.544 = 10.590 mm

Y = U/U' = -0.5/-132.557 = 0.00377; $\quad$ y' = 72(0.00377) = 0.272 inch.

A similar but more lengthy calculation is necessary to find the "equivalent thin mirror" that forms the fourth Purkinje image.

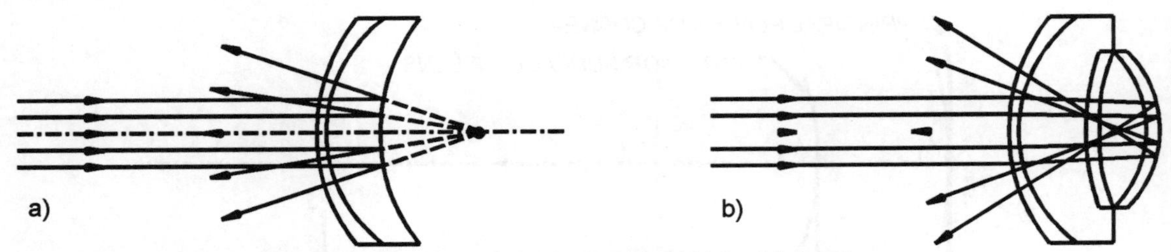

Figure 11.15 Exact ray tracing of a) Purkinje III and b) Purkinje IV.

REVIEW QUESTIONS

1. Define front and back focal lengths.
2. Define front and back vertex power.
3. How does increasing the lens thickness affect the positions of the principal planes?
4. How does lens shape affect the positions of the principal planes?
5. Cite two applications in photography of shifting the principal planes.
6. Why is vertex power used for specifying ophthalmic lenses?
7. Why do presbyopes allow their reading glasses to slide down their noses when reading?
8. Which Purkinje images are virtual and erect?
9. How are effective and vertex powers related?
10. Which ocular surfaces produce Purkinje images I-IV?

PROBLEMS

11.1 Find (a) the equivalent power, (b) positions of the principal and focal points, (c) the front and back vertex powers, (d) the separation between the principal points, and draw a clearly labeled diagram for the following lenses:

1a. An equiconvex lens of index 1.5, with radii of 15 cm, and 3 cm thickness.

Ans.: (a) $F = 6.444$ D, (b) $A_1H = 1.034$ cm, $A_2H' = -1.034$ cm, $A_1F = -14.483$ cm, $A_2F' = 14.483$ cm, (c) $F_v = 6.905$ D and $F_v' = 6.905$ D, (d) $HH' = 0.931$ cm.

1b. A meniscus lens of index 1.5, $r_1 = +15$ cm, $r_2 = +10$ cm, and 3 cm thickness.

Ans.: (a) $F = -1.333$ D, (b) $A_1H = 7.5$ cm, $A_2H' = 5$ cm, $A_1F = 82.5$ cm, $A_2F' = -70.0$ cm, (c) $F_v = -1.212$ D and $F_v' = -1.428$ D, (d) $HH' = 0.5$ cm.

1c. A meniscus lens of index 1.5, $r_1 = +10$ cm, $r_2 = +15$ cm, and 3 cm thickness.

Ans.: (a) $F = 2.00$ D, (b) $A_1H = -3.333$ cm, $A_2H' = -5$ cm, $A_1F = -53.333$ cm, $A_2F' = 45$ cm, (c) $F_v = 1.875$ D and $F_v' = 2.222$ D, (d) $HH' = 1.333$ cm.

1d. An equiconcave lens of index 1.5, with radii of 15 cm, and 3 cm thickness.

Ans.: (a) F = -6.889 D, (b) A_1H = 0.968 cm, A_2H' = -0.968 cm, A_1F = 15.484 cm, A_2F' = -15.484 cm (c) F_v = -6.458 D and F_v' = -6.458 D, (d) HH' = 1.065 cm.

11.2. Given a real object 5cm high, located 25 cm from the front surface of each of the above lenses, find (a) the distance of the image from the second surface of the lens, (b) the size of the image.

2.1 Ans.: (a) A_2M' = +37.377 cm, (b) y' = -7.377 cm.

2.2 Ans.: (a) A_2M' = -17.674 cm, (b) y' = +3.488 cm.

2.3 Ans.: (a) A_2M' =-43.235 cm, (b) y' = +8.824 cm.

2.4 Ans.: (a) A_2M' =-10.279 cm, (b) y' = +1.793 cm.

11.3. A concentric lens of index 1.5 has radii of +5 and -1 cm. Find (a) the equivalent power, (b) positions of the principal and focal points, (c) the front and back vertex powers, and draw a clearly labeled diagram.

Ans.: (a) F = 40. D, (b) A_1H = 5.00 cm, A_2H' = -1.00 cm, A_1F = 2.50 cm, A_2F' = 1.50 cm (c) F_v = -40.00 D and F_v' = 66.667 D.

11.4. A meniscus concentric lens of index 1.5 has radii of +7.5 and +3.0 cm. Find (a) the equivalent power, (b) positions of the principal and focal points, (c) the front and back vertex powers, and draw a clearly labeled diagram.

Ans.: (a) F = -6.667 D, (b) A_1H = 7.50 cm, A_2H' = 3.00 cm, A_1F = 22.50 cm, A_2F' = -12.00 cm (c) F_v = -4.444 D and F_v' = -8.333 D.

11.5. The equivalent power of a concentric lens is -5 D. The radius of the second surface is +4 cm. How thick is the lens, if its index is 1.5? Ans.: d = 6 cm.

11.6. A glass ball of index 1.5 is 2 cm in diameter. Find (a) the equivalent power, (b) positions of the principal and focal points, (c) the front and back vertex powers, and draw a clearly labeled diagram.

Ans.: (a) F = 66.667 D, (b) A_1H = 1.00 cm, A_2H' = -1.00 cm, A_1F = -0.50 cm, A_2F' = 0.50 cm (c) F_v = 200.00 D and F_v' = 200.00 D.

11.7. A glass ball of index 2.0 is 2 cm in diameter. Find (a) the equivalent power, (b) positions of the principal and focal points, (c) the front and back vertex powers, and draw a clearly labeled diagram.

Ans.: (a) F = 100.00 D, (b) A_1H = 1.00 cm, A_2H' = -1.00 cm, A_1F = 0.00 cm, A_2F' = 0.00 cm (c) F_v = ∞ D and F_v' = ∞ D.

11.8. The glass ball in Problem 7 is immersed in water. Find (a) the equivalent power, (b) positions of the principal and focal points, (c) the front and back vertex powers, and draw a clearly labeled diagram.

Ans.: (a) F = 88.889 D, (b) A_1H = 1.00 cm, A_2H' = -1.00 cm, A_1F = -0.5 cm, A_2F' = 0.5 cm (c) F_v = 266.67 D and F_v' = 266.67 D.

11.9. How far in front of the ball in Problems 6, 7 and 8, must an object be placed to form a real inverted image of the same size? Ans.: Prob.6: -2 cm; Prob 7: -1 cm; Prob 8: -2 cm.

11.10. A 20 D equiconvex lens is 2 cm thick and of index 1.5 glass. What are the radii of curvature?
Ans.: $r_1 = -r_2$ = 4.641 cm, or 0.359 cm.

11.11 A telephoto lens is made of a +20 D and a -30 D lens separated by 25 mm. Find (a) the equivalent focal length, (b) positions of the principal and focal points, (c) the front and back focal lengths, and (d) the telephoto effect. Draw a clearly labeled diagram.

Ans.: (a) f = 200 mm, (b) A_1H = -150 mm, A_2H' = -100 mm; (c) A_1F = FFL = f_v = -350 mm; A_2F' = BFL = f_v' = 100 mm; (d) f/OVL = 1.6.

11.12. Reverse the above lens to form a wide-angle lens. Find (a) the equivalent focal length, (b) positions of the principal and focal points, (c) the front and back focal lengths, and (d) the ratio of focal length to overall length. Draw a clearly labeled diagram.

Ans.: (a) f = 200 mm, (b) A_1H = 100 mm, A_2H' = 150 mm; (c) A_1F = FFL = f_v = -100mm; A_2F' = BFL = f_v' = 350 mm; (d) f/OVL = 0.533.

11.13. The telephoto lens in Problem 11 is used to photograph the sun. The sun subtends 32 minutes of arc. How large is its image? Ans.: 1.86 mm

11.14. The lenses in Problems 11 and 12, initially focused for infinity, are refocused to photograph a scene 3 meters away from the front lens. By how much and in which direction must the film plane move?
Ans.: Prob. 11: +15.09 mm; Prob. 12: +13.79 mm.

11.15. Using the data for the Gullstrand eye, except that it is emmetropic (24 mm long), how large is the retinal image of the moon. The moon subtends 0.5°. Ans.: 0.149 mm

11.16. In Chapter 8, Problem 9 two concave mirrors face each other and are 75 mm apart. Their radii are 30 and 60 mm, respectively. An object 2 mm high is placed 20 mm from the first mirror. Find (a) the equivalent focal length, (b) positions of the principal and focal points, (c) the position and size of the image. Draw a clearly labeled diagram.

Ans.: (a) f = -15 mm, (b) A_1H = -37.5 mm, A_2H' = 75 mm; (b) A_1F = -22.5 mm; A_2F' = 60 mm; (c) A_2M' = -30.0 mm; y' = -12 mm

11.17. A 5 mm thick positive meniscus lens of index 1.5 has surface powers F_1 = +10 D and F_2 = -5 D. The back surface of the lens is 15 mm in front of the cornea of the eye. Find (a) the equivalent power; (b) the back vertex power and (c) the effective power of the lens at the cornea.
Ans.: (a) +5.17 D; (b) +5.34 D; (c) +5.81 D.

11.18. A 5 mm thick negative meniscus lens of index 1.5 has surface powers F_1 = +5 D and F_2 = -10 D.

The back surface of the lens is 15 mm in front of the cornea of the eye. Find (a) the equivalent power; (b) the back vertex power and (c) the effective power of the lens at the cornea.

Ans.: (a) -4.83 D; (b) -4.92 D; (c) -4.58 D.

11.19. Using the data for the Gullstrand eye, find (a) the far point (the point in space to which the retina is conjugate), and (b) the refractive error of the eye, if the length of the eye is 25.667 mm.

Ans.: (a) HM = 319.64 mm; (b) 3.13 D myopia.

11.20. Repeat Problem 19 for an eye length of 22.8 mm.

Ans.: (a) HM = 228.4 mm; (b) 4.38 D hyperopia.

11.21. Using the data from Section 11.8.1 Calculation of Purkinje Image I, draw a scale diagram of the cornea and lens, and show the position of the image.

11.22. Using the data from Section 11.8.2 Calculation of Purkinje Image II, draw a scale diagram of the cornea and lens. Show the position and center of curvature of the equivalent "mirror" and the position of the image.

11.23. Using the data from Section 11.8.3 Calculation of Purkinje Image III, draw a scale diagram of the cornea and lens. Show the position and center of curvature of the equivalent "mirror" and the position of the image.

11.24. A thin equi-convex lens, with surface powers of +5 D, is in the spectacle plane at a vertex distance from the cornea of 15 mm. The lens corrects the refractive error of the eye. Find the power of the equivalent contact lens.

Ans.: +11.76 D.

11.25. A thin meniscus lens has powers of F_1 = +10 D and F_2 =-15 D. It is 15 mm from the cornea. Find the equivalent contact lens.

Ans.: -4.65 D.

11.26. The patient pushes the eyeglass frame in which the above lens is mounted closer to the eye. The vertex distance becomes 10 mm. Find the effective power at the cornea.

Ans.: -4.76 D.

11.27. A thin biconvex lens, with surface powers of +10 D and +5 D, is at a vertex distance of 10 mm. Find its effective power at the cornea.

Ans.: +17.65 D.

11.28. The patient with the above Rx, to examine a fine watch mechanism, slides the glasses down along his nose until it is 20 mm from the cornea. How much of an add has he created?

Ans.: +3.78 D.

11.29. A 15 mm thick lens (n=1.5) has surface powers of +10 D and -11.1111 D. Find: a) the effective power of the front surface at the back surface, and b) the back vertex power of the lens.

Ans.: a) +11.111 D; b) 0 D.

11.30. An aphakic eye is 25 mm long. The power of the cornea, as a single spherical refracting surface, is +20 D. A lens is to be implanted at a distance of 10 mm from the cornea. Find; a) the power of the implanted lens in the ocular medium of index 4/3; b) the equivalent power of the eye; c) the back vertex power of the eye. Assume the lens is thin.

Ans.: a) +65.36 D; b) +75.56 D; c) +88.89 D.

11.31. Suppose in Problem 30, that the implanted lens power is +60 D. (therefore incorrect). Find: a) the distance to the far point (HM); b) the residual ametropia.

Ans.: a)+260.662 mm; b) 3.84 D hyperopia.

11.32. Suppose in Problem 30, that the implanted lens power is +70 D. Find: a) the distance to the far point (HM); b) the residual ametropia. **Ans.: a)-296.104 mm; b) 3.38 D myopia.**

11.33. Show that distance HN = (n'-n)/F.

11.34. Given F_2=-6.00 D,and c = 0.01, design a lens with a BVP = +10 D.

Ans.: F_1 = +13.793 D; F = +8.621 D.

CHAPTER 12

STOPS, PUPILS AND PORTS

12.1 TYPES OF STOPS

In paraxial optics all rays from an object point are assumed to converge on the conjugate image point. No restriction is placed on the diameter of the lens system or the size of the object. Obviously, only rays that pass through the lenses and stops of an optical system can reach the image plane. As shown in Figure 12.1, the rays through **F** and **F'**, which locate **Q'**, are not transmitted through the lens, they are merely construction rays. Only the rays in the shaded regions are transmitted. The lens aperture is a stop. Stops in optical systems are physical apertures, such as lens rims, iris diaphragms and rings. Stops have various functions. An *aperture stop* limits how much light from each point of the object reaches the conjugate image point. The iris diaphragm of a camera lens is an aperture stop. The duration of the film exposure depends upon its diameter, and its position in the lens serves to minimize aberrations by blocking aberrated rays. A *field stop* limits the *field of view* (FOV) i.e., the extent of the object to be imaged by the system. The ideal field stop produces a sharply demarcated image. A *baffle* prevents glare or scattered light reflections off the inside walls of the system from reaching the image plane. In Figure 12.2.a an optical system, designed to collect light from a distant axial object and to transmit it through a pinhole, due to internal reflection, also transmits stray light from an off-axis point. The ring shaped baffle in Fig. 12.2.*b* blocks the reflected rays from reaching the pinhole.

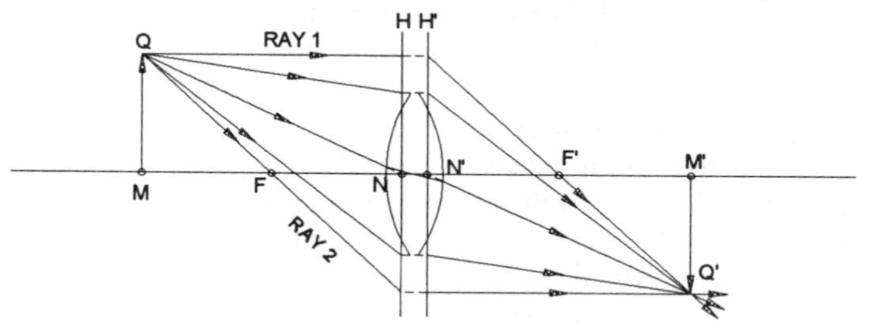

Figure 12.1 The lens is an aperture stop. The lens diameter limits the pencil of rays that is transmitted. Rays 1 and 2 are merely for constructing the image.

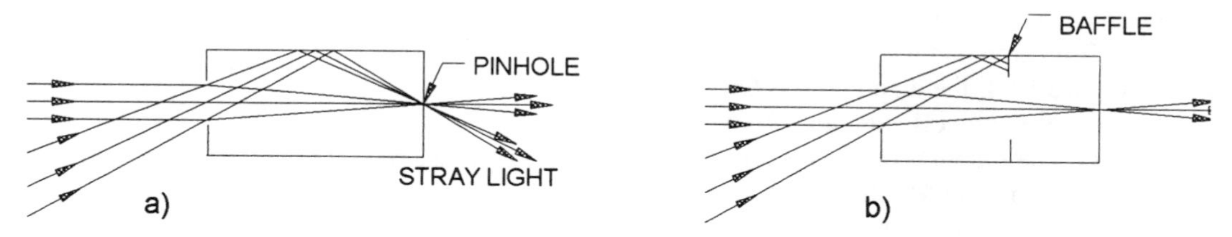

Figure 12.2 a) Stray light is transmitted through pinhole. b) Baffle blocks stray light.

12.2 PUPILS

All optical systems, from a simple lens to the most complex, contain an aperture stop. The rim of a simple lens is the aperture stop. The aperture stop can be a diaphragm in front of a lens, behind the lens, or between two or more lenses, i.e., a front, a rear or an interior stop, respectively. *Pupils* are images of aperture stops. See Figure 12.3. The *entrance pupil* is the image of the aperture stop produced by all lenses in front of it. The *exit pupil* is the image of the aperture stop formed by all lenses that follow it.

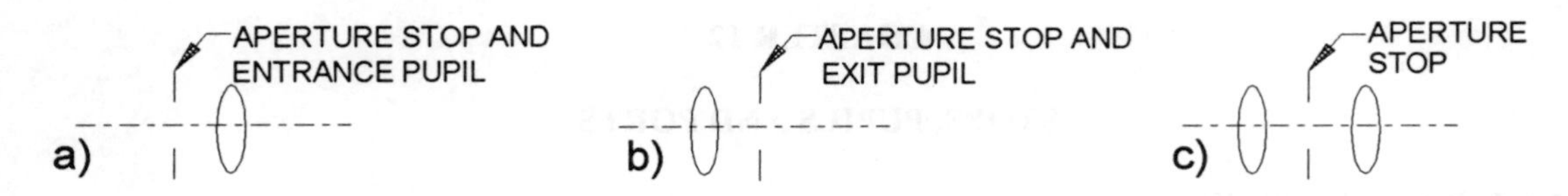

Figure 12.3 a) Front Stop; b) rear stop; c) interior stop.

The entrance pupil coincides with the aperture stop if it is a front stop. The exit pupil coincides with the aperture stop if it is a rear stop. Both an entrance and an exit pupil will be formed when the aperture stop lies between lenses. The entrance and exit pupils are conjugate to the aperture stop and to each other. Although pupils are images, they effectively block rays from being transmitted through the system, because each point in the pupil corresponds to a point in the aperture stop. Consequently, only rays through points in the pupil will get through the aperture stop.

The entrance pupil belongs to object space. It is found by placing your eye at the object position **M** and looking into the optical system. The smallest visible opening is the entrance pupil. If that smallest opening is an image of a component, that component is the aperture stop. Rays are drawn as straight lines from the object point to the opposite edges of the entrance pupil. The solid angle formed by these rays shows the light gathering efficiency of the system. Except front stops, the rays merely appear to go through the entrance pupil. They actually are refracted through the aperture stop.

Similarly, the exit pupil is the smallest opening seen from the image point **M'**. The exit pupil belongs to image space. The solid angle subtended by rays drawn from its edges to the image determines the illumination in the image.

Because an object may be at any distance from an optical system it is evident that the smallest opening seen from **M** will depend on where **M** is with respect to the various apertures or aperture images of the system. For example, in a simple box camera containing a front stop and a simple lens, as shown in Figure 12.4, the diaphragm is the aperture stop for objects ranging from infinity to $\mathbf{M}_3$. The diaphragm and the lens subtend the same angle at $\mathbf{M}_3$, thus, both act as the aperture stop. Between $\mathbf{M}_3$ and $\mathbf{M}_4$ the lens is the aperture stop.

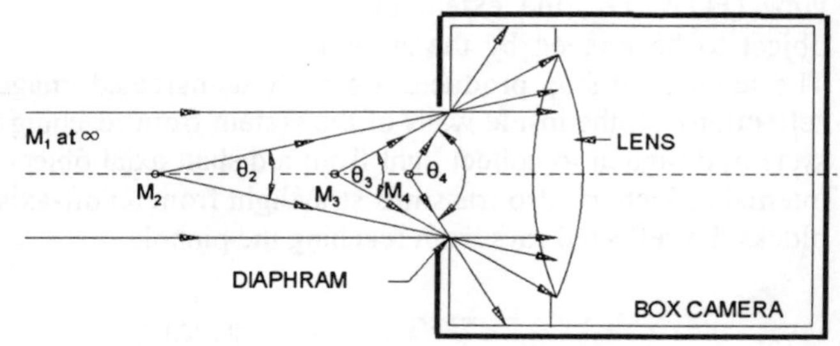

Figure 12.4 The entrance pupil is determined by the object distance.

12.3 THE FIELD STOP AND THE ENTRANCE AND EXIT PORTS

Once the aperture stop of a system has been identified the *field stop* must be found, to establish the field of view **FOV**. Ideally, the field stop is in an image plane. For example, in a camera consisting of a simple lens and a film plane, the lens will be the aperture stop and the frame that borders the film is the field stop. In a complex lens system another component, such as a diaphragm or lens element, may be the field stop.

The *entrance port* is the image of the field stop formed by the lenses in front of it. It is the smallest visible opening seen from the center of the entrance pupil and belongs to object space. Correspondingly, the *exit port* is the image of the field stop formed by the lenses that follow it and it belongs to image space

12.4 FIELD OF VIEW AND VIGNETTING

Suppose that an optical system consists of a diaphragm placed a short distance in front of a simple lens, and an object and image plane, as shown in Figure 12.5. As shown, the diaphragm is the aperture stop and the lens is the field stop. The amount of light transmitted through the two stops from a point in the object plane depends on the height of the object point. Rays obstructed by the stops are *vignetted.* Vignetting determines the field of view.

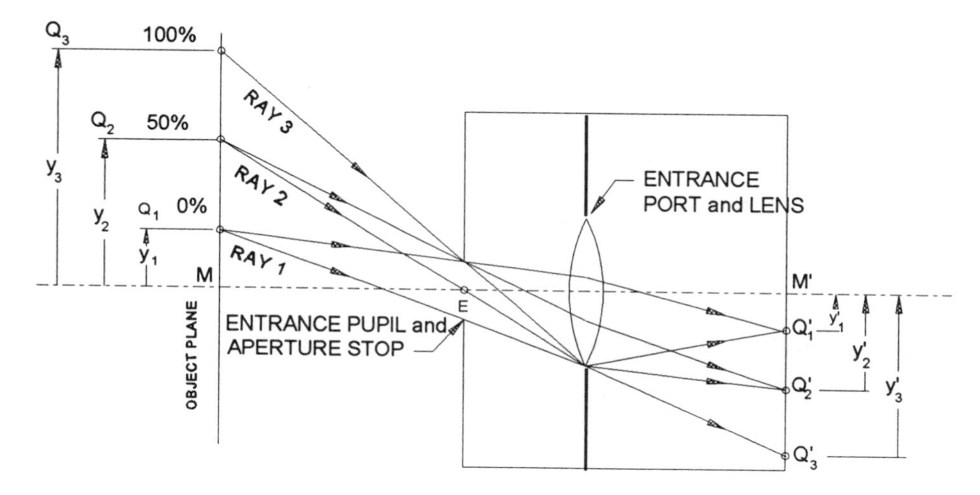

Figure 12.5 Vignetting. Points in the object from M to Q_1 (y_1) undergo zero vignetting. Vignetting increases to 50% for the points from Q_1 to Q_2. Ray 2 is the chief ray. Vignetting increases to 100% at Q_3.

Rays from all object points between **M** and $\mathbf{Q}_1$ *fill* the entrance pupil, and are transmitted by the field stop (the lens). Rays from this central zone of the object undergo zero vignetting. The boundary for *zero vignetting is defined by the ray (No. 1) that grazes corresponding edges of the entrance port and entrance pupil.*

Rays from object points beyond $\mathbf{Q}_1$ become increasingly vignetted. The ray (No. 2) *that grazes the field stop and goes through the center of the aperture stop determines the 50% vignetting zone in the field.* This ray originates at object $\mathbf{Q}_2$.

Finally, Ray 3, *which grazes opposite edges of the field stop and aperture stop, corresponds to the 100% vignetting zone.* It is the only ray from point $\mathbf{Q}_3$ to be transmitted. In effect, the image has a central disk of maximum illumination then gradually grows dimmer until it disappears.

The field of view is understood to be to the 50% vignetting zone, because it is independent of the size of the aperture stop. Ray 2 from object Q_2 is a *chief ray*. It will be transmitted even though the aperture stop is reduced to a pin hole. The angular field of view in object space is equal to the angle subtended by object y_2 at the center of the entrance pupil, i.e.,

$$\tan\theta = \frac{y_2}{ME}$$

The total *angular* field of view at 50% vignetting is 2θ. Alternatively, the total *linear* FOV in object space is $2y_2$.

12.5 PROCEDURE FOR FINDING THE STOPS, PUPILS, PORTS AND FIELD OF VIEW

The best way to explain the procedure for analyzing the pupils, ports and field of view of a system is by example. Consider an optical system consisting of two lenses and an interior stop, as illustrated in Figure 12.6. The object is 200 mm in front of the first lens, and the system data are tabulated below:

	Focal length	Diameter	Separation	Index
Object			$MA_1 = 200$ mm	air
Element 1	$f_1 = 100$ mm	$D_1 = 20$ mm	$A_1A_2 = 200$ mm	air
Element 2	$f_2 = \infty$	$D_2 = 5$mm	$A_2A_3 = 100$ mm	air
Element 3	$f_3 = 50$ mm	$D_3 = 18$ mm		

Step 1. Find image position and size. Figure 12.6.

1st lens. $$\frac{1}{u_1'} = \frac{1}{-200} + \frac{1}{100}$$

Thus, $u_1' = +200 = A_1M_1' = A_1M_2$

2 n d lens. $$\frac{1}{u_2'} = \frac{1}{-100} + \frac{1}{50}$$

Thus, $u_2 = u_1' - A_1A_3 = 200 - 300 = -100$ mm
Therefore, $u_2' = +100 = A_3M_2' =$ distance of image from lens No. 2.

The lateral magnification is

$$Y = \frac{u_1'}{u_1} \cdot \frac{u_2'}{u_2} = \frac{200}{-200} \cdot \frac{100}{-100} = +1$$

Note: Steps 2-4 determine the positions and sizes of the several apertures or openings seen from the object point.

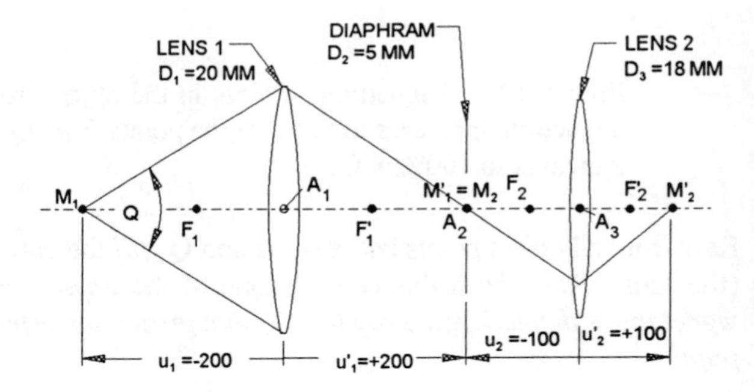

Figure 12.6 Optical system schematic from object to image surface.

Step 2. Find the angle subtended by the first lens at M. See Figure 12.6

$$\theta_1 = \frac{D_1}{MA_1} = \frac{20}{200} = 0.1 \textit{ radian}$$

Note: The diaphragm and the 2nd lens are not directly seen from point M, therefore, find the positions and sizes of their images, as formed by the 1st lens.

Step 3. Find the image of the diaphragm, where u = -200, and f_1 =100. Note: light travels from *right to left*. See Figure 12.7.

$$\frac{1}{u'} = \frac{1}{-200} + \frac{1}{100}$$

thus, u' = +200 mm = distance from lens No. 1

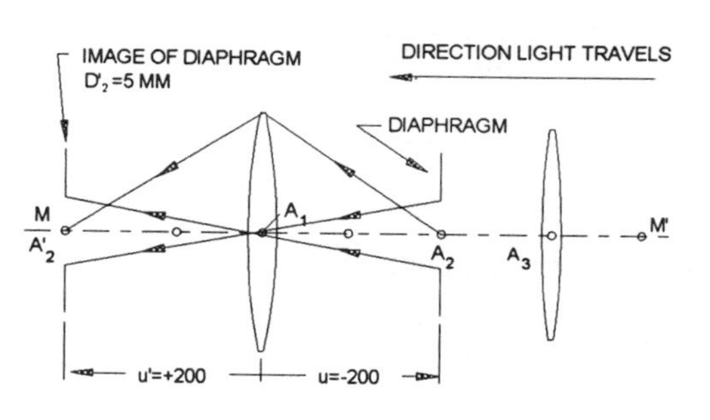

Figure 12.7 Location and size of the image of the diaphragm in object space. This step is required for finding the entrance pupil.

The image of the diaphragm is at the object plane. It is inverted and its diameter D_2' is

$$D_2' = D_2 \cdot \frac{u'}{u} = 5 \cdot \frac{200}{-200} = -5\ mm$$

What angle does this image subtend at M?

$$\theta_2 = \frac{D_2'}{MA_2'} = -\frac{5}{0} = \infty\ radians$$

Step 4. Find the image of lens No. 2, given u = -300, and f_1 = 100. See Figure 12.8.

Result: u' = +150 mm = distance from lens No. 1, but distance to M = 50 mm. The image size is:

$$D_3' = D_3 \cdot \frac{u'}{u} = 18 \cdot \frac{150}{-300} = -9.0\ mm$$

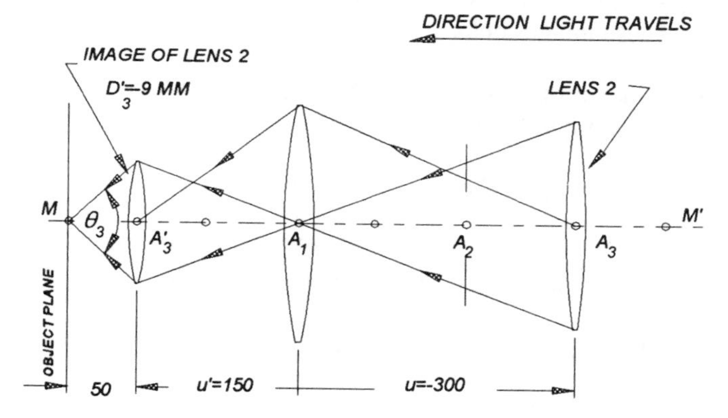

Figure 12.8 Location and size of the image of Element 3 lens in object space. This step is required for finding the entrance pupil.

Find the angle subtended by the image at M.

$$\theta_3 = \frac{D_3'}{MA_3'} = \frac{9.0}{50} = 0.18\ radians$$

Step 5. Identify the entrance pupil and aperture stop. Lens No. 1 subtends the smallest angle. Therefore, it is the entrance pupil and aperture stop.

Step 6. Find the exit pupil. See Figure 12.9. The exit pupil is the image of the aperture stop (lens No. 1) formed by lens No. 2. u = -300 mm, f_3 = 50 mm.

$$\frac{1}{u'} = \frac{1}{-300} + \frac{1}{50}$$

Therefore, u' = +60 mm (distance from lens No. 2).

Find the size of exit pupil D_1':

$$D_1' = D_1 \cdot \frac{u'}{u} = 20 \cdot \frac{60}{-300} = -4 \ mm$$

The field stop will be either Lens No. 2 or the diaphragm. *To find the entrance port, determine which of the object space openings that correspond to these elements appears to subtend the smallest angle at the entrance pupil. That element is the entrance port.* In this example, both elements were imaged by lens No. 1. These images belong to object space.

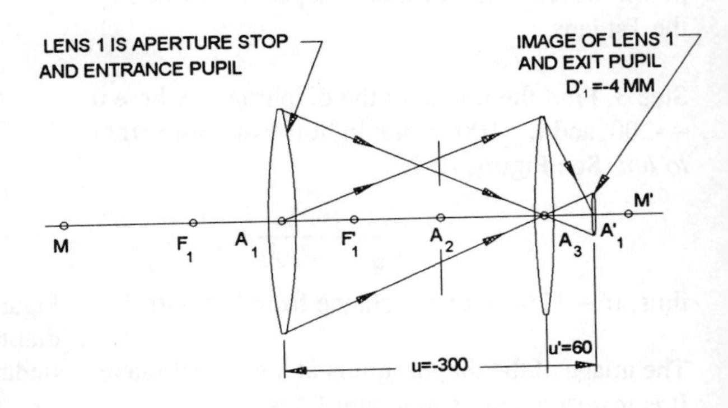

Figure 12.9 Location and size of exit pupil.

Step 7. Find entrance port, by comparing the angles subtended by images D_2' and D_3' at A_1, the center of the entrance pupil. Figure 12.10

The image of the diaphragm subtends $\theta_D = \frac{D_2'}{A_2'A_1} = \frac{5}{200} = 0.025 \ radian$

The image of Lens No. 2 subtends $\theta_L = \frac{D_3'}{A_3'A_1} = \frac{9.0}{150} = 0.060 \ radian$

The image of the diaphragm is the entrance port, therefore, the diaphragm is the field stop.

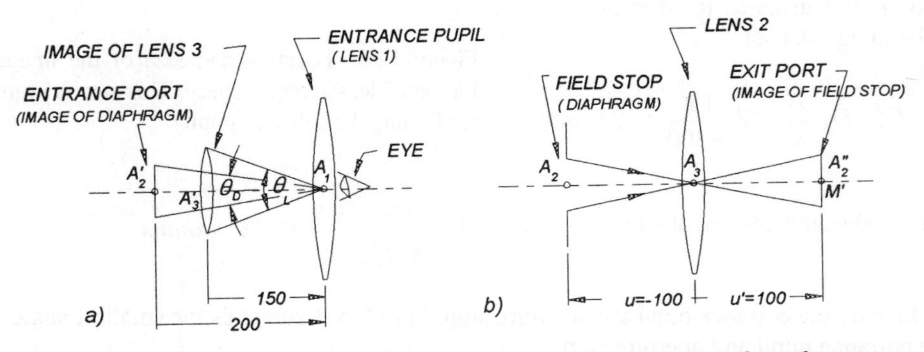

Figure 12.10 (a) The entrance port is the smallest opening in object space seen from the center of the entrance pupil. (B) The exit port is the image of the field stop produced by lens No. 2.

Step 8. Find the exit port. See Figure 12.11.
Note: The image of the field stop formed by all lenses that follow it is the exit port. u = -100 mm, f_3 = 50.

$$\frac{1}{u'} = \frac{1}{-100} + \frac{1}{50} \qquad \text{and} \qquad u' = +100 \text{ mm (distance from lens No. 2)}$$

The exit port coincides with the image plane. Why? Y = -1, therefore, the diameter of the exit port = -5 mm.

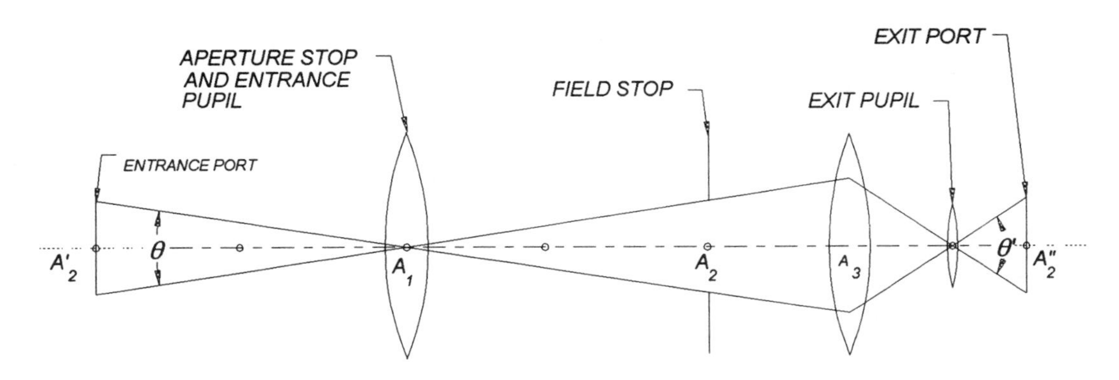

Figure 12.11 The fields of view are determined by the chief rays through the center of the aperture stop and edge of the field stop.

Step 9. Find angular field of view for 50% vignetting. Figure 12.12.
In object space, this is the ray that appears to graze the entrance port and go through the center of the entrance pupil. Actually, the ray grazes the field stop and goes through the center of the aperture stop.

$$\tan \theta/2 = 2.5/200 = 0.0125 \qquad \text{and} \quad \theta/2 = 0.7162°$$

Total field of view in object space is $\theta = 1.432°$

The field of view in image space is found from the slope of the ray through the center of the exit pupil to the edge of the exit port.

$$\tan \theta'/2 = 2.5/40 = 0.0625 \qquad \text{and} \quad \theta'/2 = 3.576°$$

The total field of view in image space is $\theta' = 7.153°$

Step 10. Find the linear field of view. Figure 12.12

Because the field stop in *this example* is in its ideal position, i.e., at the intermediate image plane, the ports coincide with the object and image planes. Therefore, the linear FOVs correspond to the diameters of the ports in *this example*: The total linear fields of view of the object the image = 5 mm.

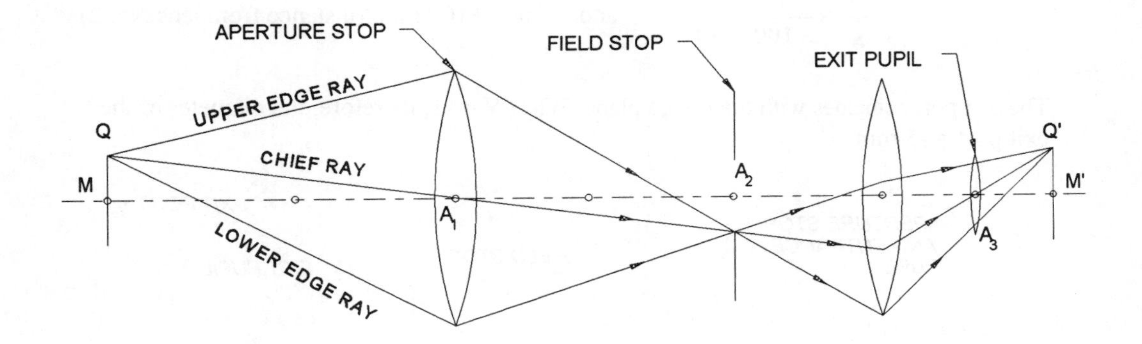

Figure 12.12 Rays from point Q, at the edge of the field of view fill the entrance and exit pupils. Note, however, that they fill only about half of the clear aperture of lens No. 2.

Field of view for zero and 100% vignetting.

Zero vignetting is determined with the ray through corresponding edges of the A.S. and F.S. (Ray 1); 50% vignetting, with the ray through the edge of the F.S. and the center of the A.S. (Ray 2); and 100% vignetting, with the ray through opposite edges of the A.S. and F.S. (Ray 3). *Because the field stop is in the intermediate image plane* the rays for zero, 50% and 100% vignetting meet at points Q and Q' in the object and image planes, and at the edge of the field stop. There is no gradual increase in vignetting by the field stop. The image is sharply circumscribed; illumination abruptly goes to zero at the edge of the exit port. See Figure 12.12. The linear fields of view equal 5 mm in object and image space. Why are they equal? .

Student exercise: Move the diaphragm 25 mm toward lens No. 1, increase the diameter of lens No. 2 to 20 mm, and repeat the calculations. Draw a diagram and show paths of the zero, 50% and 100% vignetting rays.

REVIEW QUESTIONS

1. What is the function of the aperture stop; the field stop; baffles?
2. What are pupils of an optical system? Must rays that go through the aperture stop also go through the pupils?
3. How can you find the entrance pupil of an optical system?
4. Where is the exit port of a camera?
5. Why is the field of view taken to be at the 50% vignetted angles?

PROBLEMS

1. A lens with a +50 mm focal length and a 25 mm diameter is 20 mm in front of a diaphragm of 20 mm diameter. The object is infinitely distant. a) identify the aperture stop, find the positions of the entrance and exit pupils; b) identify the field stop, find the positions of the entrance and exit port; c) find the total angular real and apparent field of view (total field at 50% vignetting), and the linear field of view in the image plane.

Ans.: a) the lens is the aperture stop, the entrance pupil and exit pupil coincides with the lens; b) the diaphragm is the field stop, the entrance port is 33.33 mm to the right of the lens, the exit port coincides with the diaphragm; c) 53.1°, 53.1°, 50 mm.

2. Repeat Problem 1 for an object plane 200 mm in front of the lens. In addition to answering the above question, find the linear field of view at the object and image planes.

Ans.: a) the lens is the aperture stop, the entrance pupil and the exit pupil coincide with the lens; b) the diaphragm is the field stop, the entrance port is 33.33 mm behind the lens and the exit port coincides with the diaphragm; c) 53.13°, 53.13°, object field of view = 200 mm, image field of view = 66.67 mm.

3. Reverse the lens and diaphragm in Problem 2. The object is 200 mm from the diaphragm. Find all quantities as above.

Ans.: a) the diaphragm is the aperture stop, the entrance pupil coincides with it, the exit pupil is 33.33 mm in front of the lens; b) the lens is the field stop, the entrance and exit ports coincide with it; c) 64°, 41.11°, object and image field of view = 250 and 73.53 mm.

4. A lens with a +50 mm focal length and a 12 mm diameter is 10 mm in front of another lens of 75 mm focal length and 10 mm diameter. The object is at ∞ a) identify the aperture stop, find the positions of the entrance and exit pupils; b) identify the field stop, find the positions of the entrance and exit port; c) find the total angular real and apparent field of view, and the linear field of view in the image plane.

Ans.: a) Lens No. 1 is the aperture stop, the entrance pupil coincides with it, the exit pupil is 11.54 mm in front of lens No. 2; b) lens No. 2 is the field stop, the entrance port is 12.50 mm to the right of the first lens, the exit port coincides with lens 2; c) 53.13°, 46.86°. Linear field of view of Image = 32.6 mm

5. Repeat Problem 4, but place the object 100 mm in front of the first lens.

Ans.: a) Lens No. 2 is the aperture stop, the entrance pupil is 12.5 mm to the right of lens No. 1; the exit pupil coincides with lens No. 2. b) lens No. 1 is the field stop, the entrance port coincides with the first lens, the exit port is 11.54 mm in front of lens 2; c) 51.28°, 61.92°. Linear field of view of object = 108 mm; Linear field of view of image = 49.1 mm.

6. A diaphragm with a 4 mm aperture is 50 mm in front of a lens of 25 mm focal length and 8 mm diameter. An object plane is 60 mm in front of the diaphragm. a) identify the aperture stop, find the positions of the entrance and exit pupils; b) identify the field stop, find the positions of the entrance and exit ports; c) find the total angular real and apparent fields of view, and d) the linear fields of view at the object

and image planes.

Ans.: a) the diaphragm is the aperture stop, the entrance pupil is at the diaphragm, the exit pupil is 50 mm behind the lens; b) the lens is the field stop, the entrance and exit ports coincide with the lens; c) 9.14°, 9.14°; d) 9.6 and 2.8 mm.

7. Repeat Problem 6 with an object plane at 40 mm from the diaphragm.

Ans.: a) the lens is the aperture stop, the entrance and exit pupils coincide with the lens; b) the diaphragm is the field stop, the entrance port coincides with the diaphragm, the exit port is 50 mm behind the lens; c) 4.58°, 4.58°; d) 7.2 and 2.76 mm.

CHAPTER 13

NUMERICAL APERTURE, f-NUMBER AND RESOLUTION

13.1 NUMERICAL APERTURE

The numerical aperture describes the light gathering power and determines the diffraction limited resolving power of an optical system. It is specified in object and image space:

$$NA_{\text{object space}} = n \sin \theta; \; NA_{\text{image space}} = n' \sin \theta' \quad 13.1$$

Where θ is the slope angle of the ray from the object to the edge of the entrance pupil, and θ' is the slope angle of the ray from the edge of the exit pupil to the image point. See Figure 13.1a. Matching the numerical apertures of optical systems arranged in series is important so they efficiently transmit light. In Figure 13.1b, the second lens has a smaller NA than the first lens, consequently, the illumination in the final image is reduced.

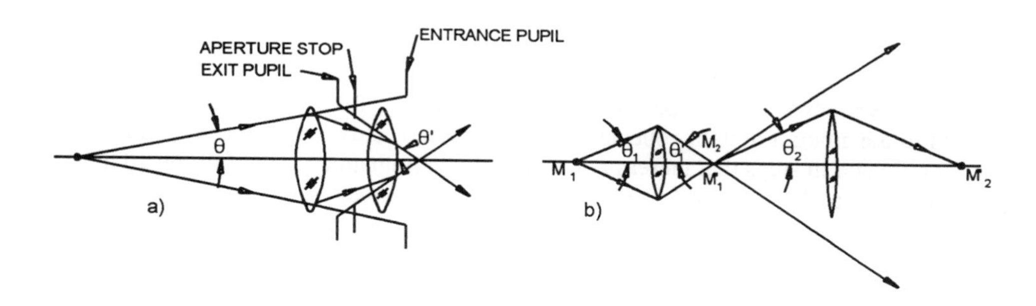

Figure 13.1 a) Numerical apertures in object and image space; b) Unmatched numerical apertures cause loss in transmitted light. The NA in object space of the second lens is smaller than the NA in image space of the first lens.

13.2 f-NUMBER

The numerical aperture is applicable to systems with finite conjugates. The light gathering power (speed) of a photographic lens, however, is defined for the lens focused for infinity. This *relative aperture* or *f-number* **(f/No)** is equal to the focal length divided by the diameter of the entrance pupil.

$$f/No = \frac{f}{D} \quad 13.2$$

The f-number is inversely related to the diameter **D** of the entrance pupil. The size of the solid cone of light that converges on the image point depends on the *area* of the pupil, and the area is proportional to D^2. For example, an f/8 lens with a 100-mm focal length will have a 12.5 mm entrance pupil diameter. At f/4, the diameter is twice as large (25 mm), but the pupil area is greater by a factor of four, as will be the illumination on the film.

According to Figure 13.2, for an *aplanatic* photographic lens, that is, a lens corrected for spherical aberration and coma, the second principal surface must be a sphere centered on F'. The sphere has a radius equal to the focal length. All rays are normals to the sphere, therefore, they all converge on F'. Consequently, the relationship between NA and f/No for a system in air is

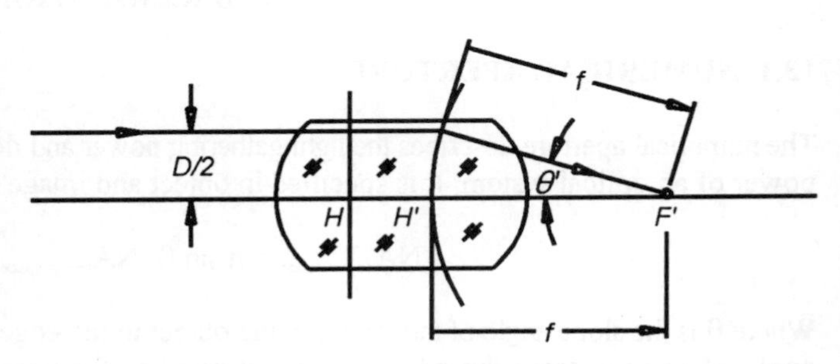

Figure 13.2 f-number.

$$\sin\theta' = \frac{D}{2f} = \frac{1}{2\,f/No} = NA$$

$$f/No = \frac{1}{2\,NA} \tag{13.3}$$

Modern camera lenses have relative apertures that range from f/1.0 to f/22. Because the f-number is inversely related to the amount of light collected, small f-numbers allow for shorter exposure times. The f-numbers are graduated in steps that increase or reduce the illumination on the film by a factor of two. Each step is called a stop. The standard f-numbers, relative pupil areas **A,** and relative exposure times **t** are:

f/No	22	16	11	8	5.6	4	2.8	2	1.4	1
Pupil Area	A	2A	4A	8A	16A	32A	64A	128A	256A	512A
Time (sec)	t	t/2	t/4	t/8	t/16	t/32	t/64	t/128	t/256	t/512

Since, the pupil area is proportional to the illumination on the film, the table above shows that the *product of illumination and exposure time is a constant.* Thus, photos taken at f/22 and a time of one second, and at f/1.4 and 1/256th of a second will be equally exposed. The one second exposure will be unsuitable for stopping action and the diffraction blur due to a small aperture is likely to be appreciable. The photo shot at f/1.4 will have little diffraction blur, but the blurs due to aberrations may limit the resolution of the photo. The relative light gathering efficiency of lenses of different f/Nos, e.g., f/a and f/b, is given by

$$\textit{Relative efficiency} = \left(\frac{a}{b}\right)^2.$$ Thus if the lenses are f/a = f/16 and f/b = f/4, the relative efficiency is 16.

In other words, the f/4 lens collects 16 times as much light as the f/16 lens.

13.3 ANGULAR RESOLUTION

The notion that light consists of rays that travel in straight lines works well for finding object and image positions and sizes. A direct experiment to isolate a ray might be to allow a pencil of rays to pass through a circular aperture and fall on a screen. As the aperture is made smaller, fewer rays are transmitted and the spot of light on the screen should constrict to a point. Although the spot on the screen initially does constrict, at a certain stage, as the aperture continues to shrink, the spot on the screen begins to spread. See Figure 13.3. This phenomenon, known as diffraction, sets a limit to the minimum size of an image. Diffraction is a bending of light as it passes near the edge of an aperture or lens rim. As a result, even a lens free from aberrations will not focus light to a point.

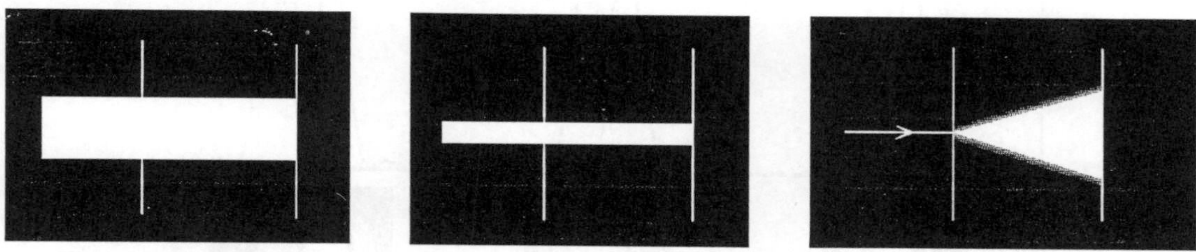

Figure 13.3 Diffraction increases with a reduction in the aperture diameter.

The diffracted image of a point source, produced by a lens with a circular aperture, is a pattern containing a central bright spot, known as the Airy disk, surrounded by a series of concentric bright and dark rings. Approximately 84% of the total energy is contained in the Airy disk. Another 7.1% is in the first bright ring. The remaining energy is distributed in the surrounding rings that become progressively fainter. Figure 13.4 illustrates the diffraction pattern and shows the relative intensity of light across the pattern. The peak energy is at the center of the disk. The angular subtense of the disk depends only on the wavelength of light and the diameter **D** of the aperture of the lens. In particular, the semi-diameter of the Airy disk subtends an angle **ω** with respect to the lens of

$$\omega = \frac{1.22\,\lambda}{D} \qquad 13.4$$

Both λ and **D** must be in the same units. The equation shows that the size of the disk increases with increasing wavelength, and decreases with increasing an aperture diameter. See Fig. 13.6. The following discussion assumes monochromatic light.

Diffraction limits the ability of optical systems that range from the eye to the telescope at Mt. Palomar, to resolve two adjacent object points. Systems with small apertures will produce two large diffraction patterns that will run together

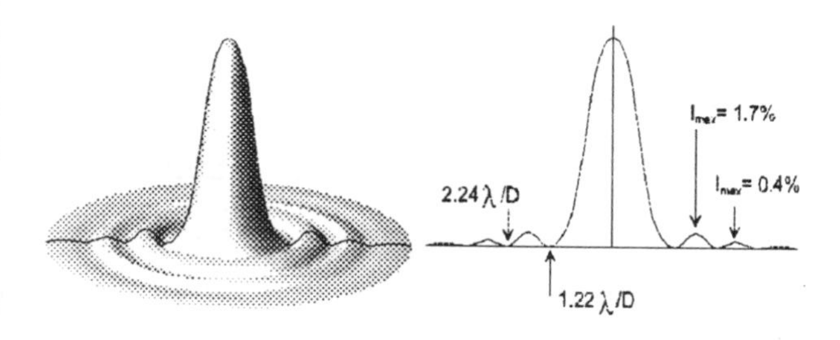

Figure 13.4 Fraunhofer diffraction by a circular aperture. The intensity distribution is shown at the right.

and appear single at relatively larger angular separations than will the two fine diffraction patterns produced by a system with a large aperture. See Figure 13.5. Conversely, for a telescope objective of a fixed aperture, the smaller the angle between two stars the closer their patterns will be imaged and the more the Airy disks will overlap. The dark region between the disks becomes increasingly brighter until the two disks seem indistinguishable. The smaller the angle at which distinguishing two patterns is no longer possible, the finer the resolution.

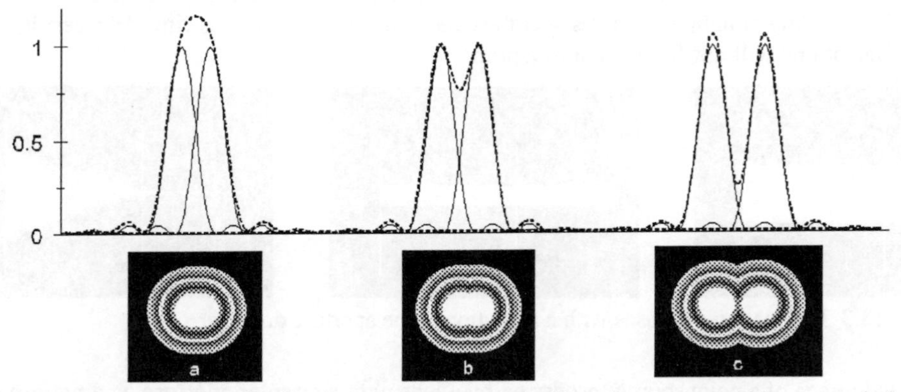

Figure 13.5 a) Intensity distributions for two unresolvable points separated by an angle less than $1.22\lambda/D$, i.e., less than the radius of the airy disk. The dashed line represents the combined intensity of the two patterns. b) Intensity distributions for two just-resolvable points separated by an angle of $1.22\lambda/D$ (Rayleigh criterion). This separation equals one disk radius; the peak of one pattern falls on the first minimum of the other. The dip in the combined intensity curve is sufficient to resolve the patterns. c) Further separation between the points makes them easily resolvable.

Lord Rayleigh found that two stars are resolvable when the peak (central maximum) of one disk was coincident with the first dark ring of the other disk. The addition of the intensities of these two overlapping patterns results in a dip in the combined light distribution (dotted line) that is perceptible. This separation is equal to the semi-diameter of the Airy disk. Thus, the angular resolution of a lens system is given by Eq. 13.4.

If λ and D are in mm, and $\lambda = 555\text{x}10^{-6}$ mm the angular resolution will be in radians.

$$\omega = \frac{1.22 \cdot 555x10^{-6}}{D} = \frac{677x10^{-6}}{D} \quad radians$$

To convert the angular resolution from radians to seconds of arc, divide by $4.85\text{x}10^{-6}$ radians/sec.

Thus, $\omega = 140/D$ seconds of arc. 13.5

For example, the theoretical or diffraction limited resolution of an eye with a 3 mm pupil is $\omega = 140/3 = 47''$. The usual resolution of the eye, based on 20/10 acuity, is 1' or 60". The eye resolves nearly at the diffraction limit.

13.4 LINEAR RESOLUTION

The linear size of the semi-diameter **s** of an Airy disk in the second focal plane of a lens, as illustrated in Figure 13.6, is **s = f tan ω = fω,** for small angles. By substituting Eq. 13.4 for angle **ω**, we obtain the minimum resolvable separation between two points.

$$s = \frac{1.22\,\lambda f}{D}$$

But, f/D = f/No. Thus,

$$s = 1.22\lambda \cdot f/No \qquad 13.6$$

The smaller the f/No, the better the resolution. For example, given an f/2 lens and a wavelength of 500 nm = 0.0005 mm, the semi-diameter of the disk s = 0.00122 mm. This is how close two diffraction images can be and still be resolved. At f/22, s = 0.0134 mm.

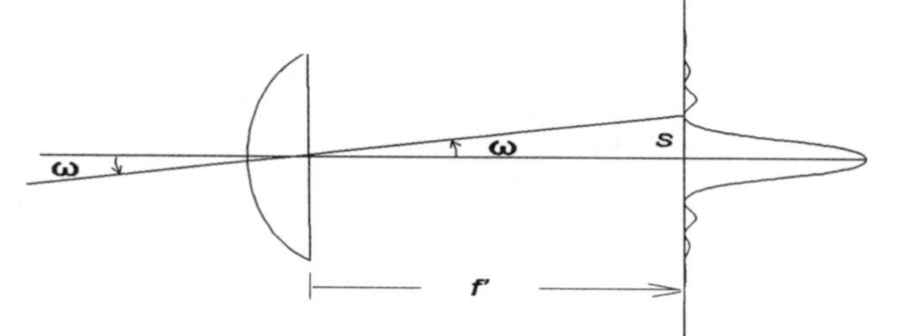

Figure 13.6 The semi-diameter of the Airy Disk (not to scale).

If the object is a pattern of equal black and white stripes, the resolution **R** can be stated as line-pairs-per-mm **l/mm** by taking the reciprocal of **s**. The resolving power of the f/2 lens is

$$R = \frac{1}{s} = \frac{1}{0.00122} \approx 820 l/mm$$

Compare this with the 75 l/mm resolution for the f/22 lens.

13.5 RESOLUTION CHARTS

Astronomers, who were the first to study the resolving power of lenses, were interested in resolving double stars. Seen unaided or through a poor quality telescope, double stars seem to be single points. With sufficient telescope magnification and image quality they are resolvable. Astronomers, therefore, considered the ability to resolve binary stars as the criterion for comparing the quality of telescope objectives. The equivalent criterion for the quality of a microscope objective lens was its ability to resolve closely spaced lines. This idea was extended in photography. Special charts, consisting of black and white lines were developed. The USAF Resolution Chart, shown in Figure 13.7a ,is an example of one such test target. The spatial frequencies in cycles/mm or **c/mm** (which is the same as the number of black and white line-pairs-per-mm) range from 0.25-228. These high contrast charts may be made with black bars on a white background, or white bars on a black background. A target element consists of three vertical and three horizontal bars. Six elements constitute a group. The groups are numbered in ascending order from -2 to +7. The spatial frequencies increase by the sixth root of two ($\sqrt[6]{2}$), or a factor 1.122, therefore, the spatial frequencies double for every sixth element. See Table 1. To test photographic lenses over a large field of view, an array of test charts is mounted on a large panel. The NBS test chart is shown in Figure 13.7b. This chart has a frequency range of 1 to 18 c/mm.

TABLE 1. US AIR FORCE RESOLUTION TEST CHART

ELEMENT NUMBER	GROUP NUMBER									
	-2	-1	0	1	2	3	4	5	6	7
1	0.250	0.500	**1.00**	2.00	4.00	8.00	16.00	32.00	64.00	128.00
2	0.281	0.561	1.12	2.24	4.49	8.98	17.96	35.92	71.84	143.68
3	0.315	0.630	1.26	2.52	5.04	10.08	20.16	40.32	80.63	161.27
4	0.354	0.707	1.41	2.83	5.66	11.31	22.63	45.25	90.51	181.02
5	0.397	0.794	1.59	3.17	6.35	12.70	25.40	50.80	101.59	203.19
6	0.445	0.891	1.78	3.56	7.13	14.25	28.51	57.02	114.04	228.07

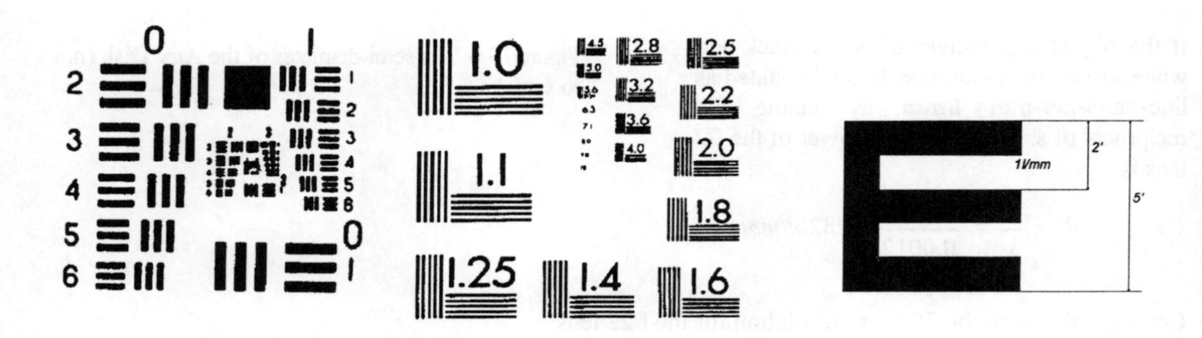

a b

Figure 13.7 Resolution charts: a) black on white USAF resolution chart; b) the NBS resolution chart.

Figure 13.8 The resolution angle of the 20/20 Snellen **E** is 2 minutes of arc.

13.6 SNELLEN CHARTS

Visual acuity, using Snellen letters, is a measure of legibility, but also of the resolving power of the eye. See Figure 13.8. The 20/20 letter **E** subtends 5'. Each black bar and white space subtends a visual angle of 1'. Visual acuity is reported as a Snellen fraction or as decimal acuity, i.e., the reciprocal of the visual angle subtended by the white space. Thus, 20/20 and 20/40 correspond to decimal acuities of 1 and 0.5, respectively. Resolving power, however, is expressed as the angle subtended by a just resolved line-pair. For a 20/20 letter, a line-pair subtends two minute of arc. One minute resolution corresponds to 20/10 acuity.

13.7 MODULATION TRANSFER FUNCTION

Resolution, classically, has been measured with high contrast (black and white) targets. At the limit of resolution the contrast in the image, between what should be black and white lines, is reduced to a barely discriminable level. Resolving power provides no information about the contrast in the images of coarser patterns that are resolvable. Photographs generally contain a wide range of detail or spatial frequencies. The trunks of trees in the woods are coarse or low spatial frequency targets that are easily resolvable, but the leaves on the trees or the bark on the trunks are higher spatial frequency targets. The quality of their photographic images depends on how much contrast they contain. It is not solely dependent on the highest spatial frequency that is just resolvable by the lens. Generally, the contrast in the image should be as high as possible at all spatial frequencies.

The modulation transfer function **MTF** describes the contrast of all spatial frequencies in the image. Ideally, the light distributions across the images of a series of bar targets should be square waves, i.e., have sharp edges, and flat peaks and valleys, despite spatial frequency. At best, the image of a point is a diffraction pattern. Lens aberrations will smear the image even more. The light distribution in the image is called a *point spread function*. See Figure 13.9. If the object is a line, i.e., an infinite number of points, the image is a *line spread function.* Modulation describes the distribution of light across a series of parallel sinusoidal line spread functions

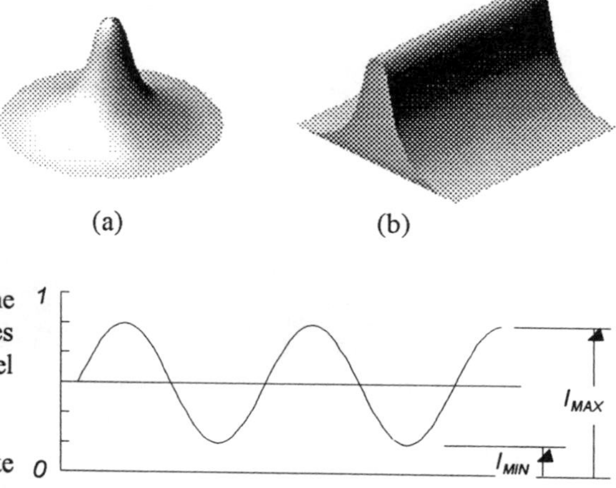

Figure 13.9 (a) Point and (b) line spread functions, (c) Modulation of a sine wave.

About two cycles of an object pattern of black and white bars is shown in Figure 13.10a. The square wave brightness of the object is shown in Figure 13.10b. The bright bars consist of an infinite number of bright lines. Each bright line is imaged as a line spread function. Overlapping line spread functions, as shown in Figure 13.10c, cause the sharp edge of the bright bars to be imaged with a rounded intensity profile, thus forming the *edge function* in Figure 13.10d. In effect, some light spills into the adjacent dark area, and the edges of the bars are blurred as indicated by Figure 13.10e.

As the spatial frequency increases, i.e., as the bright and dark stripes become narrower, the edge functions overlap. The peak intensity in the bright stripes begins to diminish and the dark stripes begin to brighten. The square wave target is imaged with a sinusoidal illumination distribution. Its amplitude, which is a measure of the contrast, decreases as spatial frequency increases until the contrast is so small that the illumination in the image looks uniform and no pattern is discernable. See Figures 13.10 f, g. The contrast in these images is expressed as a modulation.

$$Modulation = \frac{I_{Max} - I_{Min}}{I_{Max} + I_{Min}} \qquad 13.7$$

Where, I_{MAX} is the intensity at the peaks of the sinusoidal-like light distribution and I_{MIN} is the intensity at the valleys. See Figure 13.9c.

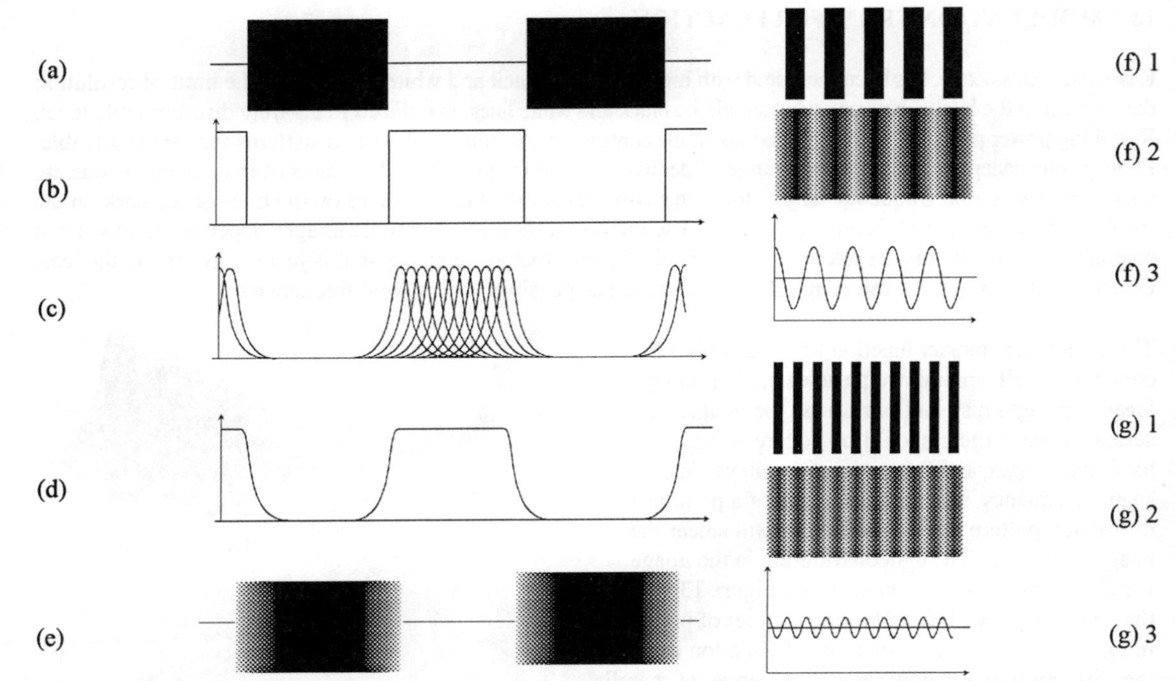

Figure 13.10 Square wave patterns and their images due to blurring produced by line spread functions: a) coarse square wave object; b) brightness distribution of object; c) line spread functions in image; d) illumination in image due to edge function; e) blurred edges of the image; f and g) higher spatial frequency square wave objects are imaged with sinusoidal illumination distributions of decreasing contrast or modulation. (After Modern Optical Engineering by W. J. Smith)

The ratio of the modulation in the image $\mathbf{M}_i$ to the modulation in the object $\mathbf{M}_o$ at any given spatial frequency is the modulation transfer function **MTF**.

$$MTF = \frac{M_i}{M_o} \qquad 13.8$$

Generally, the object is a high contrast black and white pattern, therefore, M_o will be 1. When the spatial frequency of the object is zero, i.e., the bars are infinitely wide, the edge function is negligible and the **MTF** will be 1 or 100%. As the spatial frequencies increase the **MTF** will be less than 100%, and ultimately drop to zero. The **MTF** curve is a plot of modulation versus spatial frequency. It shows the contrast in the image at all spatial frequencies. *Resolving power is the end point on this curve, where contrast is nearly zero.*

The **MTF** curve due to diffraction by a perfect parabolic mirror is shown by the △ symbols in Figure 13.11. This diffraction limited curve corresponds to a mirror with a focal length of 50 mm and a relative aperture of f/2.5. The size of the diffraction blur is 0.0018 mm. This figure also shows the **MTF** when the parabola is deformed to introduce spherical aberration corresponding to peak optical path differences of $\lambda/4$, $\lambda/2$ and $3\lambda/4$. Obviously, the **MTF** provides much more information about the quality of optical systems. For example, lens A in Fig. 13.12 produces higher contrast images of coarser patterns than lens B. The latter lens, however, has a higher resolving power. Lens A may provide a pleasingly soft portrait photograph; lens B will provide greater detail.

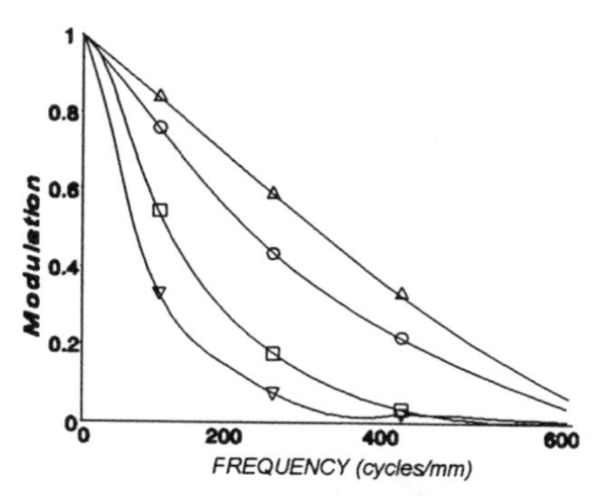

Figure 13.11 **MTF** curves for a 50mm, f/2.5 mirror with peak OPD errors of 0 (△), λ/4 (⊙), λ/2 (□), and 3λ/4 (▽) wavelengths.

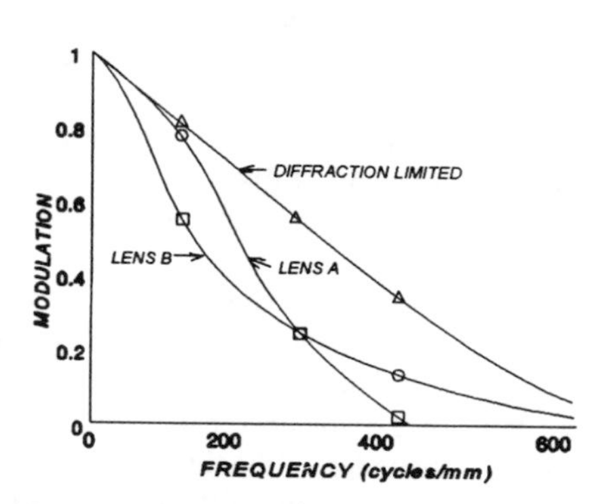

Figure 13.12 **MTF** curves for two lenses. Lens A offers high contrast for coarse objects, lens B has greater resolving power.

13.8 THE NUMERICAL APERTURE OF A FIBER OPTIC

Rays that enter at one end of an optical fiber will be transmitted or piped to the far end by a succession of total internal reflections at the wall of the fiber, as shown in Figure 13.13. Rays 1, 2, 3 and 4 have increasingly large angles of incidence at the end of the fiber. Only rays 1, 2 and 3 make angles of incidence at the fiber wall that exceed the critical angle. These rays are piped down the fiber. Ray 4 fails to be piped because it refracts through the wall of the fiber. Figure 13-14 shows the condition for finding the maximum angle of incidence of a ray which will undergo total internal reflection. This ray will be incident at the critical angle α_c at point B in the meridional section of a fiber. From Snell's law, $\sin \alpha_c = n_0/n_f$, where n_0 is the index surrounding the fiber. In the right triangle ABC, the angle of refraction $\alpha' = 90-\alpha_c$. Thus, the maximum angle of incidence α at the end of the fiber, also called the *acceptance angle*, is given by

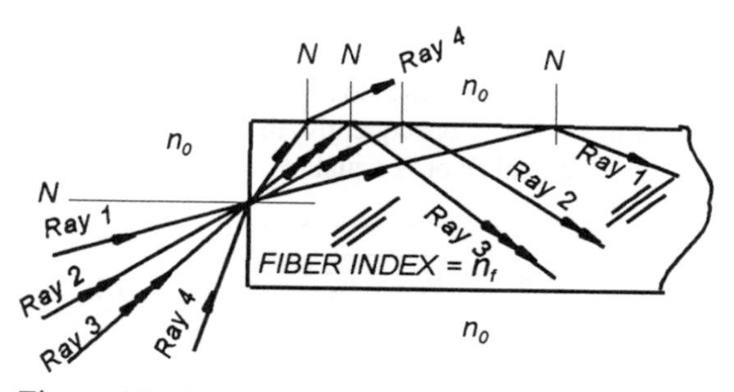

Figure 13.13 Not all rays undergo total internal reflection.

$$n_0 \sin \alpha = n_f \sin \alpha' = n_f \sin(90-\alpha_c)$$

$$\therefore n_0 \sin \alpha = n_f \cos \alpha_c \qquad 13.9$$

Bundles of randomly arranged bare fibers simply transfer light from one end to the other. *Coherent* fiber bundles have a one-to-one correspondence of fiber positions at both ends. These bundles transfer images from one end to the other. When bare fibers in a bundle are in contact, as in Figure 13.15, the rays in one fiber can cross over to the others. This *crosstalk* destroys the image transfer in coherent bundles. To prevent crosstalk each fiber is coated. The index of the cladding $\mathbf{n_c}$ is lower than the fiber index so that rays that enter one end of fiber are confined to that fiber and coherence is maintained. See Figure 13.16.

In general, then, three media or refractive indices ($\mathbf{n_o}$, $\mathbf{n_f}$ and $\mathbf{n_c}$) are germane to finding the maximum acceptance angle. The index n_0 corresponds to the medium at the ends of the fiber; n_f is the index of the fiber; and n_c is the index of the thin film or cladding in contact with the fiber wall. If there is no cladding, $\mathbf{n_c = n_0}$. The critical angle now is $\mathbf{\sin \alpha_c = n_c / n_f}$. Applying the Pythagorean Theorem to triangle ABC will give us the $\mathbf{\cos \alpha_c}$.

$$\cos\alpha_c = \frac{\sqrt{n_f^2 - n_c^2}}{n_f}$$

Substitute into Eq. 13.9, $n_0 \sin\alpha = \sqrt{n_f^2 - n_c^2}$

Because the angle α is also the slope θ of the ray, $n_o \sin \alpha = n_o \sin \theta = \text{NA}$. Consequently, the numerical aperture of an optical fiber is equal to

$$NA = \sqrt{n_f^2 - n_c^2} \tag{13.1}$$

This equation shows the maximum acceptance angle of rays. If a single fiber is surrounded by air, i.e., $n_o = n_c = 1$, the maximum value of NA is 1. Therefore, the right-hand term cannot exceed 1, and the fiber index cannot exceed $\sqrt{2}$. If $n_f = \sqrt{2}$, the acceptance angle will be ±90°, i.e., all rays with less than grazing incidence at the end of the fiber will be transmitted. For the grazing ray, the angle of refraction α' equals $45° = \alpha_c$, the critical angle.

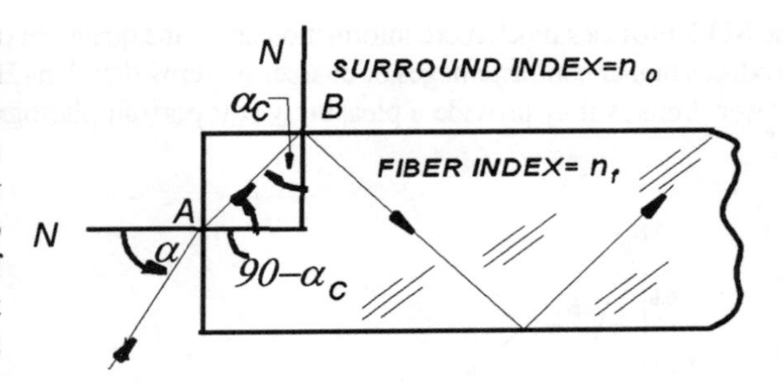

Figure 13.14 A fiber with a NA = 1 transmits all rays.

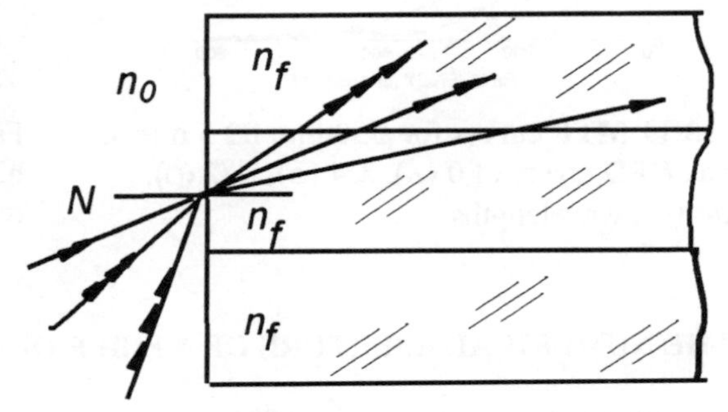

Figure 13.15 Three unclad fibers in contact. Rays are no longer totally internally reflected. Crosstalk results.

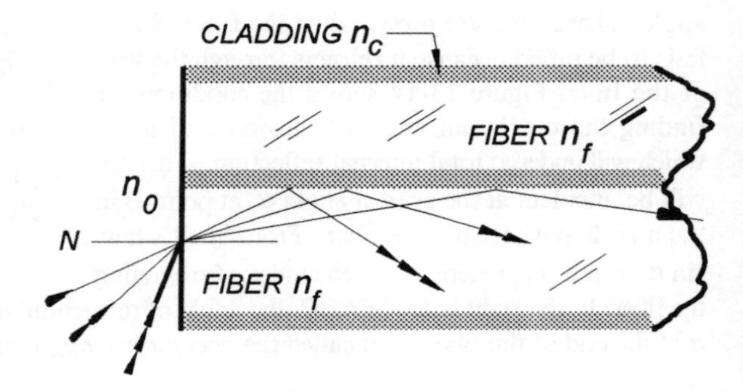

Figure 13.16 A thin layer of cladding on the fiber restores total internal reflection.

Example: A coherent fiber bundle will be made of glass fibers with an index of 1.60, and coated with a cladding of index 1.4. Find the NA and the maximum angle incidence.

$NA = \sqrt{1.6^2 - 1.4^2} = 0.7746$ and $\sin\theta = 0.7746$ and therefore $\theta = \alpha = 50.769°$

Fiber bundles may be flexible or rigid. Flexible bundles contain loose fibers encased in a rubber tube or other casing. Rigid bundles are made by heat fusing or bonding the fibers together. Non-coherent fiber bundles are used to detect which holes are punched, and transmit light, in punch cards for computers, to provide "cold light," in operating rooms, high intensity visual instruments, and in explosive atmospheres where the heat generated by the light source is undesirable. Coherent fiber bundles are used to transfer images in medical endoscopes, borescopes and through remote and convoluted optical paths that would require many lenses, mirrors and prisms in accurate alignment. A very important application of fiber optics is to replace copper wire cables in telephone systems. The sounds to be transmitted are transduced to modulate light sent through the fiber. A transducer at the receiver converts the light impulses to sound. A tapered coherent fiber bundle may be made by heating part of a rigid coherent bundle of fibers and stretching it, as is done with Bunsen burners and glass tubing in chemistry or biology labs. An image at the small end will be magnified at the large end and vice-versa, in proportion to the diameters of the ends.

PROBLEMS

1. The numerical aperture in object space (air) of a lens is 0.5. Find the half-angle of the cone of light collected by the lens.
Ans.: 30 deg

2. Given a half-angle of 30 ° and a NA of 0.75, find the index of the medium.
Ans.: 1.5

3. A 100 mm focal length lens is set at f/8. What is the diameter of the entrance pupil?
Ans.: 12.5 mm

4. What is the NA of the above lens?
Ans: 0.0625

5. An f/16 camera lens properly exposes the film at a 1 second exposure time. How long should the exposure be when the lens is stopped to f/4?
Ans.: 0.0625 seconds

6. What is the angular subtense of the semi-diameter of the Airy disk, at 555 nm, produced by a perfect lens with a diameter of 25.4 mm?
Ans.: 0.000026 radians

7. Convert the above answer to seconds of arc.
Ans.: 5.5 sec.

8. If the focal length of the above lens is 100 mm, find a) the linear *diameter* of the Airy disk in the focal plane and b) how many l/mm can be resolved according to Rayleigh?
Ans.: a) 0.0053 mm; b) 375

9. A fiber optic of index 1.75 is clad with an index of 1.5. Find a) NA; b) the slope of the limiting ray; c) the critical angle.
Ans.: a) 0.90; b) 64.34 °; c) 59 °.

10. A fiber optic of index 1.523 is clad with an index of 1.5. Find a) NA; b) the slope of the limiting ray; c) the critical angle.
Ans.: a) 0.26; b) 15.29 °; c) 80 °.

11. A pinhole camera is 400 mm long. What is the optimum diameter of the pinhole for starlight of 550 nm. (Object at infinity. Hint: Let camera length equal focal length and set geometric blur diameter equal to diffraction blur diameter.
Ans.: 0.73 mm.

12. Paul Revere was told that *two* lanterns would indicate that the Redcoats were coming by sea. What was the maximum distance at which he could resolve two lanterns if they were placed 500 mm apart, he had a 3 mm pupil diameter and λ = 550 nm?
Ans.: 2.2 km.

13. What is the linear separation of two points on the surface of the moon that are just resolved by a telescope lens with a 1 meter diameter. The moon is 380,000 km away and λ = 555 nm?
Ans.: 260 meters.

CHAPTER 14

MAGNIFIERS AND MICROSCOPES

14.1 INTRODUCTION

Magnifiers extend vision. They are used by watchmakers to enlarge the appearance of the fine gears and springs of a watch, or enable the surgeon to make exquisitely fine sutures. When spectacles fail, magnifiers often restore the ability to read to people with low vision. *Simple magnifiers* are constructed of positive lenses. They may be hand held, as in the clue-searching Sherlock Holmes pose, mounted in spectacle frames, or built into stands that permit them to sit on a page of print. *Microscopes* are compound magnifiers that are much more powerful than simple magnifiers. We will explore how both types of magnifiers work.

14.2 VISUAL ANGLE AND ANGULAR MAGNIFICATION

The *visual angle* is the angle subtended by an object of height **y= MQ** at distance **z = NM** from the nodal point **N** of the eye. The exact position of the nodal point, within the eye, is not important if the object is distant. Visual angle corresponds to the *apparent size* of an object. See Figure 14.1.

$$\tan \omega = \frac{y}{z} \qquad 14.1$$

Magnifiers and microscopes provide *angular magnification*. It is the ratio of the visual angle of the image seen through the instrument to the visual angle of the object seen directly at some defined distance.

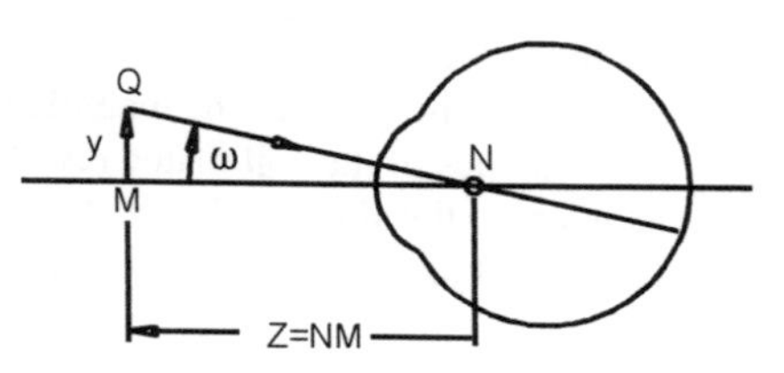

Figure 14.1 The apparent size of an object or visual angle ω.

14.3 THE MAGNIFIER OR SIMPLE MICROSCOPE

When we want to examine fine detail or read small print we bring the object of interest close to the eye. Obviously, bringing the object from a normal reading distance of 40 cm to 10 cm from the eye, for example, will increase the visual angle of the detail by a factor of four. However, accommodation must increase from 2.5D to 10D to see the detail clearly. An inability to accommodate this much will preclude the gain of four times the visual angle, and even those able to focus at 10 cm may experience discomfort in sustaining such an effort. Consequently, a nearest distance for comfortable sustained seeing has been adopted by convention to serve as a reference distance for comparing visual angles. This *least distance of distinct vision* **lddv** is equal to 250 mm, or 10 inches.

A positive lens permits an object to be brought close to the eye without requiring the eye to accommodate or without requiring excessive accommodation, because it enables the eye to view an image that is more distant than the object. When used this way the lens is a magnifier or, for the watchmaker, a loupe.

The magnification of a magnifier is given by the ratio of the visual angle subtended by the image ω' to the visual angle subtended by the object at the least distance of distinct vision ω.

$$M_a = \frac{\tan\omega'}{\tan\omega} \qquad 14.2$$

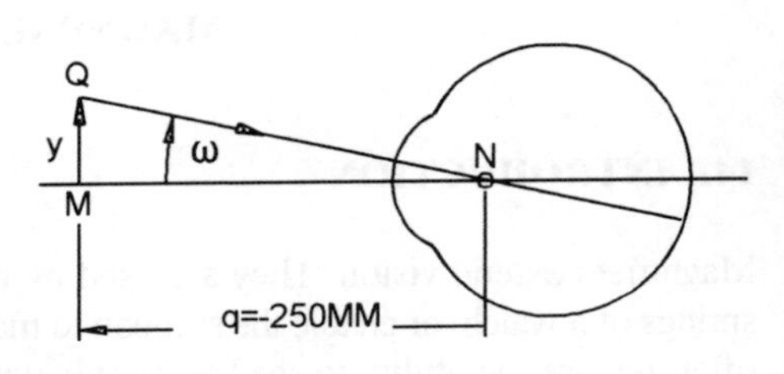

Figure 14.2 The reference visual angle for the least distance of distinct vision,

In Figure 14.2 the object subtends the *reference* visual angle: **tan ω = y/q**, where q = lddv = NM = -250 mm.

14.3.1 NOMINAL MAGNIFICATION

In nominal or standard magnification the object is placed at the first focal point **F** of the magnifier. The image is seen at infinity and subtends the same visual angle ω' as the object subtends at the center of the magnifier lens. That is, the slopes of the collimated rays from **Q'** are identical to the slope of the chief ray from point **Q** in the object. See Figure 14.3. Thus,

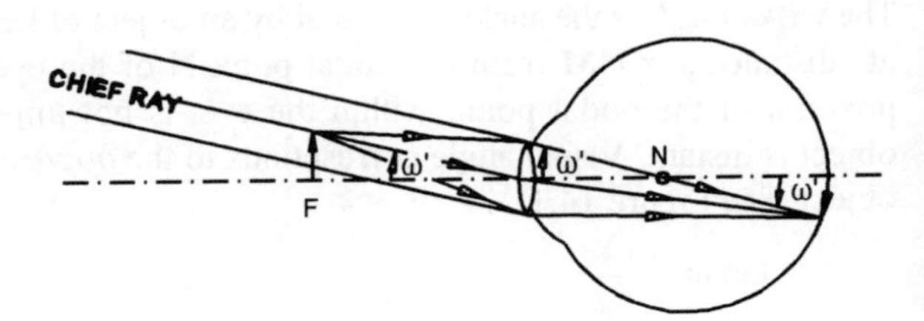

Figure 14.3 The visual angle subtended by an image formed at infinity.

$$\tan\omega' = y/-f$$

Nominal magnification $\mathbf{M_N}$ is given by

$$M_N = \frac{\tan\omega'}{\tan\omega} = -\frac{q}{f} \qquad 14.3$$

Since q = -250 mm, or -25 cm, or -10 inches:

$$M_N = \frac{250}{f\ mm} = \frac{25}{f\ cm} = \frac{10}{f\ inch} \qquad 14.4$$

Eq. 14.4 may be transformed to dioptric power, since $\boldsymbol{F} = 1/f$ and q = -1/4 meter.

$$M_N = \frac{F}{4} \qquad 14.5$$

Because the image is seen at infinity, the distance of the eye from the magnifier will not affect the angular magnification. It will affect the field of view, however. Why?

14.3.2 MAXIMUM MAGNIFICATION

By moving the object to less than a focal length away from the magnifier a virtual image may be formed that is nearer than infinity. In fact, magnifier users commonly do this because, being aware that they are holding the object close to the lens and their eye, they often accommodate. If the image is formed at the nearest acceptable distance from the eye, namely, the lddv, and that the eye is up against the magnifier, we may obtain the maximum magnification from the magnifier. This condition is illustrated in Figure 14.4. The object at distance **u**, is imaged at distance **u'**. Both the object and image subtend angle **ω'** at the lens and at the eye if the distance between lens and eye is neglected. From the figure we see that u' = q, therefore, the angular and lateral magnifications are equal.

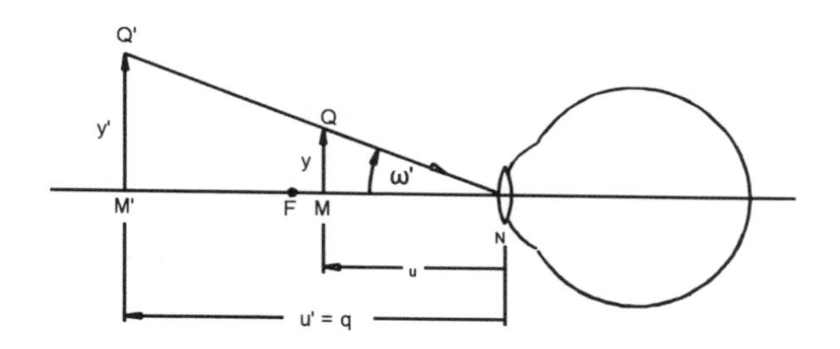

Figure 14.4 The condition for the maximum magnification - image formed at LDDV.

$$U' = \frac{1}{q} \tag{14.6}$$

$$\tan\omega' = \frac{y}{u} = yU \tag{14.7}$$

$$but,\ U = U' - F = \frac{1}{q} - F \tag{14.8}$$

Substitute Eq. 14.8 into 14.7.

$$\tan\omega' = \frac{y}{q} - yF \tag{14.9}$$

Maximum magnification $\mathbf{M_M}$ = tan ω'/tan ω = **1 - q*F***
or

$$M_M = \frac{F}{4} + 1 \tag{14.10}$$

Thus, by forming the image at -250 mm, the magnification is increased by one. For example, a +10 **D** lens has a nominal magnification of 2.5x, and a maximum magnification of +3.5x. Manufacturers of magnifiers frequently inflate the power of their lenses by quoting maximum magnification values.

The conditions for maximum magnification included two assumptions: 1) the image is formed at -250 mm from the lens, and 2) the eye is in contact with the lens, i.e., **d** = **0**. If the distance of the eye from the lens is increased, its distance from the image is also increased. The visual angle subtended by the image will be reduced, consequently, the magnification will be reduced. To determine the magnification when neither assumption is fulfilled we will consider a general equation for magnification.

14.3.3 A GENERAL EQUATION FOR MAGNIFICATION

A general set of conditions for using a magnifier is illustrated in Figure 14.5. The object is at a distance u<f. (Why not u>f?). The eye is at a distance **d = AN** behind the lens, and the image is at a distance **z = NM' = u'-d** from the eye. The reference angle is given by Eq. 14.2: **tan ω = y/q**. The visual angle of the image is

$$\tan\omega' = \frac{y'}{u'-d}$$

$$\therefore\ M = \frac{\tan\omega'}{\tan\omega} = \frac{y'q}{y(u'-d)}$$

but, y'/y = U/U', and u' = 1/U'. After substitution we find the general magnification equation

$$\boxed{M = \frac{qU}{1-dU'}} \qquad 14.11$$

Because U' is always negative (the image is always virtual and erect) the denominator of Eq. 14.11 increases as d increases, and magnification decreases. Therefore, the eye should be as close as possible to the lens when the image is formed at a finite distance. The eye distance ***d*** does not affect the nominal magnification, but it does affect the size of the field of view for all magnification conditions.

When the object is at the first focal point, U = - *F*, and U' =0, the general magnification equation reduces to the nominal magnification equation. Similarly, when the image is formed at the lddv, and d = 0, Eq. 14.11 yields the maximum magnification equation (Eq. 14.10).

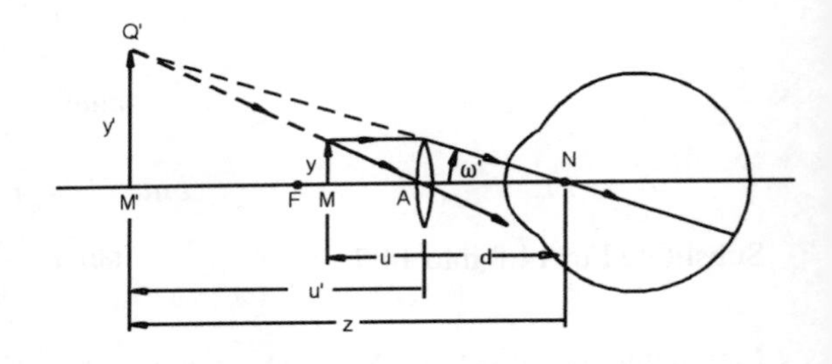

Figure 14.5 General conditions for determining magnification.

Example 1: An emmetropic eye, 5 cm behind a thin +8 D lens, views the image of an object 5 cm in front of the lens. Find the angular magnification. The eye position is represented by its nodal point **N**.

Given: q = -1/4 m, *F* = +8 D, u = -5 cm, or U = -20 D, d = AN = 5 cm.
thus, U' = -12 D, or u' = -8.33 cm.

Substitution into Eq.14.11 results in a magnification of **M = 3.125x**.

$$M = \frac{(-0.25)(-20)}{1-0.05(-12)} = 3.125x$$

This result may be obtained from first principles (Eq. 14.2). Suppose that the object is 1 cm high. The *lateral* magnification Y = U/U' = 1.67. Therefore, the image height y'= 1.67 cm. This image is at a distance from the eye of M'N = 13.33 cm. Thus, tan ω' = 1.67/13.33 = 0.125; tan ω = 1/25 = 0.04, and M = 0.125/0.04 = 3.125x

14.3.4 THE ACCOMMODATION REQUIRED TO VIEW THE IMAGE

To see the magnified image clearly, the eye must accommodate for the image distance. The image distance is NM'. In the above example the image distance NM' = NA+AM'= u'-d = -8.33-5.0 = -13.33 cm. Thus, the vergence at the eye is U_a = 1/NM' = -7.5 D. The total amount of accommodation is the sum of the uncorrected ametropia (**A**) and the image vergence (**U_a**). For an emmetrope, **A** = 0. A hyperopic eye has insufficient power, i.e., must additionally accommodate to overcome its ametropia, therefore, **A** is negative. In myopia, on the other hand, the power of the eye is excessive. **A** is positive. Accommodation required to focus on the image is reduced. The necessary accommodation for viewing the magnified image is

$$\text{Accom.} = -(U_a + A) \qquad 14.12$$

The amounts of accommodation required of an emmetrope, a one diopter hyperope and a one diopter myope, in the above example are: 7.5, 8.5 and 6.5 diopters.

14.3.5 THE FIELD OF VIEW OF MAGNIFIERS

If it is assumed that the eye is the aperture stop and the magnifier is the field stop, as shown in Figure 14.6, the first step toward finding the fields of view in object and image space is to locate the entrance pupil. The half-field angle subtended by the object is shown by the projected ray BE, from the edge of the entrance port to the center of the entrance pupil. This ray originates at the point Q in the object plane. Thus, MQ is half the linear field of view in the object. The ray BE' indicates the half-field angle subtended by the image. Its projection to the point Q' determines half the linear field of view in the image plane.

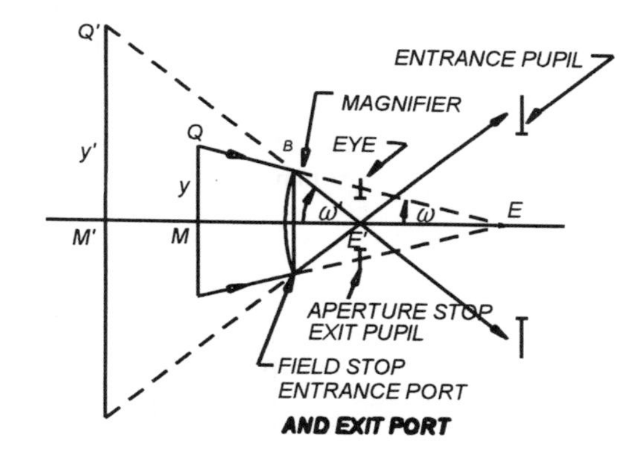

Figure 14.6 The field of view of a magnifier.

Example 2 : Given a lens diameter D = 40 mm, find the angular and linear FOVs of the above magnifier.

Step one: Treat the eye (aperture stop) as an object at distance u_e = -50 mm, and find its image (entrance pupil).Given: U_e = -20 D; F = +8 D

Thus, U'_e = -12 D, and u'_e = -83.33 mm. The entrance pupil (E) is 83.33 mm to the right of the lens

Step 2: Find the angle ω subtended at E (the center of the entrance pupil) by half the entrance port (lens).
tan ω = 20/83.33 = 0.24; thus, ω = 13.5°. This is the half-field angle in object space.

Step 3: Find MQ. MQ = ME(tan ω) = (133.33)0.24 = 32 mm. Thus, the magnifier covers 64 mm of the object.

Step 4: Find the angle ω' subtended at E' (the center of the exit pupil) by half the exit port (lens).

$\tan \omega' = 20/50 = 0.40$; thus, $\omega' = 21.8°$. This is the half-field angle in image space.

Step 5: Find M'Q'. $M'Q' = M'E'(\tan \omega') = (133.33)0.40 = 53.33$ mm.

14.3.6 TYPES OF MAGNIFIERS

A convex lens made with spherical surfaces is the simplest magnifier. See Figure 14.7. It is used in watchmakers' loupes and cheap reading glasses. It has low power (2 or 3x), and covers a field of view of about 1/5 of its focal length before the aberrations become excessive. Higher magnifications (4 or 5x) can be obtained by using aspherical surfaces. Air spaced doublets are better able to correct aberrations and have powers up to about 15x. They are used as eyepieces in low cost microscopes and telescopes. The Coddington magnifier has even higher power (20x). It is a lens cut from a sphere, with an equatorial groove that acts as an aperture stop to control spherical aberration. The spherical shape, with a stop at the center, essentially eliminates such off-axis aberrations as coma and astigmatism. Also commonly available up to 20x is the Hasting's triplet, made of three cemented lenses.

The distance between the object and the magnifier is called the *working distance*. The stronger the magnifier: the shorter its focal length and working distance; the smaller the lens diameter and linear field of view; and the closer object and lens must be from the eye. Typical values for the Hasting triplet are:

Magnification	10x	15x	20x
Working Distance	26 mm	16 mm	11 mm
Field of View	23 mm	13 mm	8 mm

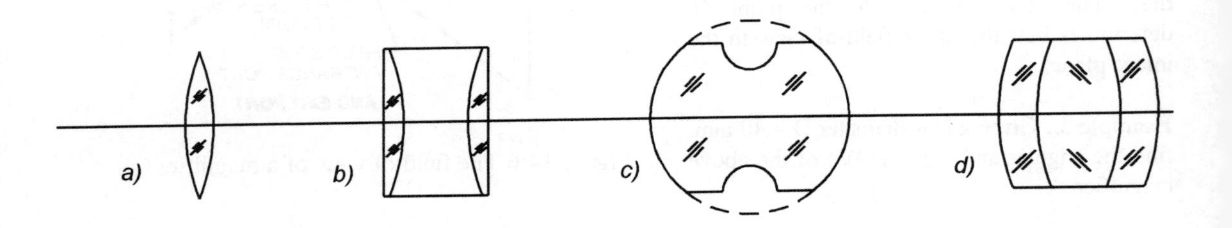

Figure 14.7 Common magnifiers: a) biconvex lens, b) air-spaced doublet, c) Coddington, d) Steinheil triplet.

14.4 THE COMPOUND MICROSCOPE

A two-stage magnification system will circumvent the limits imposed by progressively shortening focal length to increase magnification. *The compound microscope provides an initial stage of lateral magnification by means of its objective lens, and the ocular lens, acting as a simple magnifier, provides a second stage of angular magnification*. The objective forms a real, inverted and laterally magnified image at $\mathbf{M}_1'$ of an object located just beyond the first focal point. When $\mathbf{M}_1'$ falls in the first focal plane of the ocular, the microscope is in *normal adjustment*. The eye will see an image at infinity, which is desirable since there will be no need to accommodate for sustained periods. See Figure 14.8. Of course, by moving the ocular to place $\mathbf{M}_1'$ less than a focal length from it, the image seen by the eye can be formed at any distance, including the least distance of distinct vision.

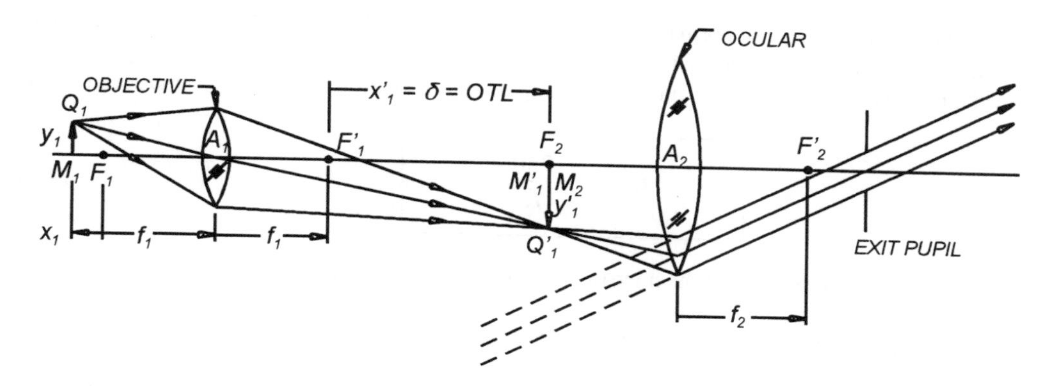

Figure 14.8 The compound microscope.

14.4.1 MAGNIFICATION BY A MICROSCOPE

The magnification of a microscope is the product of the lateral magnification by the objective lens and the angular magnification by the ocular. The Newtonian equation $\mathbf{Y} = \mathbf{x'}/\mathbf{f'}$ is the most appropriate form to use for finding the lateral magnification of the objective lens, because $\mathbf{x'} = \mathbf{F}_1'\mathbf{M}_1' = \mathbf{F}_1'\mathbf{F}_2 = \boldsymbol{\delta}$ will be equal to the *optical* tube *length* OTL. This instrument constant has been standardized at 160 mm. Thus, the first stage of magnification is

$$Y_1 = -\frac{x_1'}{f_1} = -\frac{OTL}{f_1} = -\frac{\delta}{f_1}$$

and the second stage is the angular magnification of the eyepiece,

$$M_N = \frac{250}{f_2}$$

The magnification of the microscope in normal adjustment is

$$M = -\frac{\delta}{f_1}\cdot\frac{250}{f_2} \qquad 14.13$$

Example 3: A microscope contains a 16 mm focal length objective lens, a 25 mm focal length eyepiece, and has an OTL of 160 mm. Find the power engraved on the objective and the power of the microscope.

The engraved power is the lateral magnification of the objective. Y = - 160/16 = -10, or 10x.

The angular magnification of the eyepiece is $M_N = 250/25 = 10x$

Therefore, the angular magnification of the microscope M = -100x.

14.4.2 MAXIMUM MAGNIFICATION BY A MICROSCOPE

The image produced by a microscope in normal adjustment is at infinity. If the ocular lens is pushed toward the objective the image can be produced at the LDDV, i.e., A_2M_2' will be -250 mm. As in the simple magnifier, this condition provides maximum magnification. Thus, the first stage of magnification is $Y_1 = -\delta/f_1$; the second stage is

$$M_{MAX} = \frac{250}{f_2} + 1 \qquad 14.14$$

Therefore, the maximum magnification of the microscope in the previous example will be -110x.

14.4.3 THE COMPOUND MICROSCOPE AS A SIMPLE MAGNIFIER

Despite the distinctions made between simple and compound microscopes the latter may be treated as a simple microscope with the Gullstrand equation, $F = F_1 + F_2 - dF_1F_2$.

In a microscope, $d = f_1 + \delta + f_2$. Substitute this into the Gullstrand equations and solve for F.

$$F = -\delta F_1 F_2$$

The magnification of the compound microscope, treated as a simple magnifier, is

$$\mathbf{M} = -\frac{\delta F_1 F_2}{4} \qquad 14.15$$

In the previous example: $\delta = 0.16$ m, $F_1 = 1000/16 = 62.5$ D, and $F2 = 1000/25 = 40$ D. Substitute into Eq. 14.15:

$$M = -\frac{(0.16)(62.5)(40)}{4} = -100x$$

This is the same magnification as found with Eq. 14.12. The minus sign means that the image is inverted.

14.4.4 THE STOPS OF A MICROSCOPE

Generally, the objective of a microscope is the aperture stop and entrance pupil. The image of the aperture stop, formed by the ocular lens, is the exit pupil. This is also called the *eye point* because the eye should be placed at the eye point to see the entire field of view of the microscope slide. The distance from the ocular to the eye point or exit pupil is called the *eye relief*. If the eye relief is too short spectacle wearers cannot see the entire field of view. If the only other element in the microscope is the ocular, it will be the field stop and exit port. The angle subtended by the exit port at the center of the exit pupil corresponds to the 50% field of view. Usually, however, a field stop is placed at the first focal point of the ocular. It produces a sharply defined field of view. The entrance port coincides with the object and the exit port is at infinity.

If the diameter of the pupil of the eye at the eye point of the microscope is less than the diameter of the exit pupil of the microscope, the full numerical aperture will not be attained. However, the apparent brightness (luminance) of the image, if light absorption and transmission losses are neglected, will be the same as the object luminance. However, if the exit pupil is smaller than the eye pupil, the image will be dim.

14.5 NUMERICAL APERTURE AND RESOLUTION OF MICROSCOPES

Numerical apertures in object space of microscopes range from 0.07 to 1.6. To obtain numerical apertures >**1** the object space is filled with an oil that matches the index of the cover slide and the first plano-convex lens element of the objective lens. Rays in the cover glass at angles > θ_a (i.e., > the critical angle) undergo total internal reflection with a dry objective, shown in the right half of Figure 14.9. The **NA** = **nsin** θa. The right half of the figure shows oil filling the space between the cover glass and the lens. All three media have the same index **n**. In effect, the object **M** is immersed in a homogeneous medium. The ray to the edge of the lens travels in a straight line. Obviously, **NA** = **n sin** θ_o. is much greater with the oil immersion objective.

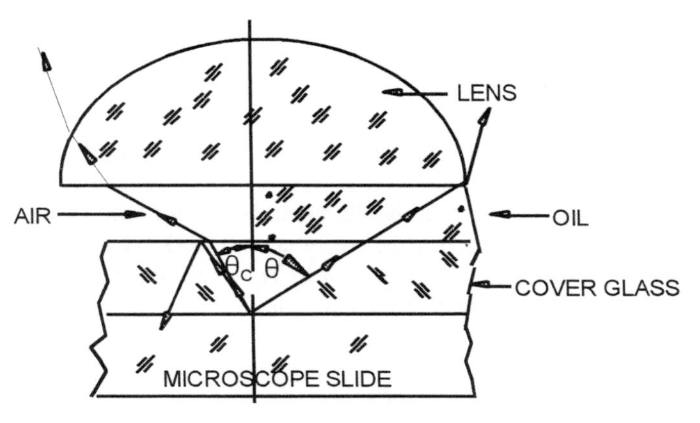

Figure 14.9 Numerical aperture of dry and oil immersion objective lenses.

Abbe measured the ability of microscopes to resolve fine gratings or closely spaced black and white lines. It turns out that the resolution of a microscope depends on the numerical aperture of the objective lens. According to Abbe, the minimum linear separation **s** of two resolvable points on a microscope slide, based on diffraction effects, is given by the following equation, where, λ = the wavelength of the light.

$$s = \frac{1.22\lambda}{2NA} \tag{14.16}$$

Better resolution is obtained in practice, so the 1.22 factor is reduced to 1, and the actual separation is

$$s = \frac{\lambda}{2NA} \tag{14.17}$$

As discussed in Chapter 13, the diffracted images of these two points, produced by a perfect objective, are patterns consisting of central Airy disks surrounded by a series of bright rings that rapidly fade away. The patterns are resolvable when the peak of one Airy disk coincides with the first dark ring of the other pattern. The image separation **s'** is equal to the radius (semi-diameter) of the Airy disk.

Actually, Abbe measured the spacings of fine gratings or closely spaced black and white lines. Eqs. 14.16 and 14.17 apply to the minimum resolvable spacing **s** of these gratings, where **s** is the width of one pair (a black and a white) of lines. If **s** is in mm then the *resolution* **R = 1/s**, i.e., line pairs per mm (L/mm), or cycles per mm (c/mm). According to Eq. 14.17 the resolution in l/mm is

$$R = \frac{2NA}{\lambda} \qquad 14.18$$

Example 4: Given an objective lens with a NA = 0.5 and light of 555 nm, find the diffraction limited resolution in line pairs per mm (L/mm).

$$R = 2(0.5)/\ 555\text{x}10^{-6} = 1802 \text{ L/mm}.$$

14.6.1 MAXIMUM USABLE MAGNIFICATION OF A MICROSCOPE

Now for the eye to resolve the diffraction patterns, the separation **s'** at the ocular must subtend a minimum angle of one minute of arc at the eye, or **ω' = 0.0003** radians. The ocular is treated as a simple magnifier with the eye in contact with it. See Figure 14.10. For small angles the angular magnification $\mathbf{M_a = \omega'/\omega}$, where **ω** is the angle subtended by the object at 250 mm. Thus,

$$\omega = s/250$$

Substitute Eq. 14.17 for s,

$$\omega = \frac{\lambda}{500NA}$$

and

$$M = \frac{0.15NA}{\lambda} \qquad 14.19$$

The eye is most sensitive to λ = 555 nm or 0.000555 mm, therefore, **M = 270NA**. (If the 1.22 factor is included **M= 221NA**). Assuming a maximum NA = 1.6, we find that the maximum usable magnification is 432x. Just as it would be uncomfortable to read 20/10 print for any period of time, so it is a strain to view slides at the limit of resolution. Consequently, in practice, using double or triple this magnification is desirable. This is called **empty magnification**, because the blurs are simply enlarged and no additional detail is seen.

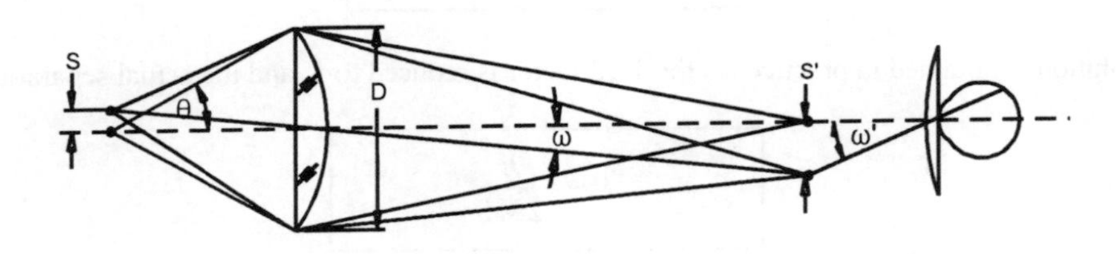

Figure 14.10 Maximum usable magnification of a microscope. See text.

14.6.2 RESOLUTION, WAVELENGTH AND THE ELECTRON MICROSCOPE

Magnets can focus electrons much as lenses focus light. The electron microscope employs this property of magnets to obtain extremely high resolving power.

According to Eq. 14.17, $s = \lambda/2NA$, the smallest resolvable separation of two points, assuming a NA of 1.5, will be equal to 1/3 the wavelength of the light, or about 1,850 Å, using visible light. Because resolution is inversely related to wavelength (Eq.14.18), higher resolving power may be obtained by using shorter wavelengths. For example, the wavelength of electrons, according to de Broglie, is given by

$$\lambda = \frac{h}{mv}$$

where, h is Planck's constant = 6.624×10^{-34} joule-sec
m is the mass of the electron = 9.108×10^{-31} kg
v is the velocity of the electron, which depends on its accelerating voltage (V) and its charge (e)
$e = 1.602 \times 10^{-19}$ coulombs.

The velocity of the electron is found with $eV = \frac{1}{2}mv^2$

Solve this for v and substitute in de Broglie's equation, to obtain the following equation for wavelength in Å.

$$\lambda = \frac{12.27}{\sqrt{V}}$$

If V = 150 eV, the wavelength of electrons = **1** Å. Thus, the resolution in lines per mm is several thousand times greater for the electron microscope compared with a visible light microscope and its maximum useful magnification exceeds 300,000x.

14.6.3 DEPTH OF FOCUS OF MICROSCOPE

It is apparent to anyone who has used a microscope that only very thin layers of the specimen are in focus at a time. Racking the microscope up or down is necessary to focus at different depths. A simple way to understand this effect is by considering the axial magnification of the objective. For example, a 10x objective has an axial magnification of 100x. That is, a thin layer of the specimen will be imaged 100x thicker at the ocular. Assume that this thick image extends on either side of the first focal plane of the ocular, and is an *axial* object for the ocular. The portion of this object that lies more than a focal length away will have converging wavefronts after refraction by the ocular, i.e., this portion will be imaged behind the eye. The eye cannot negatively accommodate, so this portion of the specimen will be blurred. Assume that the eye can clearly see images, produced by the ocular, that range from infinity to the lddv (-250 mm). We will work this example backwards, from the image seen by the eye to the specimen on the slide. First we will find the conjugate positions of an axial object that correspond to ∞ and -250 mm, for an ocular with a focal length f_2 = 25 mm. See Figure 14.11.

For the final image to be at infinity ($u_R' = \infty$) the image formed by the objective lens must coincide with F_2, or the conjugate distance u_R = -25 mm. For a final image distance of u_P' = -250 mm, the lens equation gives u_P = -22.7273 mm. Thus, x' = 25 - 22.7273 = 2.2727 mm equals the axial thickness of the object at the ocular that will be seen clearly, provided the eye accommodates 4 D. But x' is the thickness of the image of the specimen produced by the 10x objective. Therefore, we must find x the corresponding thickness of the specimen. The axial magnification X of the 10x objective is

$$X = 100 = \frac{x'}{x} = \frac{2.27}{x}$$

$$\therefore x = 0.0227\ mm$$

Thus, the thickness of the specimen that is in focus, for an eye accommodating from zero to 4 D, is a little less than one thousandth of an inch.

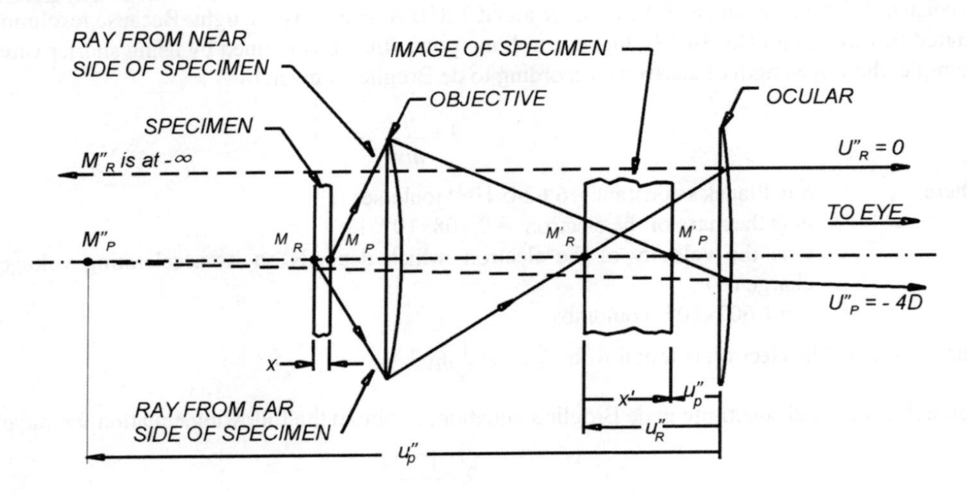

Figure 14.11 Thickness of specimen that corresponds to four diopters of accommodation.

REVIEW QUESTIONS

1. Define visual angle and apparent size.
2. Define angular magnification of a magnifier. Why does it reference the least distance of distinct vision?
3. Why doesn't the position of the eye behind the magnifier affect the nominal magnification?
4. Why does the position of the eye behind the magnifier affect the field of view?
5. What ultimately limits the magnification obtainable with simple magnifiers?
6. Identify the two stages of magnification of a compound microscope.
7. Define optical tube length. To what other quantity is it identical?
8. Define eye point and eye relief.
9. What class of viewers is affected by a short eye relief, and how are they affected?
10. What properties of a microscope are affected by the numerical aperture?

PROBLEMS

1. Find the nominal and maximum magnification of a 2 inch focal length lens. Ans.: 5x; 6x.

2. What is the refracting power of a 5x magnifier? Ans.: +20 D.

3. Show that the linear and angular magnifications are equal when angular magnification is maximum.

4. What is the working distance for a 50-mm focal length lens to obtain maximum magnification? Ans.: 41.67 mm.

5. An object is 100 mm from the eye. A +8 D magnifier, 40 mm in diameter, is placed midway.
   a) Find the magnification. Ans.: 3.125x.
   b) Find the total real and apparent angular FOVs. Ans.: 27°, 43.6°
   c) Find the total real and apparent linear FOVs. Ans.: 64 mm, 106.7 mm.
   d) How much accommodation is required to see the image; by an emmetrope, a 1 D hyperope and a 1 D myope? Ans.: 7.5, 8.5 and 6.5 D.

6. A 20 D magnifier lens, 50 mm in diameter, is used to read the telephone directory. The page is 40 mm in front of the lens and the eye is 15 mm behind the lens.

   a) Find the nominal and maximum magnifications, and the magnification for the given conditions. Ans.: 5, 6 and 5.8x
   b) Find the total real and apparent angular FOVs. Ans.:98.8°, 118°
   c) Find the total real and apparent linear FOVs. Ans.: 143.3 mm, 716.6 mm.
   d) How much accommodation is required of an emmetrope, a 2 D hyperope and a 2 D myope? Ans.: 4.65, 6.65 and 2.65 D

7. A microscope has an 8 mm focal length objective and a 12.5 mm focal length ocular lens. The separation between the two lenses is 180.5 mm. Find a) the lateral magnification by the objective; b) the angular magnification by the ocular, c) the angular magnification of the microscope, d) the distance of the slide from the objective lens. Assume normal adjustment.

   Ans.: a) -20; b) 20; c) -400; d) -8.4 mm.

8. A 200 power microscope with an OTL = 160 mm has a 40 times objective. Find the focal length of the ocular. Ans.: 50 mm.

9. The separation between the objective (f=4mm) and ocular lenses (f=50mm) of a microscope is 254 mm. Find a) the lateral magnification by the objective; b) the angular magnification by the ocular, c) the angular magnification of the microscope, d) the distance of the slide from the objective lens. Assume normal adjustment. Ans.: a) -50; b) 5; c) -250; d) -4.08 mm.

10. According to the example in Sect. 14.4.1, a microscope contains a 16 mm focal length objective lens, a 25 mm focal length eyepiece, and has an OTL of 160 mm. The magnification of the microscope was -100x. a) Find the magnification of this microscope when the final image is at the lddv. Ans.: -110x.

11. In Prob. 10 the object slide is 17.6 mm from the objective lens. The position of the ocular lens is altered to produce the final image at the LDDV, i.e., the system is not in normal adjustment. Find a) the object distance for

the ocular; b) separation between the lenses A_1A_2; c) the lateral magnification by the objective lens; d) the lateral magnification by the ocular lens; e) the lateral magnification of the microscope; f) why are the lateral and angular magnifications equal?

Ans.: a) $u_2 = -22.7272$; b)$A_1A_2 = 198.7272$; c) $Y_1 = -10$; d) $Y_2 = +11$; e) $Y = -110$.

12. A microscope has a 16mm objective lens with an aperture diameter $D_1 = 4$ mm , and a 25 mm ocular lens with an aperture diameter $D_2 = 8$mm. Find a) the working distance, b) the position and size of the aperture stop, entrance and exit pupil, c) the position and size of the field stop, entrance and exit ports, d) the linear FOV on the slide at 50% vignetting and e) compare the angular FOV at 50% vignetting at the eye to the angle subtended by the linear FOV (found in part d) at 250 mm, f) Draw a diagram showing slide, pupils, ports and FOV's.

Answers: a) 17.6 mm; b) the objective is the aperture stop and entrance pupil, of 4 mm diameter, the exit pupil is at an eye relief of 28.55 mm and has a diam of -0.5682 mm; c) the ocular is the field stop and exit port of 8 mm diameter, the entrance port is 17.38 mm in front of the objective lens; d) diameter of linear FOV on slide = 0.7005 mm; e) 8° vs. 0.08° or (-)100x.

13. A microscope with an OTL of 160 mm contains an objective lens with a focal length of 16 mm and a diameter of 4 mm. a) What is the NA of this lens; b) the resolvable separation of two points on a slide due to diffraction theory; c) the practical resolvable separation; d) the theoretically resolvable number of line pairs/mm; e) the practical resolvable L/mm. Assume a wavelength of 555 nm.

Ans.: a) 0.1136; b) 0.00298 mm; c) 0.00244 mm; d) 335.66 L/mm; e) 409.5 L/mm.

14. Repeat part e), in Problem 13, for wavelengths of 400 and 700 nm.

Ans.: 568 L/mm; 324 L/mm

15. A fine grid is engraved with 500 L/mm. Find the smallest NA required to resolve these. Assume a wavelength of 500 nm.

Ans.: Theor. NA= 0.1525; practical NA = 0.125.

CHAPTER 15

TELESCOPES

15.1 INTRODUCTION

Telescopes ordinarily are afocal instruments. Their focal points and principal points are at infinity. Light from distant objects is collimated on entering and exiting the telescope. Because the objects and images are distant, we specify their apparent size or angular subtense rather than their linear size. Telescopes provide *angular magnification* **M**, which is *the ratio of the angle subtended by the image* ω' *to the angle subtended by the object* ω.

$$M = \frac{\omega'}{\omega} \qquad 15.1$$

Telescopes for visual use consist of an objective, which may be a refracting or reflecting component and an ocular lens. The three classes of telescopes are refracting, catoptric and catadioptric.

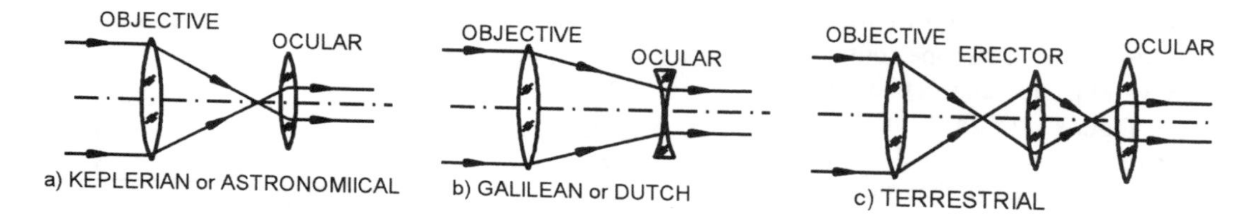

Figure 15.1.1 Refracting telescopes.

1. *Refracting* telescopes are constructed entirely of lenses. They include a) astronomical (Keplerian) telescopes that produce inverted images, b) Dutch (Galilean) telescopes that produce erect images and c) terrestrial telescopes that erect the image produced by the astronomical telescopes.

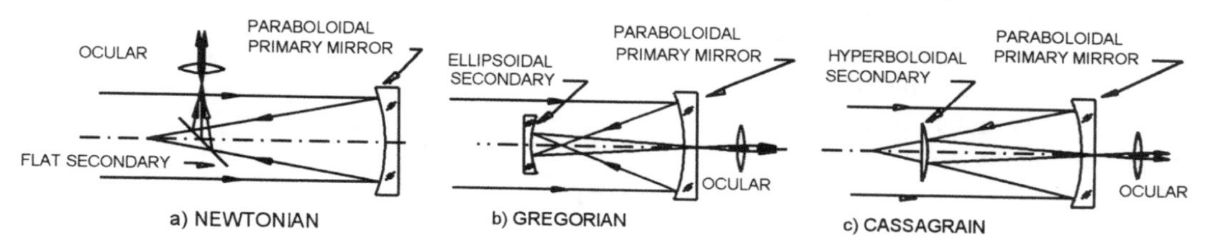

Figure 15.1.2 Catoptric telescopes.

2. *Catoptric* telescopes have mirrors for objectives. The mirrors are generally aspherical (conics) to avoid spherical aberration. Newtonian telescopes have a paraboloidal primary mirror to collect light and bring it to a prime focus, and a flat secondary mirror to fold the ray path out of the telescope tube so that the image may be viewed through the ocular. The objective of the Gregorian telescope comprises a paraboloidal primary and an ellipsoidal secondary mirror. The image is made accessible through a hole in

the primary mirror. Cassagrainian telescopes use a paraboloidal primary and a hyperboloidal secondary mirror. Access to the image is also via a hole in the primary.

3. *Catadioptric* telescopes use a combination of reflectors and refractors to produce the prime focus image. The Schmidt camera uses a spherical primary mirror, but a very special lens or corrector plate eliminates the spherical aberration of the mirror.

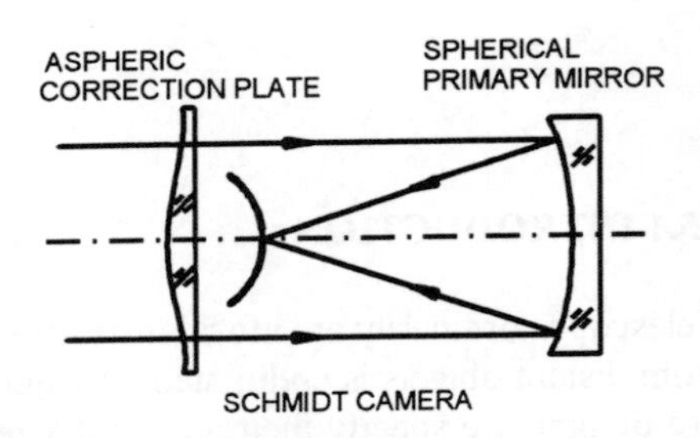

Figure 15.1.3 Catadioptric telescope; the Schmidt Camera.

15.2 REFRACTING TELESCOPES

15.2.1 THE ASTRONOMICAL OR KEPLERIAN TELESCOPE

15.2.1.1 Angular Magnification

The astronomical telescope consists of positive objective and ocular lenses. See Figure 15.2a. In afocal adjustment, also called *normal adjustment,* it is set so that the second focal point of the objective coincides with the first focal point of the ocular. An inverted image **y'**, formed by the objective lens, is viewed through the ocular. See Figure 15.2b. The angle **ω** subtended by the object is defined by the chief ray, consequently, $\omega = y'/f_1$. Since, **Q'** is in the first focal plane of the ocular all rays from **Q'** will leave collimated. The slope of these rays will be determined by the chief ray through the ocular. Thus, $\omega' = -y'/f_2$.

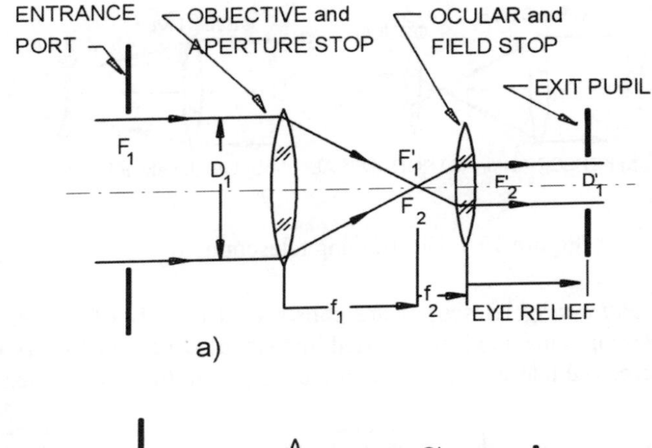

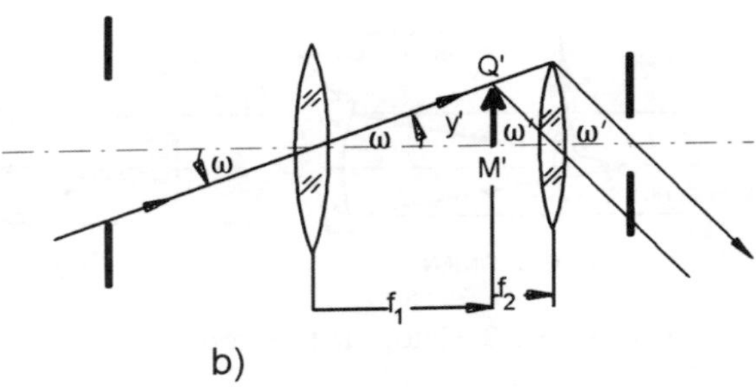

Figure 15.2 The Keplerian or astronomical telescope.

The angular magnification of the astronomical telescope is

$$M = \frac{\tan \omega'}{\tan \omega} = -\frac{y' f_1}{f_2 y'}$$

or,

$$M = -\frac{f_1}{f_2} \qquad 15.2$$

The minus sign means that the image is inverted.

EXAMPLE 1. An objective and an ocular lens are to be mounted at opposite ends of a 110 mm long tube to make a 10 power astronomical telescope. Find their necessary focal lengths.

The basic relationships are $M = -f_1/f_2 = -10$, and $f_1 + f_2 = 110$
by substitution, $10f_2 + f_2 = 11f_2 = 110$
thus, $f_1 = 100$ mm and $f_2 = 10$ mm

The objective lens (diameter = $\mathbf{D_1}$) of the astronomical telescope is usually the aperture stop of the system. The exit pupil or *Ramsden circle* is a real inverted image of the aperture stop and has a diameter $\mathbf{D'_1}$. The similar triangles in Figure 15.2a show that $f_1/f_2 = D_1/D'_1$, therefore, angular magnification is also given by

$$M = -\frac{|D_1|}{|D'_1|} \tag{15.3}$$

Telescopes appear to bring objects closer in direct proportion to their power. To observe objects that are closer than infinity the telescope must be refocused by backing the ocular out of the tube until F_2 coincides with the internal image $\mathbf{M'_1}$.

As the pupil of the eye constricts the retinal illumination is reduced and a viewed scene grows dimmer. If an eye with a given pupil diameter views a scene through a telescope that has a smaller exit pupil than the eye pupil the telescope, in effect, is constricting the pupil of the eye. The size of the exit pupil is inversely related to the magnification, according to Eq. 15.3. Consequently, increasing the magnification of the telescope by increasing the focal length of the objective without also increasing its diameter ultimately constricts the exit pupil until it just matches the eye pupil.

The magnification that just matches the diameter of the exit pupil of the telescope with the eye pupil is called *normal magnification. It is the maximum magnification that does not cause the image seen through the telescope to appear dimmer than the object would appear to the unaided eye. It is the minimum magnification required to obtain the full resolving power of the instrument.* This latter condition is imposed by diffraction. Diffraction blur is inversely related to the aperture diameter of the objective, but magnification is directly related to aperture diameter. The maximum aperture and, therefore, the magnification that will minimize the diffraction blur occurs when the exit pupil just matches the pupil of the eye. Any further increase in exit pupil diameter will not reduce the diffraction blur in the eye because the pupil of the eye now determines the diffraction blur.

Equation 15.2 shows that magnification can be altered by varying either the focal length of the objective or the ocular. For example, let a 10x telescope have an objective lens with a 100 mm focal length and 30 mm diameter. The exit pupil diameter will be 3 mm. If the eye pupil diameter is also 3 mm, the telescope satisfies the conditions for normal magnification. The view through the telescope appears as bright as the unaided view, and the diffraction blur corresponds to the full 30 mm diameter of the objective. If a 200 mm focal length lens with a 30 mm diameter is substituted, the power increases to 20x. The diffraction blur is unchanged. However, because the exit pupil is reduced to 1.5mm, i.e., less than the eye pupil diameter, the scene appears dimmer. Thus, normal magnification, i.e., 10x is the maximum magnification that will not result in a diminution in brightness.

On the other hand, if the power of the telescope is reduced to 5x, i.e., the focal length of the objective is reduced to 50 mm, but its diameter remains 30 mm, then the diameter of the exit pupil will increase to 6mm. Overfilling the eye pupil does not increase image brightness because only light within the 3mm pupil illuminates the retina. However, the eye pupil, in the exit pupil plane of the telescope, is conjugate to the objective lens. Its image at the lens will be 15 mm in diameter. In effect, the diffraction blur will increase because the effective diameter of the objective lens is reduced. Here, the eye pupil is the aperture stop and its image at the objective lens is the entrance pupil. Thus, normal magnification, i.e., 10x, is the minimum magnification that will allow the full aperture of the objective to determine the diffraction blur.

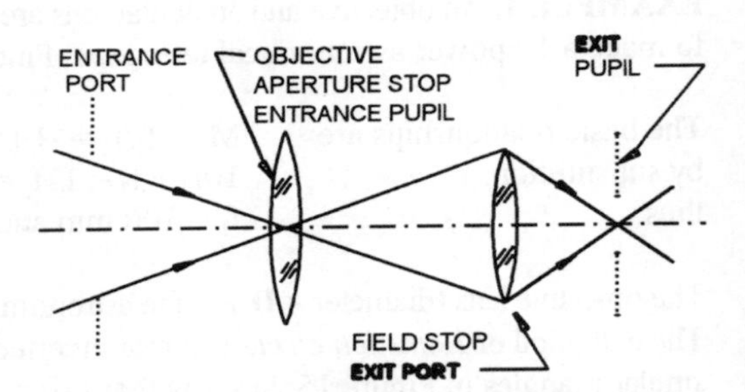

Figure 15.3 The field of view of astronomical telescopes.

15.2.1.2 The Field of View of the Astronomical Telescope

In a properly designed Astronomical (Keplerian) telescope the objective lens is the aperture stop, consequently, the entrance pupil of the telescope coincides with the objective lens. With no internal diaphragm to act as the field stop, the ocular is the field stop and exit port. See Figure 15.3. The entrance port will be a real inverted image. The field of view for 50% vignetting in object space is given by the ray that does or appears to graze the edge of the entrance port and go through the center of the entrance pupil. The ray actually grazes the edge of the field stop and goes through the center of the aperture stop. In image space, the ray grazes the exit port and goes through the center of the exit pupil. The ray pencils for 0, 50 and 100% vignetting are shown in Figures 15.4. Zero vignetting is given by the ray that goes through *corresponding* edges of the ports and pupils, 100% vignetting, by the ray that goes through *opposite* edges of the ports and pupils.

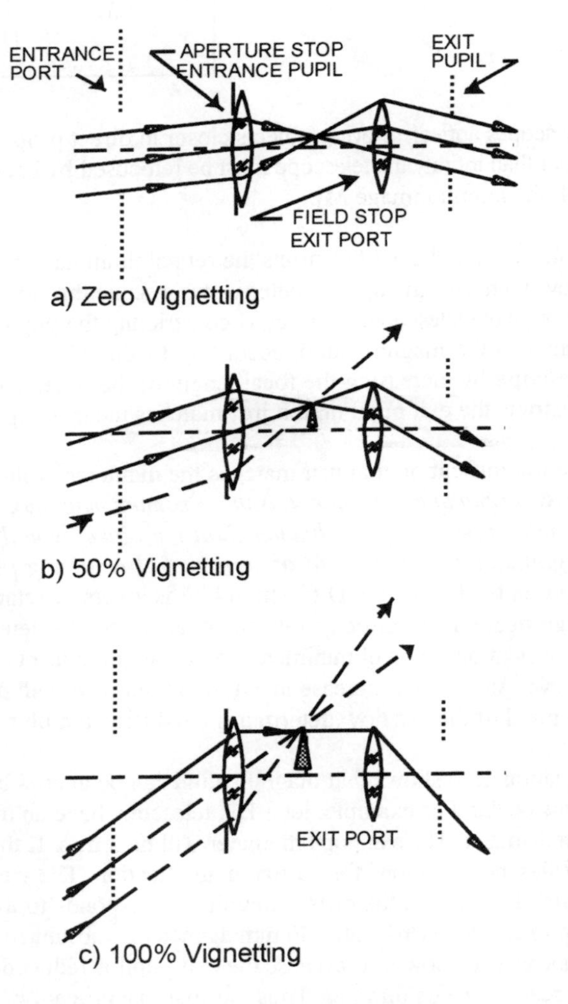

Figure 15.4 Vignetting limits the field of view of telescopes.

Student assignment

The student should follow the paths of the reference rays through the pupils, ports and stops in Figure 15.4, paying particular attention to inverted pupils and ports. Does the ray through corresponding edges of pupil and port necessarily go through corresponding edges of the aperture and field stop?

15.2.1.3 Field Lenses

Figure 15.4 shows that the field of view is limited by the diameter of the ocular. At 100% vignetting all rays miss the ocular. A field lens placed in the internal image plane will increase the field of view without the need to increase the diameter of the ocular or affecting the power of the telescope. An example of the action of a field lens is shown in Figure 15.5.a. The focal length of this lens is such that it forms an inverted image of the objective on the ocular. The exit pupil coincides with the ocular, i.e., the eye relief is zero. Rays 1-3, from the 100% vignetted field angle, through points BAC of the objective are refracted to their conjugate points in the exit pupil B'A'C', and are collimated by the ocular. Field lenses increase the field of view, but decrease the eye relief. By increasing the focal length of the field lens a smaller field increase and an increased eye relief may be obtained. See Figure 15.5.b. In this diagram, Rays 1-3 from the 100% vignetted field angle are bent by the field lens just enough to cause the rays between 1 and 2 to go through the ocular and exit pupil. The field angle is now seen with 50% vignetting. The eye relief is not as great as it was without the field lens, because the exit pupil (the image of the

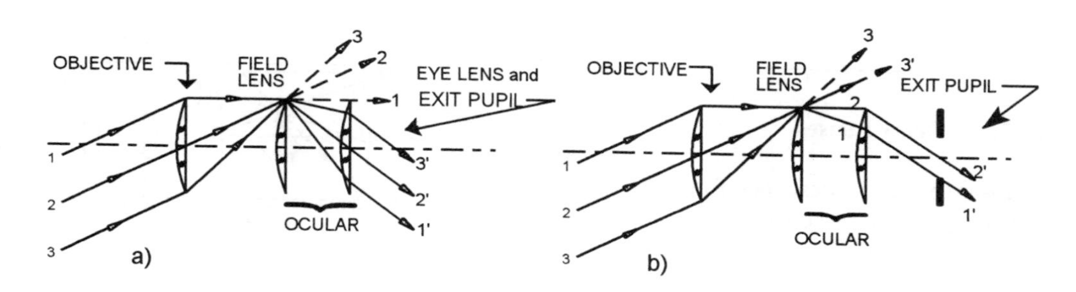

Figure 15.5 Field lenses increase the field of view, but decrease the eye relief of the astronomical telescope.

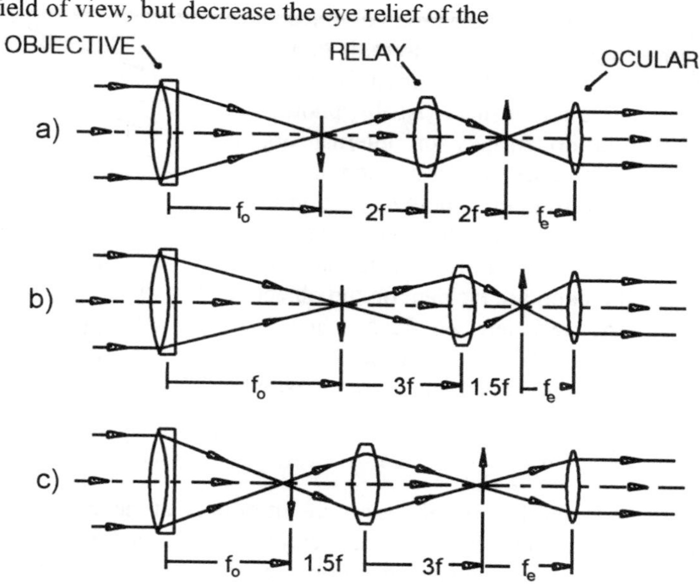

Figure 15.6 Terrestrial telescopes contain relay lenses to erect the image. They provide lateral magnification stages of a) -1; b) -½; and c) -2.

objective) is formed by the combined powers of the field and ocular lenses.

15.2.2 THE TERRESTRIAL TELESCOPE

A *relay* lens inserted between the objective and ocular of an astronomical telescope will erect the image, thus, converting it to a terrestrial telescope. This is the type used for rifle scopes. As shown in Figure 15.6, the relay can be positioned with various conjugates to provide any desired lateral magnification. The figure illustrates three such sets of conjugates that will modify the overall length of the telescope. The angular magnification of the terrestrial telescope is the product of this lateral magnification and the angular magnification of the astronomical telescope.

$$M = -\frac{u'f_1}{uf_2} = -Y\frac{f_1}{f_2} \qquad 15.4$$

15.2.2.1 Relay Lenses

Endoscopes and borescopes formerly were designed with a series of field and relay lenses to transfer an image through a long rigid narrow tube. Today fiber optics is used to transfer images, through flexible endoscopes. Endoscopes are inserted into body openings, such as throat, urethra and rectum to inspect visually, photograph and do surgery on internal organs. Borescopes allow inspection of the bore of a rifle, or remote inspection of the interior of a radioactive chamber. In Figure 15.7a, rays from an off-axis direction are incident on the objective. The rays between Rays 2 and 3 are vignetted at the relay lens. The rays between Rays 1 and 2 are vignetted at the ocular. Vignetting is 100% from this field angle.

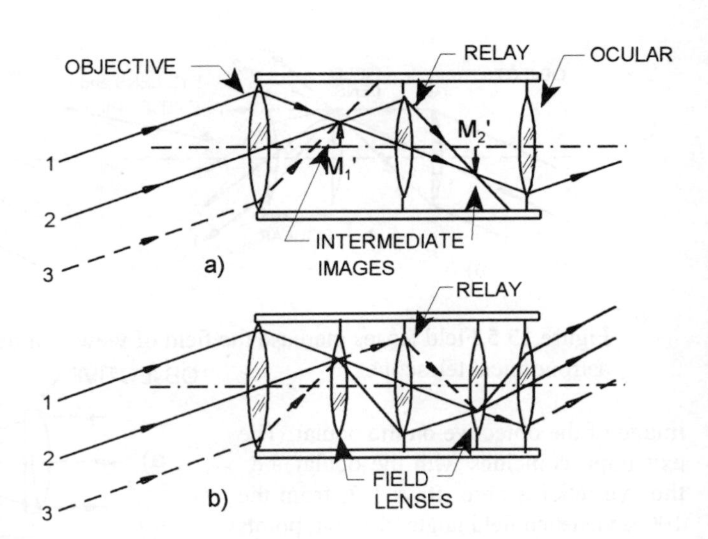

Figure 15.7 a) A terrestrial telescope in a long narrow tube seriously vignettes an image. b) Vignetting is reduced by inserting field lenses at the intermediate image planes on either side of the relay lens, thereby, increasing the field of view.

In Figure 15.7.b, field lenses are introduced at the intermediate image planes. The first field lens forms an image of the objective on the relay lens; the second field lens forms an image of the relay lens on the ocular. There now is no vignetting from the original field angle. Adding additional relay and field lenses will extend the tube length. The image on the final field lens serves as the object for the ocular. If each image stage is relayed with a lateral magnification m = -1, then the spacings between image planes will be 4f.

15.2.3 The Galilean Telescope

The Galilean (Dutch) telescope has a negative ocular. In normal adjustment the second focal point of the objective coincides with the first focal point of the ocular, as for the Keplerian telescope. The angular magnification, as shown in Figure 15.8, is the same as for the Keplerian telescope. However, f_2 is negative, so the image is erect. No real intermediate image is formed, therefore, crosshairs and reticles cannot be inserted.

$$M = \frac{\tan w'}{\tan w} = -\frac{y' f_1}{f_2 y'} \quad \text{or}$$

$$M = -\frac{f_1}{f_2} \qquad 15.5$$

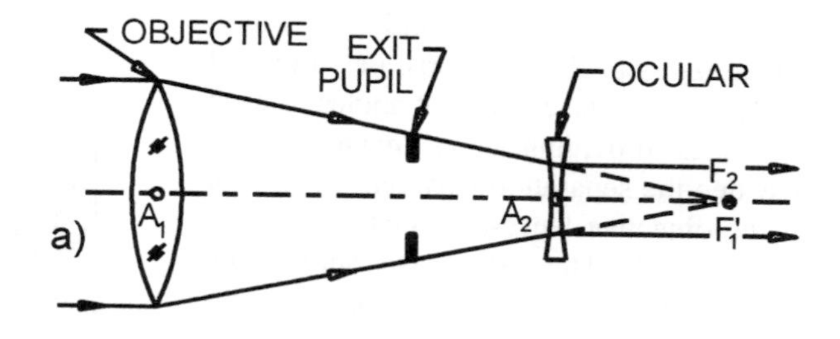

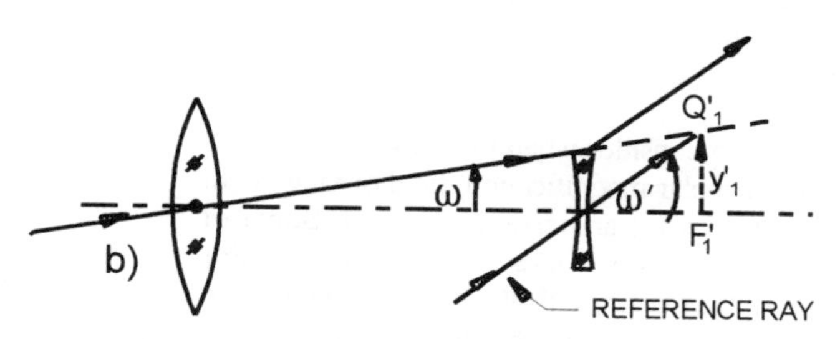

Figure 15.8 a) The Galilean or Dutch telescope; b) angular magnification = ω'/ω.

Furthermore, if we were to treat the objective lens as the aperture stop, the exit pupil would be virtual and inaccessible. Instead, the telescope is designed so that the pupil of the eye is the aperture stop and exit pupil, and the pupil of the eye is considered in contact with the ocular. Consequently, the objective will be the field stop and entrance port. The entrance pupil will be a virtual image of the pupil of the eye. Figure 15.9 illustrates the real and apparent fields of view. The diameter of the objective lens must be increased to increase the field of view. Galilean telescopes generally have low power and small fields of view.

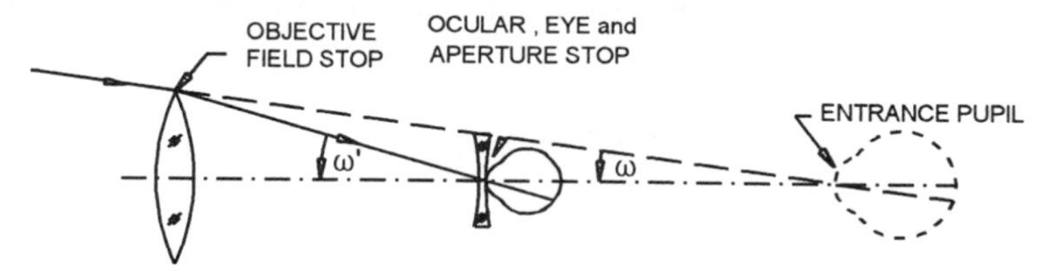

Figure 15.9 Field of view of the Galilean telescope.

15.2.3.1 Field and Opera Glasses

Field and opera glasses comprise pairs of Galilean telescopes with parallel optical axes to provide binocular viewing. Although they are usually only about 2 or 3 power and have relatively small fields of view, they are inexpensive and compact. They provide bright images because they contain as few as four air/glass interfaces at which light reflection losses may occur.

15.2.4 THE PRISM BINOCULAR

Prism binoculars consist of a pair of Keplerian telescopes. The images are commonly re-inverted and re-reverted with *Porro* prisms, i.e., a pair of right-angle prisms, placed at right angles to each other. Besides providing erect images, Porro prisms widen the separation between the optical axes of the objective lenses, by that, enhancing stereoscopic acuity. The prisms also shorten the length of the binoculars by folding the light path back and forth. The separation between the oculars can be adjusted to match the pupillary separation. See Figure 15.10.

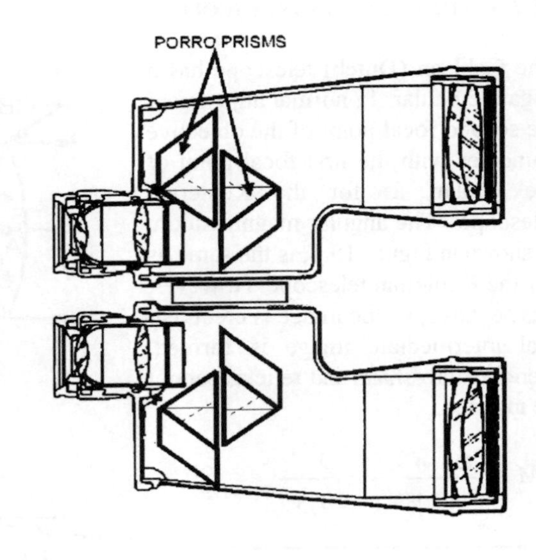

Figure 15.10 The prism binocular.

Features to consider when purchasing binoculars include: angular magnification, field of view, light gathering power, and size and weight. Some of these features are marked on the housing. For example, the markings, 8x30, 360' indicate that the angular magnification is 8x, the diameter of the objective lens is 30 mm, and the width of the linear field of view that can be seen at 1000 yards is 360 ft. Some binoculars mark the field angle in degrees. This is a field of view in object space. Generally, the field of view in object space varies inversely with power for a particular type of eyepiece.

The 8x30 markings immediately tell us that the diameter of the exit pupil is 3.75 mm. See Eq. 15.3. Exit pupil size is important. If the exit pupil is smaller than the pupil of the eye the view will be dim compared with the view seen with the unaided eye and the potential resolving power will not be obtained. The pupil of the eye dilates at low light levels, and may become as large as 8 mm in diameter. Consequently, an 8x30 binocular is not a good choice for twilight use. The relative brightness of binoculars is given by

$$RelativeBrightness = \frac{D_1^2}{M^2} = D_1'^2$$

A 7x50 binocular has twice the relative brightness of a 7x35 binocular.

Some common binoculars are 6x20, 6x30, 7x35, 7x50, 8x20, 8x30, 8x40, 10x50 and 15x60. Binoculars of less than 10x are generally hand-held. Holding binoculars of more than 10x steady enough is difficult, since hand tremors also are magnified. Binoculars ordinarily have 9 mm eye reliefs. Special designs have long eye reliefs (17 mm) for spectacle wearers.

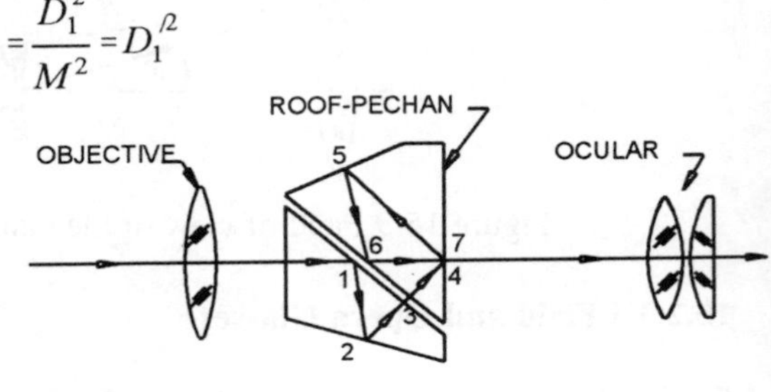

Figure 15.11 The direct-vision monocular employs a Roof-Pechan or Schmidt prism.

Compact straight-tube monoculars and binoculars use roof-Schmidt or roof-Pechan prisms. The prism consists of two elements separated by a thin airspace See Figure 15.11. These prisms use three ordinary reflections and three total internal reflections to fold the path. The even number of reflections results in an erect image and greatly shortens the tube length. The ray through an ordinary Pechan prism undergoes total internal reflection at points 1, 4 and 6. It is transmitted across the airspace at point 3. Mirror coatings produce reflection at points 2 and 5. The ordinary Pechan produces only five reflections, and would result in a left-handed image. To get six reflections, the prism face at 5 is made as a roof, therefore the name, roof-Pechan.

15.2.4.1 Stereoscopic Range Through Binoculars

Binocular vision is the basis for depth or stereoscopic perception. Depth discrimination or stereoscopic resolution depends on binocular parallax, that is, a difference in the convergence angle of the eyes to fixate on objects at different distances or ranges. A parallax angle ≥ 30" of arc is sufficient for stereoscopic resolution. However, trained observers under ideal conditions have stereoscopic resolution as small as 2" arc. Range and interpupillary separation determine the convergence angle.

Given an interpupillary separation **pd** and a stereoscopic resolution θ, the unaided range $\mathbf{R}_o$, based on a triangulation calculation is

$$R_o = \frac{pd}{\theta_o} \qquad 15.6$$

where, **pd** is the base, $\mathbf{R}_o$ is the altitude of a triangle and θ_o is the angle of convergence of the eyes. See Figure 15.12.a. θ is a small angle and is expressed in radians.

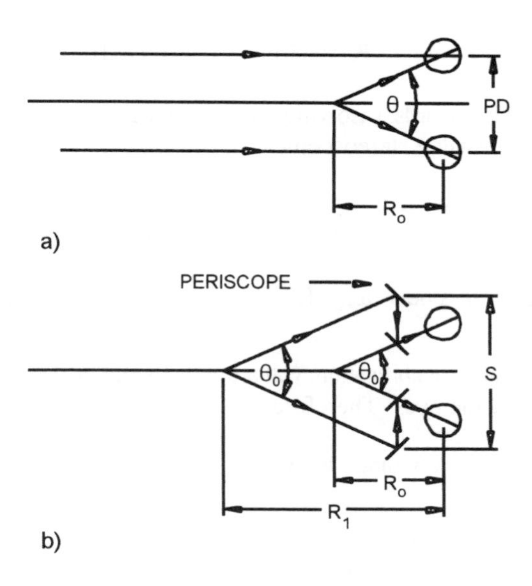

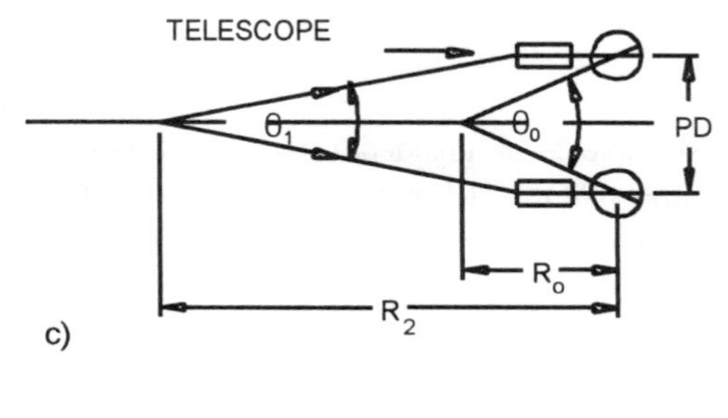

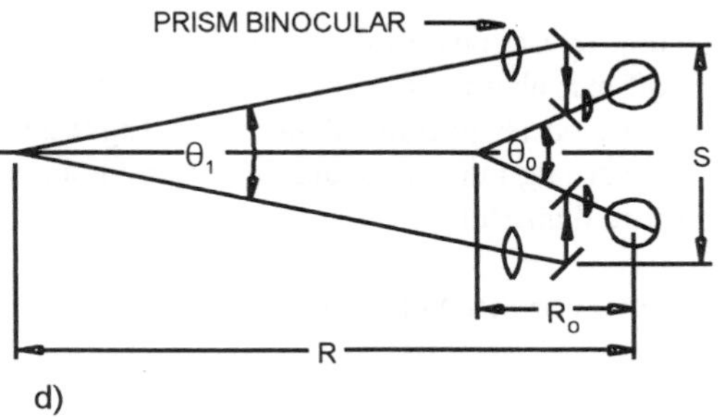

Figure 15.12 a) Stereoscopic range of unaided eyes; b) stereoscopic range is increased by increasing the base factor **S/pd;** and c) by introducing an angular magnification factor θ_o/θ_1; ; or d) by both factors.

EXAMPLE 2: Given a **pd** = 65 mm and θ_o = 30", the unaided range according to Eq. 15.6 is

$$R_o = 65/\tan 30'' = 4.47\text{x}10^5 \text{ mm} = 447 \text{ m} \approx 500 \text{ yards.}$$

In other words, resolving a point light source at a distance greater than 500 yards from a star at infinity is not possible stereoscopically.

Binoculars will increase this range by two factors: 1) the increase in separation of the centers of the objectives S, and 2) the angular magnification. Substitution of the separation **S** for **pd** in Eq. 15.6 will result in a new range R_1, as shown in Figure 15.12b. Where, $R_1 = \dfrac{S}{\theta_o}$

Angular magnification will amplify a smaller parallax difference in object space θ_1 to the necessary angle θ_o at the eyes. Therefore, we can substitute θ for the convergence angle in Eq. 15.6. Figure 15.12.c shows a new range R_2 where, $R_2 = \dfrac{pd}{\theta_1}$.

Since, $\mathbf{M} = \theta_o/\theta_1$, we can substitute $\theta_1 = \theta_o/\mathbf{M}$, and replace **pd** with **S** in the above equation to obtain the stereoscopic range with binoculars.

$$\boxed{R = \frac{MS}{\theta_o}} \qquad 15.7$$

In Example 2, for unaided vision, M = 1 and S = pd. If a binocular supplies S = 130 mm, the range R will become 894 m ≈ 1,000 yds. That is, by doubling the separation, the range is also doubled. If the angular magnification M = 6x, the binocular will increase the stereoscopic range to 6,000 yds.

15.3 REFLECTING TELESCOPES

Reflecting telescopes produce no chromatic aberration but have very narrow real fields of view due to severe off-axis aberrations. They are commonly used for astronomical observations because large diameter mirrors are easier to make than lenses. Light gathering power depends on the aperture diameter, and is critical for seeing faint stars and deep into the universe. Furthermore, the resolving power is proportional to aperture diameter.

15.3.1 NEWTONIAN TELESCOPE

Isaac Newton's telescope uses a mirror in place of the objective lens. Newton erroneously concluded that the chromatic aberration of lenses could not be eliminated. A concave paraboloidal mirror is usually used in the Newtonian telescope to avoid the spherical aberration of the axial image point. The off-axis images are aberrated. As seen in Figure 15.1.2 a, a small flat diagonal mirror is used to fold the image plane out of the tube, where the image is observed with an eyepiece. The diagonal mirror blocks the central zone, or about 25% of the surface area of the primary mirror. The angular magnification is equal to the ratio of focal length of the mirror to that of the eyepiece.

15.3.2 GREGORIAN TELESCOPE

A paraboloidal primary mirror is retained, but the flat secondary mirror is replaced with a concave ellipsoidal mirror in the Gregorian telescope. See Figure 15.1.2b. The focal point of the paraboloidal mirror coincides with one of the two foci of the ellipsoidal secondary mirror. After reflection from this mirror the rays converge on the second focus. The final image is reinverted and free of spherical aberration. To allow a straight tube design, a hole is cut in the primary mirror. Off-axis images are severely aberrated.

15.3.3 CASSAGRAINIAN TELESCOPE

This telescope, illustrated in Figure 15.1.2c, uses a convex hyperboloidal secondary mirror. The focus of the paraboloidal primary mirror coincides virtually with the first focus of the secondary mirror. Again the final image at the secondary focus of the hyperboloid is free of spherical aberration, but off-axis images are aberrated.

The Hubble space telescope is a Cassagrain type known as the Ritchey-Chretien design. Both the primary and secondary mirrors are hyperboloidal. It has zero coma and so covers a wider field than the traditional design.

15.4 CATADIOPTRIC TELESCOPES

Lenses are introduced to reduce the off-axis aberrations and to increase the field of view of reflecting telescopes. Only one such design will be treated here. The **Schmidt camera** uses a special aspherical corrector plate at the center of curvature of the spherical primary mirror. It is the aperture stop of the system. The aspheric corrects the spherical aberration. Since all chief rays through the center of the corrector are also radii of the primary mirror there are no off-axis aberrations. This system will cover large fields of view. It is used to photograph galaxies. See Figure 15.1.3.

15.5 THE MAXIMUM USEFUL MAGNIFICATION OF A TELESCOPE

The magnification of a telescope, according to Eq. 15.1, is $M = \dfrac{\omega'}{\omega}$

Let the angle ω correspond to the angular separation between two stars that are at the limit of resolution. According to Chapter 13, Eq. 13.5 this angle of resolution for a telescope objective with a diameter D_o in mm is

$$\omega = 140/D_o \qquad \text{seconds of arc.} \qquad 15.8a$$

or, if D_o is in inches, this equation becomes

$$\omega = 5.5/D_o \qquad \text{seconds of arc.} \qquad 15.8b$$

The diffraction images of a visual telescope also must be resolved by the eye looking through the ocular. The eye resolves an angle of 60 arc seconds. Thus, the minimum value of angle $\omega' = 60''$. Therefore, the magnification required to resolve two stars visually and, therefore, the maximum useful magnification, as a function of aperture diameter in inches, is given by ω'/ω. If D_o is in inches, M_{MAX} is

$$M_{MAX} = \frac{60\,D_o}{5.5} = 11\,D_o \qquad 15.9$$

For example, a telescope with a 2 inch diameter objective lens will resolve two stars separated by 2.75 seconds of arc. If its power is 22x the eye will visually resolve the stars. Additional power is called *empty magnification* and will serve only to enlarge blurs. Since examining image detail at the limit of visual resolution is uncomfortable, it is like reading a book printed with 20/10 letters, providing 3 or 4 times the maximum useful power is usual, namely, 80x.

Empirical measurements of telescope resolution, indicate that Rayleigh's criterion is too conservative. The Airy discs may be less than a semi-diameter apart. The separation at which no dip occurs between the peaks of the diffraction patterns is sufficient visually to resolve the two patterns. Consequently, what is called the Sparrow criterion for the angle of resolution for stars is

$$\omega = \frac{\lambda}{D} \qquad 15.10$$

Using the Sparrow criterion for the resolution in seconds of arc, Eq.15.8b becomes $4.5/D_o$, and maximum useful magnification (Eq. 15.9) equals $13.3D_o$.

PROBLEMS

1. A distant hot air balloon subtends 0.25° at the objective lens of an afocal astronomical telescope. The image subtends 1.5°. The objective and simple ocular are in a 35 cm long tube. Find the focal length of the ocular lens.
Ans.: 5 cm.

2. Repeat Problem 1 for a Galilean telescope in a 35 cm long tube. Ans.: -7 cm.

3. If the diameter of the objective lens in Problem 1 is 10 cm, find a) the diameter of the exit pupil and b) the eye relief.
Ans.: a)1.67 cm; b)5.83 cm.

4. The focal lengths of the objective and ocular lenses of a telescope are 300 mm and 30 mm, respectively. The telescope is to be designed so that the exit pupil fills a 5 mm eye pupil. Find the diameter of the objective lens.
Ans.: 50 mm.

5. a) Find the focal lengths of the objective and ocular lenses of a six power Keplerian telescope that will fit into a 35 cm long tube. b) If the inside diameter of the tube is 2 cm, find the diameter of the entrance and exit pupils. c) Find the diameters of the entrance and exit ports. d) Find the total real and apparent fields of view at 50% vignetting.
Ans.: a) 30 and 5 cm; b) 2 and 0.333 cm; c) 12 and 2 cm; d) 3.27° and 19.46°.12

6 a) Find the focal lengths of the objective and ocular lenses of a six power Galilean telescope that will fit into a 25 cm long tube. The eye is not part of this system.
   b) If the inside diameter of the tube is 2 cm, find the diameter of the entrance and exit pupils.
   c) Find the diameters of the entrance and exit ports.
   d) Find the total real and apparent fields of view at 50% vignetting.
Ans.: a) 30 and -5 cm; b) 2 and 0.333 cm; c) 12 and 2 cm; d) 4.58° and 27 °.

7. Repeat Problem 6, but place a 2.5 mm diameter eye pupil in contact with the ocular lens.
Ans.: a) 30 and -5 cm; b) 1.5 and 0.25 cm; c) 2 and 0.333 cm; d) 0.76° and 4.58°.

8. A ten power Keplerian telescope has 100 mm focal length objective lens. This lens is the aperture stop. Find the eye relief. Ans.: 11 mm.

9. Convert the ten power Keplerian into a ten power terrestrial telescope with a 10 mm focal length relay lens. a) How long is the tube? Assuming that the objective is still the aperture stop, find the eye relief. Hint: Find the image of the objective lens formed by the relay lens, and treat this image as the object for the eyelens. Ans.: a) 150 mm; b) 21 mm.

10. Convert the ten power Keplerian (with f_1 = 100 mm) into a ten power Galilean telescope. If the objective is still the aperture stop, find the eye relief.
Ans.: -9 mm.

11. A 10x50 binocular has a 50 mm focal length objective and a 3mm diameter field stop in the second focal plane of the objective. What are the total real and apparent angular fields of view?
Ans.: 3.437° and 33.39°

12. A 10x50 binocular with a 50 mm focal length objective lens covers 80 meters at 1000 meters. What is the diameter of the field stop? Ans.: 4 mm.

13. A person has a 60 mm pd, and a stereoscopic range of 275 meters. With a pair of 10x50 binoculars whose objective lenses are separated by 180 mm: a) what it his aided range? b) his unaided stereo-acuity?
Ans.:8,250.6 m; 45".

14. A person with a 64 mm pd uses a 5x30 binocular with a lens separation of 128 mm. Her stereoscopic range is 13,200.9 meters. What is a) her unaided range? b) her stereo-acuity?
Ans.: a) 1320.1 m; b) 10 secs.

15. A Newtonian telescope has an f/10 primary mirror with a 1,000 mm focal length. a) What is the axial separation between the primary and the plano secondary mirror so that the prime focus, folded out at 90° from the optical axis, just clears the incident collimated bundle of light; b) how large is the secondary mirror? c) How great is the obscuration by the secondary?
Ans.: a) 950 mm; b) 5x7.07 mm; c) 0.25%

16. In Prob. 15, the plano secondary mirror obstructs 6.25% of the aperture of the primary mirror. What is their separation? Ans.: 750 mm.

NOTES:

CHAPTER 16

CAMERAS AND PROJECTORS

16.1 PINHOLE CAMERA AND CAMERA OBSCURA

The pinhole camera is cited as a device that displays the rectilinear propagation of light. It is an example of one of the earliest types of cameras. Similar to the pinhole camera, but on a larger scale, is the camera obscura. This is a light tight room or structure with a small hole in one wall through which a sunny landscape or scene is projected, with the proper perspective, onto a translucent sheet of paper. See Figure 16.1. Canaletto painted intricately detailed buildings in Venice with the aid of the camera obscura. Eventually, simple lenses were substituted for the hole. Their larger apertures and focusing properties provided brighter and sharper images.

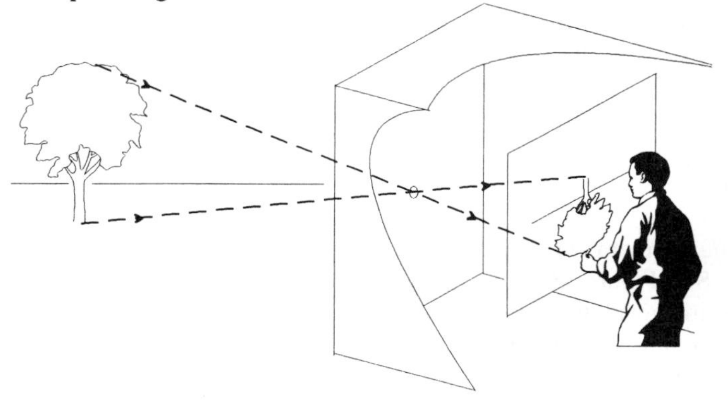

Figure 16.1 Camera Obscura.

16.2 TYPES OF CAMERAS

With the development of photosensitive materials a permanent record or photograph could be made. The camera obscura was transformed into the box camera of the early 19th century. The box camera contained a landscape lens, that is, a simple meniscus lens with either a front or rear aperture stop. Chromatic aberration was uncorrected. Although the camera covered a field of about 55°, at f/14 to f/16 it required long exposures times. Subsequently, an achromatic doublet replaced the simple lens, and the aperture became adjustable for sunny and overcast lighting. Because the camera was fixed-focused for far, it could not be used for close-ups.

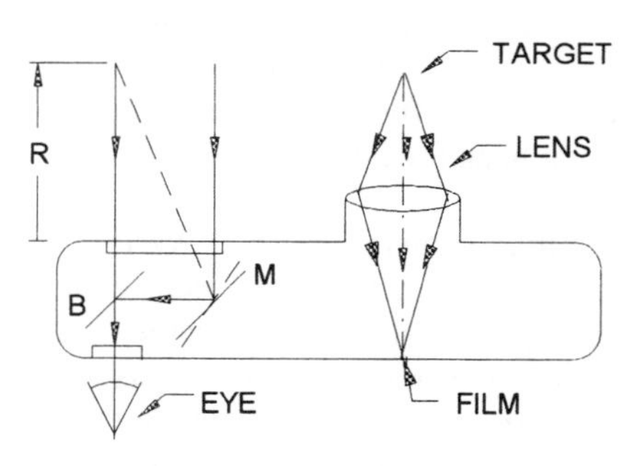

Figure 16.2 Rangefinder Camera.

Adjustable focus cameras are common today. The *range finder* camera uses a separate viewing window and a split image to focus the lens accurately. When mirror M and beamsplitter B are parallel the focus is at infinity. A closeup target requires that mirror M be rotated to eliminate double images. The requisite rotation corresponds to the range. See Figure 16.2.

The *twin-lens reflex* camera shown in Figure 16.3 uses a lens for taking the picture. The taking lens forms an image on the film. A viewing lens forms an image on a ground glass screen conjugate to the film plane via reflection from a fixed mirror. This image is viewed through the ocular. Both lenses move together as the image on the ground glass is focused. Parallax errors in close-up work, resulting from the separation between the two lenses, will produce a slightly different view on the ground glass from what is imaged on the film.

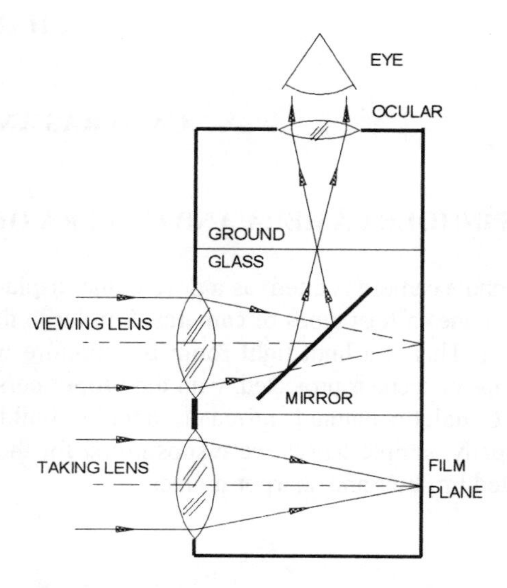

Figure 16.3 Twin-Lens Reflex Camera.

The *single-lens reflex* camera (SLR) eliminates the parallax error. See Figure 16.4. A hinged mirror reflects the image to the ground glass screen. This reflection plus three reflections in a Roof-penta prism provides an erect image for viewing, composing and focusing the picture. When the shutter is snapped, the mirror swing up against the screen and the image falls on the film. The SLR camera features interchangeable lenses that may range from 6 mm focal length wide-angle retrofocus lenses to >1000 mm focal length telephotos. To avoid interference with the pivoting mirror, short focal length lenses must have long back focal lengths.

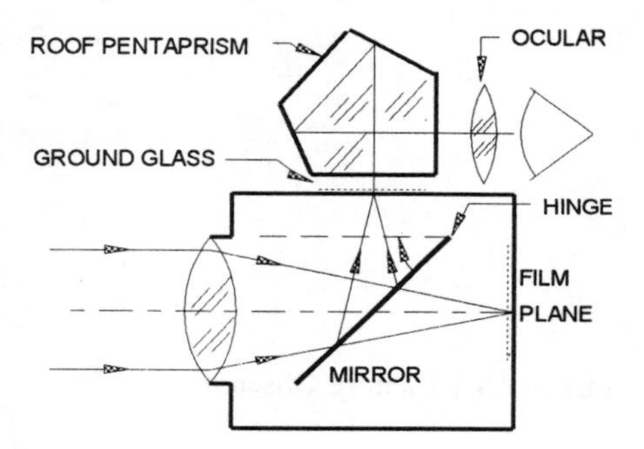

Figure 16.4 Single Lens Reflex Camera.

Many other cameras cover a variety of film formats. However, the *panoramic* camera is a particularly interesting special camera. It covers extremely wide angles by pivoting its lens about a vertical axis through its *second nodal point*. The film is on a curved platen. A vertical slit nearly in contact with the film moves with the lens along the film exposing successive portions of the film. See Figure 16.5.

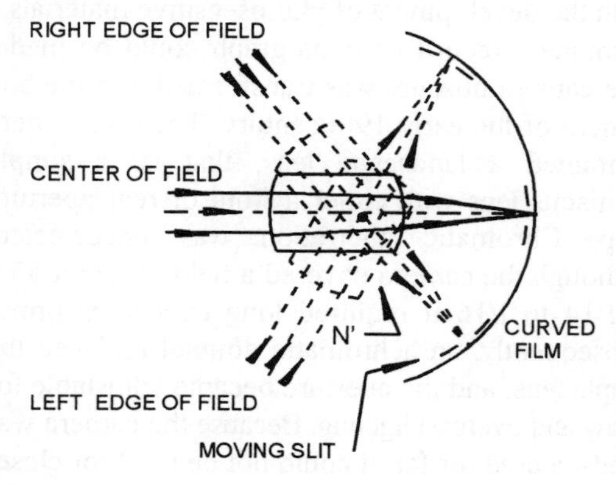

Figure 16.5 Panoramic Camera.

16.3 CAMERA LENSES

The simple landscape lens shown in Figure 16.6 was slow (f/14) and did not produce high quality photos. Fixtures to hold the head steady during exposure were required for portraits. To make the lenses faster and form sharp images over larger fields correcting the several aberrations that degraded the image was necessary. An achromatized meniscus lens (f/9) allowed for better correction of spherical aberration and coma. This, in turn, led to the design of increasingly complex lenses. Among the better known types are the Rapid rectilinear (f/8), Cooke triplet (f/6.3), Tessar (f/6.3), Petzval (f/3), and the Double Gauss (f/1.4). Fast lenses froze action and worked with available light. Many specialized lenses were designed, such as, the fish-eye for wide-angle photography, long focal length telephoto and mirror lenses for enlarged images, and zoom lenses with variable focal lengths. Usually, the wider the field of view and faster the lens the more complex is its design.

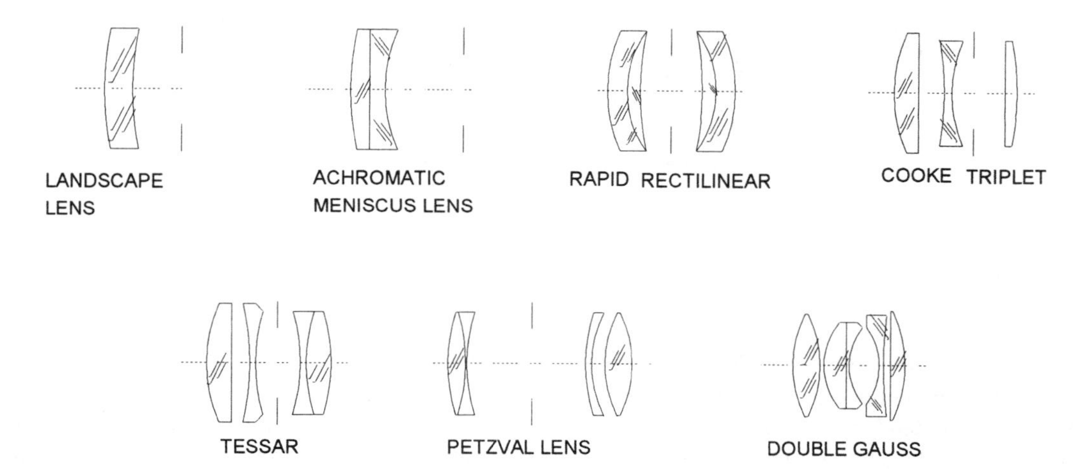

Figure 16.6 Examples of photographic lenses.

16.3.1 FIELD OF VIEW OF CAMERAS

The field of view of a camera depends on the focal length of the lens and the film format. A 35 mm SLR camera has a rectangular format of 24x36 mm, with a diagonal of 43.27 mm. A normal lens has a focal length of about 50 mm and covers about 46°, measured diagonally across the format. Wide-angle lenses have shorter focal lengths, from 35 to 6 mm, and cover fields up to a hemisphere; telephoto lens focal lengths range to more than a meter and have fields as small as a couple of degrees. Wide-angle lenses can be focused as close as several inches away, whereas, a 1000 mm telephoto lens will focus not much closer than 100 feet.

16.3.2 DEPTH OF FIELD AND FOCUS

Suppose a camera is focused to take a photograph of an object **M**, in the *focus plane*, at a distance **u,** and a sharp image **M'** is formed in the *film plane*, at distance **u'**, as shown in Figure 16.7a. If the object M is moved away from the focus plane through a distance $\delta_1 = \mathbf{MM_1}$, a blur will be formed at the film plane because the conjugate image $\mathbf{M'_1}$ will be displaced through a distance $\delta'_1 = \mathbf{M'M'_1}$. Similarly, if the object is moved toward the lens through a distance $\delta_2 = \mathbf{MM_2}$, the image displacement from the film plane will be $\delta_2' = \mathbf{M'M_2'}$, and again there will be a blurred image at the film plane.

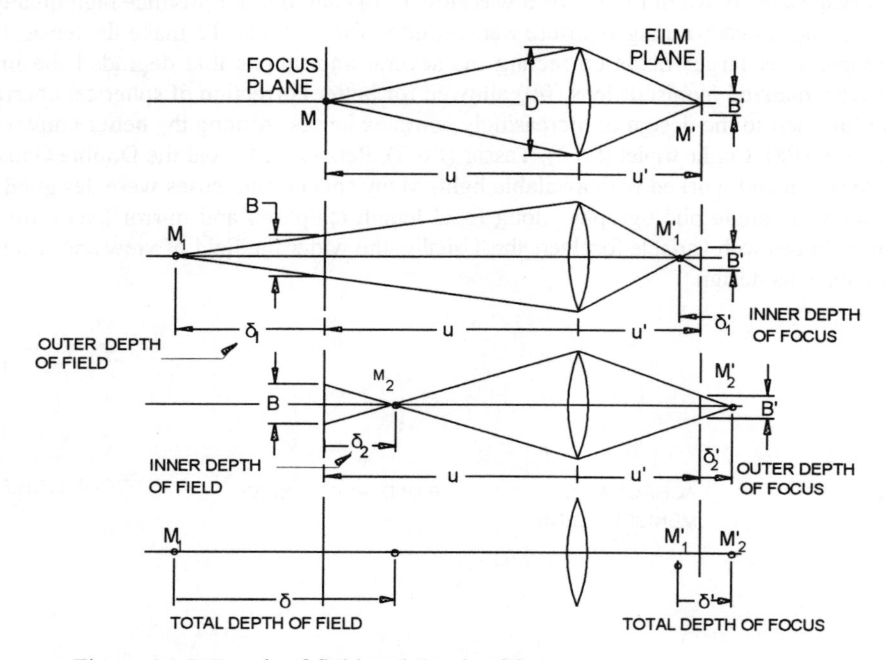

Figure 16.7 Depth of field and depth of focus.

Now a certain amount of blur will not noticeably degrade the image on the film. For instance, a blur **B'** in a picture that subtends up to one minute of arc will appear sharp to the eye. The displacements from the focus plane δ_1 and δ_2, and from the image plane δ_1' and δ_2', that correspond to the limits of acceptable blur are called *depth of field* δ and *depth of focus* δ', where; $\delta = \delta_1 + \delta_2 = \mathbf{M_1M_2}$ and $\delta' = \delta_1' + \delta_2' = \mathbf{M_1'M_2'}$.

The acceptable blur diameter B' determines the magnitude of δ and δ'. If the angular blur ß is set at one minute of arc (0.0003 radians), and the photograph is viewed at a distance d, the value of B' is given by

$$\mathbf{ß = B'/d}$$

Thus if the photograph is to be viewed at a distance of 10 inches, and ß = 0.0003 radians

$$B' = 0.0003 \times 10 = 0.003 \text{ inches.}$$

The inner and outer depths of field and focus may be derived from the similar triangles shown in Figure 16.7.

$$\textit{Outer depth of field} = \delta_1 = \frac{Bu}{D-B}\;; \qquad \textit{Inner depth of field} = \delta_2 = \frac{Bu}{D+B} \qquad 16.1$$

$$\textit{Inner depth of focus} = \delta_1' = \frac{B'u'}{D+B'}\;; \qquad \textit{Outer depth of focus} = \delta_2' = \frac{B'u'}{D-B'} \qquad 16.2$$

$$\textit{Total depth of field} = \delta = -\frac{2BuD}{D^2-B^2}\;; \qquad \textit{Total depth of focus} = \delta' = -\frac{2B'u'D}{D^2-B'^2} = \frac{2B'u'}{D} \qquad 16.3$$

The simplification of the equation for total depth of focus is based on eliminating $\mathbf{B}'^2$ in the denominator. The blur **B'** is very small compared with **D**, and is negligible when squared. For example, the diameter of a one minute blur patch **B'**, for a picture viewed at the least distance of distinct vision is $B'^2 = 0.003^2 = 0.00009$ sq. inches. The denominator = $\mathbf{D}^2\text{-}\mathbf{B}'^2$ is negligibly different from $\mathbf{D}^2$.

EXAMPLE 1: Find the depth of field and focus for an f/6 lens of 12 inches focal length, focused at 30 ft. Assume a blur tolerance of 0.0003 radians and a viewing distance of 10 inches (least distance of distinct vision). Given: f = 12 inches; D = 2 inches; u = -360 inches; B' = 0.003 inches.

Step 1. Use thin lens equation to find u' = +12.4138 inches.
Step 2. Find lateral magnification: Y = -0.03448
Step 3. Find depth of focus.

$$\delta_1' = M'M_1' = -\frac{0.003(12.4138)}{2+0.003} = -0.01859\ \textit{inches}$$

$$\delta_2' = M'M_2' = \frac{0.003(12.4138)}{2-0.003} = +0.01865\ \textit{inches}$$

$$\delta' = M_1'M_2' = \frac{2(0.003)(12.4138)}{2} = 0.03724\ \textit{inches}$$

Step 4. Find B with lateral magnification equation: B = -0.003/0.03448 = -0.0870 inches.
Step 5. Find depth of field.

$$\delta_1 = MM_1 = \frac{0.087(-360)}{2-0.087} = -16.372\ \textit{inches}$$

$$\delta_2 = MM_2 = -\frac{0.087(-360)}{2+0.087} = +15.007\ \textit{inches}$$

$$\delta = M_1M_2 = -\frac{2(0.087)(-360)(2)}{2^2-0.087^2} = 31.38\ \textit{inches}$$

Thus, the lens, when focused at 360 inches, will produce sharp images of objects within a range from 376.4 to 345.0 inches.

The depth of field and the depth of focus may be considered axial lengths of an object and image. Therefore, if we know one depth we can obtain the other through the axial magnification equation,

$$\mathbf{X} = \mathbf{Y}^2 = \delta'/\delta$$

According to Step 2, Y = -0.0345, therefore, X = 0.00119; and δ' = 0.0372 inches. Thus, δ = 31.26 inches. Compare this with 31.38 found above.

16.3.3 Hyperfocal Distance

The hyperfocal distance $\mathbf{u}_h$ is the closest distance that will be in focus when a camera or other optical system is focused at infinity. From the similar triangles in Figure 16.8, obviously

$$\boldsymbol{u_h} = -\frac{fD}{B'} \qquad 16.4$$

To maximize the depth of field of a camera it should be focused at the hyperfocal distance, i.e., let $u = u_h$.

Substitute Eq. 16.4 into the equation for outer depth of field δ_1,

$$Outer\ depth\ of\ field = \delta_1 = -\frac{BfD}{B'(D-B)}$$

However, **D** = **B**, therefore the denominator is zero and the outer depth of field δ_1 extends from the hyperfocal point to infinity.

Substitution of Eq. 16.4, and **D** = **B** into the equation for the inner depth of field δ_2, results in

$$Inner\ depth\ of\ field = \delta_2 = +\frac{BfD}{B'(D+B)} = +\frac{fD}{2B'} = -\frac{u_h}{2} \qquad 16.5$$

Thus, the inner depth of field extends from the focus plane to one-half the hyperfocal distance. *In sum, by focusing at the hyperfocal distance everything from infinity to one-half the hyperfocal distance will be in focus.*

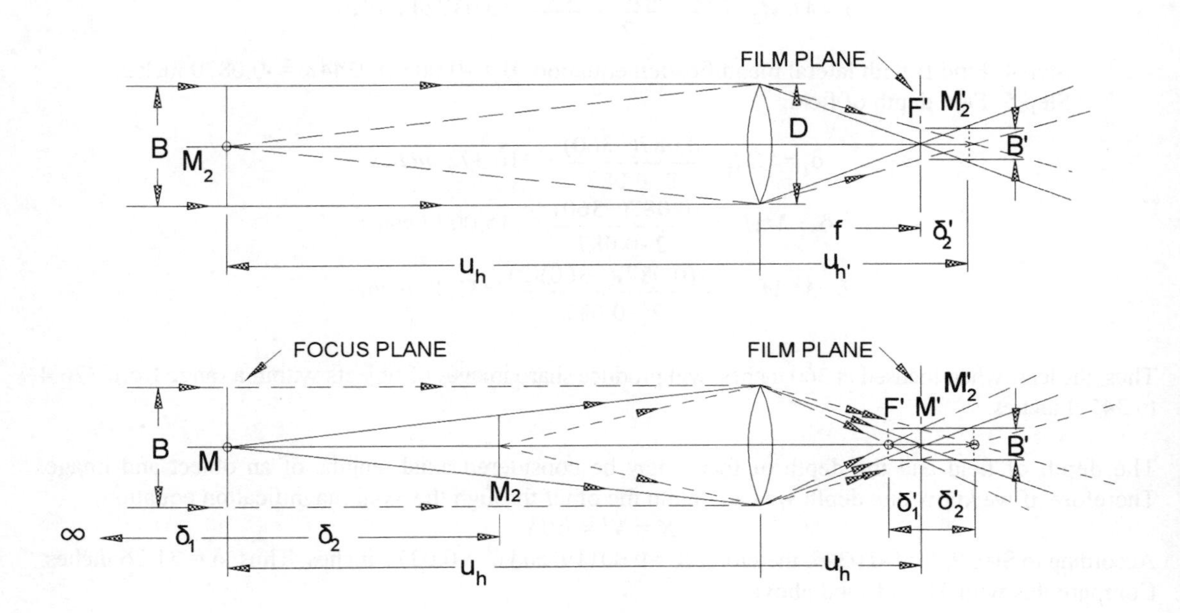

Figure 16.8 a) Lens focused at infinity; b) lens focused at hyperfocal distance.

EXAMPLE 2:

1. Find the hyperfocal distance of the f/6 lens. According to Eq. 16.4
   u_h =-12(2)/0.003=-8000 inches or -666.7 ft
2. Find the maximum depth of field. According to Eq. 16.5 the maximum depth of field is from infinity to $u_h/2$ = -333.33 feet.

16.4 THE PARAXIAL DESIGN OF A ZOOM LENS

A zoom lens has a variable focal length. For practical photographic applications the position of the second principal plane of the lens from the film plane also must vary to keep the image on the film. The simplest zoom lens is composed of a positive and negative singlet lens. The variation in focal length is obtained by adjusting the separation of the lenses, and a sharp focus is maintained by varying the distances of the lenses from the film plane. See Figure 16.9.

The focal length of the zoom lens depends on the Gullstrand equation because the equivalent power of two airspaced lenses varies with their separation d: $F = F_1 + F_2 - dF_1F_2$ 16.6

The distances of the lenses from the film plane depend on the variation in back focal length A_2F' with separation **d**, as given by

$$A_2F' = f_v' = \frac{1-dF_1}{F} \qquad 16.7$$

Eq. 16.7 shows how far the negative lens must be from the film plane. The positive lens must be an additional distance **d** from the film plane: $A_1F' = d + f_v'$ 16.8

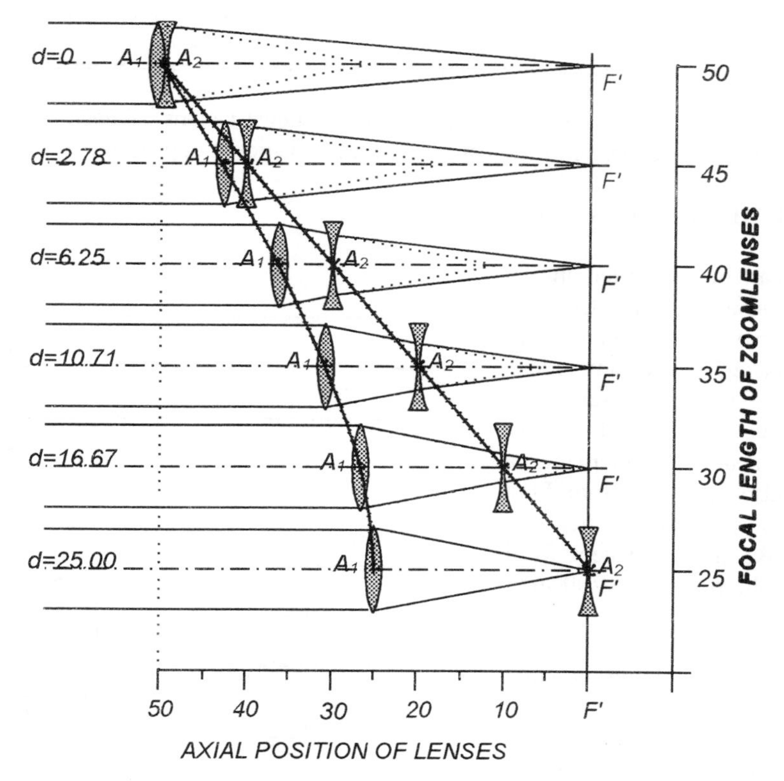

Figure 16.9 Two element zoom lens.(After Kinglake).

Suppose we want to design a two-power zoom lens that varies in focal length from 25 to 50 mm. Let the first lens will be +40 D (f_1 = 25 mm) and the second lens -20 D (f_2 = -50mm). Eq. 16.6 shows that the equivalent power will be least when d = zero or the lenses are in contact, i.e., F = +20 D or f = 50 mm. The maximum power is obtained when d = f_1 = 25 mm, accordingly, $F = F_1$ = +40 D or f = 25 mm. This corresponds to moving the negative lens to the second focal plane of the positive lens, and into coincidence with the film plane. The following table summarizes the results of Eqs. 16.6-16.8 as the equivalent focal length is varied from 50 to 25 mm in 5 mm steps.

Focal length (mm)	Separation (mm)	F=Power (diopters)	A_1F'= Overall length (mm)	A_2F'= BFL (mm)
50.00	0.00	20.0	50.00	50.00
45.00	2.78	22.2	42.78	40.00
40.00	6.25	25.0	36.25	30.00
35.00	10.71	28.6	30.71	20.00
30.00	16.67	33.3	26.67	10.00
25.00	25.00	40.0	25.00	0.00

These data are plotted in Figure 16.9. The movement of the negative lens is linear with respect to focal length. However, the positive lens requires a special cam to position it appropriately.

16.5 PARAXIAL DESIGN OF A TELEPHOTO LENS

A camera lens can be converted into a telephoto lens by inserting a negative lens behind it. Suppose we impose the condition that the back focal length of the camera lens and of the telephoto (A_2F') be the same. Let the camera lens have a focal length f_1 = 50 mm and, assuming thin lens elements, a back focal length also equal to 50 mm. Find the focal length and position of a negative lens that will make the equivalent focal length of the telephoto lens f = 200 mm. See Figure 16.10.

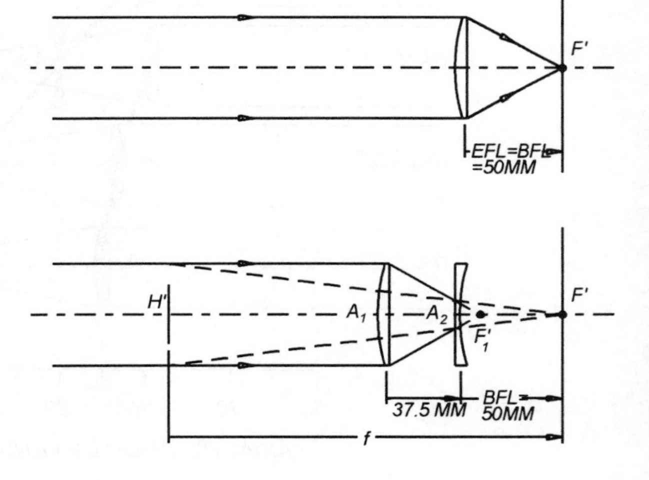

Figure 16.10 Paraxial design of a telephoto lens.

Given the specifications of the standard lens, namely, f_1 = 50 mm, or F_1 = 20 D, and A_2F'=50mm=f_1= $1/F_1$, and the requirements for the telephoto lens, f = 200 mm, or F = 5 D, find d and F_2.

The BFL of the telephoto equals $1/F_1$.

Thus,

$$A_2F' = \frac{1-dF_1}{F} = \frac{1}{F_1} \qquad 16.9$$

Solve Eq. 16.9 for d

$$d = \frac{1}{F_1} - \frac{F}{F_1^2} \qquad 16.10$$

Solve the Gullstrand equation for F_2

$$F_2 = \frac{F-F_1}{1-dF_1} \qquad 16.11$$

We will find d = 37.5 mm, F_2 = -60 D and the focal length of the negative lens, f_2 = -16.67 mm. The *telephoto ratio* **k** is the distance from the first lens to the film plane divided by the focal length of the telephoto system

$$k = \frac{A_1F'}{f} \qquad 16.12$$

The telephoto ratio **k** = 87.5/200 = 0.438

16.6 OPTICAL PROJECTION SYSTEMS

Optical projection is used to throw enlarged images of transparencies and opaque objects onto a screen. Optical comparators employ optical projection to obtain enlarged images to facilitate the measurement of the dimensions of machined parts. Besides forming sharp images, projectors must efficiently collect light so that the screen image is bright.

The crudest projector might consist of a frosted lamp behind a transparent slide, and a positive lens to form an image of the slide on a screen. Because light from each point on the frosted lamp radiates into a hemisphere but only small solid angles of light within these hemispheres are collected by the lens, this system is very inefficient, consequently, the screen image is dim.

The ordinary slide projector uses a *Kohler* illumination system to improve the screen brightness. A condenser lens system is inserted between a tungsten source and the slide. The condenser, with a very large numerical aperture, forms an image of the filament in the aperture of the projection lens. Efficiency is enhanced with a concave spherical mirror. It reflects light that radiates to the rear, back toward the condenser and projection lens. The transparency is close to the condenser lenses and uniformly illuminated. See Figure 16.11.

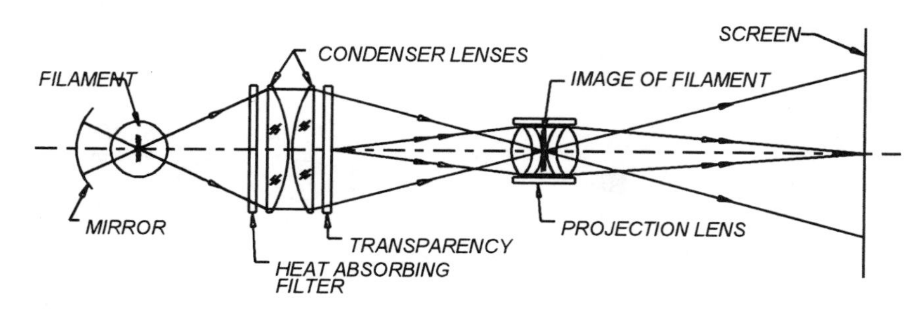

Figure 16.11 Slide Projector - Kohler Illumination.

Motion picture projectors use *Abbe* illumination. Usually an arc lamp is the source. It is imaged by the condensers onto a frame of film. The projection lens throws this brightly illuminated frame onto the screen. If the film transport mechanism jams, however, the film will melt or burn. The source in *Abbe* illumination systems must have uniform luminance if the screen brightness is to appear uniform. See Figure 16.12.

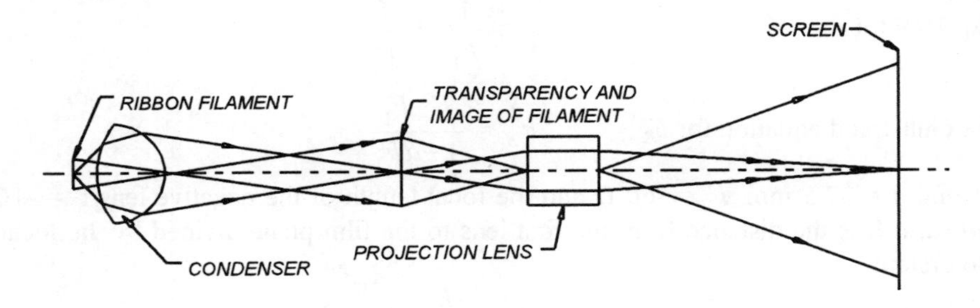

Figure 16.12 Abbe Illumination.

The *overhead projector* is similar to the slide projector. It uses Kohler illumination. The condenser is a thin Fresnel lens. A 45° mirror is incorporated into the projection lens to deviate the projection to the screen. See Figure 16.13.

The *optical comparator*, used by toolmakers, projects a magnified shadow of a mechanical part, such as a gear, onto a translucent (rear) projection screen. The enlarged contour or profile can be compared with a master drawing that has tolerance limits. Light from the source is collimated by the condenser lens. Chief rays passing around the work under examination are refracted by the projection lens through a telecentric stop at its second focal point. The size of the shadow on the screen is constant despite small changes in the position of the work.

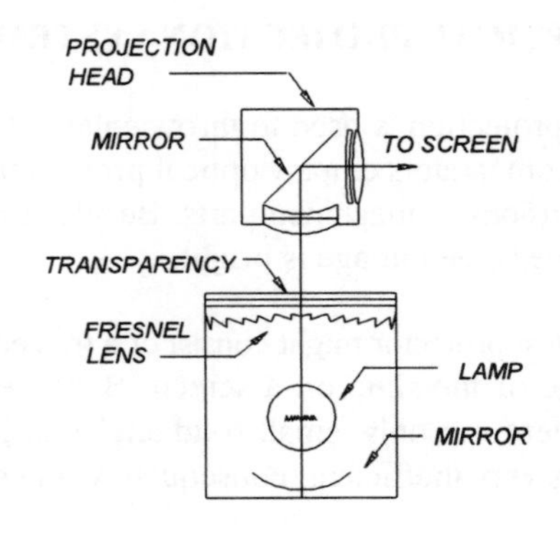

Figure 16.13 The overhead projector.

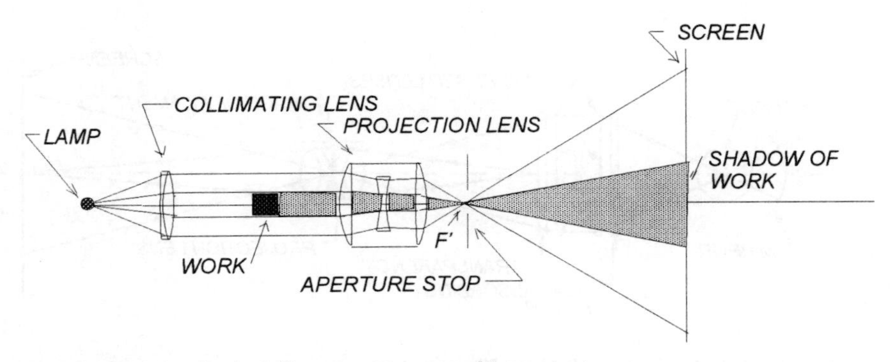

Figure 16.14 An Optical Comparator or profile projector.

PROBLEMS

1. A camera with a 20-inch focal length f/10 lens is focused on a target at 1000 inches. Find: a) outer and inner depth of field; b) distances to the furthest and nearest points that are in focus.
Assume an acceptable blur on the film of one minute of arc at ten inches.

Ans.: a) -79.3" and 68.5"; b) 1079.3" and 931.5"

2. The camera in Problem 1 was initially focused for infinity. a) how far must the film plane be moved to make it conjugate to the 1000" distance; b) with what accuracy must it be placed in the conjugate plane to receive sharp images of the target?

Ans.: a) 0.408" away from the lens; b) ±0.031"

3. Find a) the hyperfocal distance of the 20-inch f/10 lens; b) how close can an object be to be sharply focused.

Ans.: a) 1111.1 ft; b) 555.56 ft.

4. A camera lens with a 2-inch focal length f/4 lens is focused on a target 10 feet away. Find a) the distances to the furthest and nearest points that are in focus; b) the hyperfocal distance.

Ans.: a) 15.48 and 7.39 ft; b) 27.78 ft.

5. Repeat Problem 4 with an f/2 lens.

Ans.: 12.15 and 8.50 ft.

6. The negative in Problem 5 is too small to view at ten inches. You decide that you will make prints with a 5-times enlargement. The allowable blur will be one minute of arc at ten inches. Find the distances to the furthest and nearest points that are in focus on the film.

Ans.: 10.37 and 9.66 ft.

7. Design a zoom lens with a range of focal lengths from 25 to 50 mm. Let d=0 for f= 50 mm. Choose F_1 = 80 D. Find: a) F_2; b) Tabulate d, A_1F' and A_2F' for f= 50, 35 and 25 mm. Note: The lenses have a maximum focal length when in contact.

Ans.: a) -60 D

b)

f mm	d mm	A_1F'	A_2F'
50	0	50	50
35	1.786	31.786	30
25	4.167	20.833	16.667

8. Design a zoom lens with a range of focal lengths from 25 to 100 mm. Let d=0 for f= 100 mm. Choose F_1 = 80 D. Find: a) F_2; b) Tabulate d, A_1F' and A_2F' for f= 100, 55 and 25 mm.

Ans.: a) -70 D

b)

f mm	d mm	A_1F'	A_2F'
100	0	100	100
55	1.461	50.032	48.571
25	5.357	19.643	14.286

9. A 25-mm camera lens is to be converted into a 200-mm telephoto by introducing a negative lens. The BFLs of the 25 mm and 200 mm lenses are to be 25 mm. Find: a) f_2; b) A_1F' and c) the telephoto ratio.

Ans.: a) -3.57 mm; b) 46.875 mm; c) 0.234.

10. A 10-mm camera lens is to be converted into a 200-mm telephoto by introducing a negative lens. The BFLs of the 10 mm and 200 mm lenses are to be 10 mm. Find: a) f_2; b) A_1F' and c) the telephoto ratio.

Ans.: a) -0.523 mm; b) 19.5 mm; c) 0.0975.

11. The condenser lens of a Kohler type slide projector forms a 10 mm high image of the lamp filament in the aperture of the projection lens. The filament is also 10 mm high. The separation between the lamp filament and the aperture of the projection lens is 40 mm. Assuming thin lenses, find the focal length of the condenser lens. Ans.: 10 mm.

12. The condenser lens of a slide projector has a focal length of 25 mm. It produces a twice magnified image of the lamp filament. How far should the projection lens be from the condenser?

Ans.: 75 mm.

13. The format of a slide transparency is 24x36 mm. The projected screen image is 1,200x1,800 mm. The transparency is 200 mm from the projection lens. Find the focal length of the projection lens.

Ans.: 196.0784 mm.

14. A Kohler type slide projector contains a 100-mm focal length projection lens. The slide (transparency) is 24x36 mm. The projection screen is 3x4.5 feet, and the projected image fills the screen. How far is the screen from the projection lens? Ans.: 3.71 meters.

CHAPTER 17

OPHTHALMIC INSTRUMENTS

17.1 INTRODUCTION

Illumination systems, microscopes and telescopes are the building blocks of optical and ophthalmic instruments and systems. Two major ophthalmic instruments are the ophthalmometer and the lensometer. The former contains a microscope and the latter a telescope. The lensometer also contains a target and lens arrangement known as an optometer. A special illumination system and a microscope make up the slit lamp or biomicroscope. We will study how these instruments work.

17.2 OPHTHALMOMETER

The ophthalmometer measures the radius of curvature of a 2 to 3-mm zone of the front surface of the cornea. It is a fundamental tool in fitting contact lenses. With a toroidal cornea, the ophthalmometer detects and measures the amount and axis location of corneal astigmatism.

17.2.1 OPTICAL PRINCIPLE

The ophthalmometer is a low power microscope adapted to measure the first Purkinje image of a target or *mire* reflected by the corneal surface. By treating the front surface of the cornea as a spherical mirror, we may obtain its radius of curvature. An illuminated target or *mire* of known size and position is presented to the cornea, and the size and position of its reflected image is determined. The size of the image depends on the size of the mire and the radius of the cornea. In one type of ophthalmometer, for instance the B&L Keratometer or AO CLC ophthalmometer, the mire size is fixed and the image size varies with the radius of the cornea. In a second type of ophthalmometer, e.g., the Javal-Schiotz, the mire size is adjusted until a certain fixed image size is obtained. The mire image in both types becomes an object for the microscopic system of the ophthalmometer. An objective lens relays the mire image to the ocular lens.

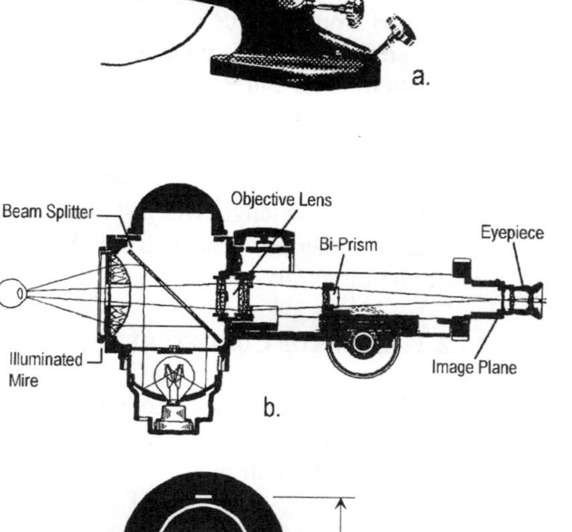

Figure 17.1 (a) a Bausch & Lamb fixed mire keratometer, (b) optical schematic of keratometer, c) illuminated mire.

The eyes normally jitter. These nystagmoid motions

are magnified by the microscope, making it nearly impossible to measure the size of the jittering image against a fixed reticle in the ocular. An elegant solution to this problem is to split the image into two and calibrate the separation due to doubling with the radius of curvature. The doubling device is usually an arrangement of thin prisms that are between the objective and the relayed image. The device produces a fixed amount of doubling in the adjustable mire type ophthalmometer. It is moved along the axis to vary the extent of the doubling in the fixed mire keratometer. The doubled image is observed through the ocular.

17.2.1.1 Fixed mire-variable image size ophthalmometers

An ophthalmometer of the fixed mire type, made by Bausch & Lomb, Inc. is shown in Figure 17.1. The patient is positioned in a chin rest at one end of the instrument. This end presents an illuminated ring-shaped mire, as shown in the figure, of known size **y** at a distance **u** from the patient's cornea. A doubled Purkinje image is seen through the eyepiece of the microscope. The doubling device is adjusted until the doubled images just touch. The centers of the two images are then one-diameter apart and, based on the calibration of the doubling device, provide the size of the Purkinje image.

17.2.1.2 Variable mire-fixed image ophthalmometers

The Haag-Streit ophthalmometer is based on the variable mire Javal-Schiotz design. Two mires are moved along an arc. See Figure 17.2. One mire is an illuminated rectangle. The other is a step pattern. The mires can be moved closer together or further apart while the examiner views their doubled images. When the two mires are separated appropriately, the step pattern and the rectangle will just abut, and the reading is obtained directly from the arc on which the mires sit.

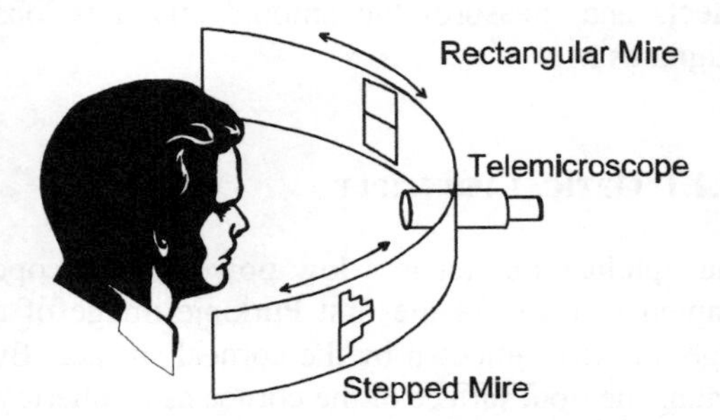

Figure 17.2 Javal - Schiotz adjustable mire ophthalmometer.

17.2.2 RADIUS OF CURVATURE

By applying the image and lateral magnification equations for spherical mirrors, the radius of the cornea is obtained. The lateral magnification of the cornea is equal to the ratio of the size of the image of the mire **y'** to the size of the mire **y**.

$$Y = \frac{y'}{y} = \frac{nu'}{n'u} \quad \text{or} \quad u' = -\frac{y'u}{y} \qquad 17.1$$

The diameter of the mire **y** and its distance **u** from the cornea are fixed by the design of the instrument. By measuring the diameter of the Purkinje image **y'** we will find the image distance **u'**. The distances **u** and **u'** are related to the radius of the corneal surface by the equation:

$$\frac{n'}{u'} = \frac{n}{u} + \frac{n'-n}{r}$$

Substituting Eq. 17.1 into this equation results in $r = -\dfrac{2uy'}{y-y'}$ 17.2

This is an exact equation for determining the radius of the cornea. However, designing instruments with linear scales and dials marked with uniform increments is desirable. Eq. 17.2 is not linear with image size **y'** because it contains **y'** in the denominator.

If the mire is at infinity the first Purkinje image will be at F', and $u' = -f = \frac{r}{2}$ 17.3

Substitution of Eq. 17.3 into Eq. 17.1 results in $r = -\frac{2uy'}{y}$ 17.4

y' has been eliminated from the denominator, consequently, **k = -2u/y** is an instrument constant, or

$$r = ky' \tag{17.5}$$

The radius now is linear with the size of the image that we measure.

The mires on ophthalmometers are not at infinity. Their location is based on making the distance **u** of the mires sufficiently large compared with the radius of the cornea so that the image is formed nearly at **F'**. The accuracy of the measurement depends on the size of the error in treating **u'** equal to **f.**

For example, if we set a 2% tolerance on the accuracy of the measurement of r then, according to Eq. 17.3, **u'** must equal r/2 (or the focal length) to within ±2%. Given a real object, a convex mirror will form a virtual image between A and F', therefore, u' must be less than r/2. That is

$$u' = \frac{0.98r}{2}$$

If the actual radius of the cornea is 8 mm, a maximum measurement error of 2% will occur if

$$u' = \frac{0.98 \times 8}{2} = 3.92\ mm$$

The conjugate mire distance is given by $-\frac{1}{3.92} = \frac{1}{u} - \frac{2}{8}$

and so, u = -196 mm = -7.7 inches. Thus, if the mire is more than 7.7 inches from the cornea the error in measuring its radius will be less than 2%.

17.2.3 Refractive Power of the Cornea

Although the ophthalmometer determines the radius of curvature of the front surface of the cornea, the refractive power is indicated by a parallel set of engraved marks on the measuring drum. Based on the Gullstrand data, the corneal index = 1.376. Given r_1 = 7.7 mm, the refractive power of the front surface of the cornea is

$$F_1 = \frac{1.376 - 1}{0.0077} = 48.83D$$

The astigmatism exhibited at the front surface of a toric cornea, as determined with index 1.376, will be much greater than the astigmatism of the entire cornea. This latter quantity has diagnostic value in assessing the refractive error of the eye. To determine the astigmatism of the cornea, the anterior and posterior surfaces are assumed to maintain a fixed relation to each other. Thus, if the anterior surface is toric, so is the posterior surface. As an example let us assume that the Gullstrand data correspond to the radii in the **y** meridian only. Then $\mathbf{r_{y1}}$ **= 7.7 mm**, and $\mathbf{r_{y2}}$ **= 6.8 mm.** Now if the radius of the anterior surface in the **z** meridian is increased by **0.1 mm**, i.e., $\mathbf{r_{z1}}$ **= 7.8 mm**, then the radius of the posterior surface in this meridian is also increased by **0.1 mm**, or $\mathbf{r_{z2} = 6.9}$ **mm**. Using, **d = .5 mm** and $\mathbf{n_1 = 1}$, **n2 = 1.376**, and $\mathbf{n_3 = 1.336}$, we obtain the following powers:

	Radius (mm)		Power (Diop)		Astigmatism (Diop)
	$\mathbf{r_y}$	$\mathbf{r_z}$	F_y	F_z	$F_y - F_z$
Anterior surface	7.7	7.8	48.83	48.21	0.62
Posterior surface	6.8	6.9	-5.88	-5.80	-0.08
Corneal power			43.05	42.51	0.54

The astigmatic error of the cornea is Fy - Fz = 0.54 D. This is less than the 0.62 D astigmatic error of the front surface alone, based on index 1.376. A fictitious index is assigned to the cornea to have the ophthalmometer indicate the lower value. The appropriate index based on the above readings is

$$0.54 = \frac{n-1}{0.0077} - \frac{n-1}{0.0078}$$

The solution of this equation is a corneal index of n = 1.324. Corneal power readings for many ophthalmometers are based on an index of 1.3375 because a standard 15 mm diameterball bearing should produce a reading of 45 D. Therefore, the accuracy of the instrument can be conveniently checked.

17.3 DOUBLING PRINCIPLE

The accuracy of the radius determination also is dependent on the accuracy with which the image diameter **y'** of the mire image is measured. Given typical values of mire diameter **y** = 100 mm, distance **u** = -200 mm and corneal radius **r** = 7.2 mm, the size **y'** of the mire image is, according to Eq. 17.4,

$$y' = -\frac{ry}{2u} = -\frac{7.2 \times 100}{-400} = 1.80mm$$

The image is rather small. An error of only 0.036 mm will result in a 2% error in measuring its diameter. A microscope is required to measure the 1.8 mm ring with sufficient precision. Ordinarily the objective lens would re-image the mire on a reticle (scale) in the focal plane of the eyepiece. The magnified image could then be measured against the reticle. See Figure 17.3. However, the eye continually exhibits nystagmoid movements. The Purkinje image bounces and jitters about. The microscope magnifies these motions, with respect to the measuring reticle and frustrates the measurement. To obviate the problem caused by nystagmus, a doubling device, such as a biprism, is introduced between the objective and the focal plane of the eyepiece.

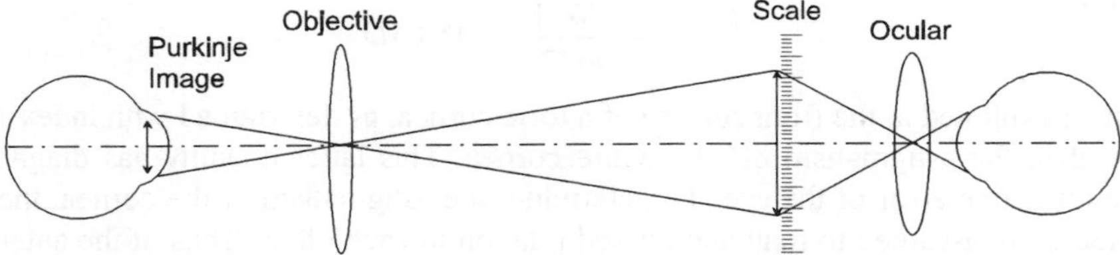

Figure 17.3 Telemicroscope relays Purkinje image to a scale in the focal plane of the ocular.

17.3.1 FIXED MIRE SIZE OPHTHALMOMETER

Positioning the Biprism

When the biprism is in the focal plane of the eyepiece, a single mire image is seen, as shown in Figure 17.4a. As the biprism is moved toward the objective lens, the mire image begins to double. See Figure 17.4b. The biprism is moved until the doubled images are tangent, as in Figure 17.4c. When they are tangent, the separation between their centers will equal the diameter of the mire image. This process is unaffected by nystagmus. The split images simply move about together, their separation does not change unless the prism is moved. The distance of the biprism from the focal plane of the eyepiece is proportional to the separation of the split images, consequently, the movement of the biprism is proportional to the radius of the cornea.

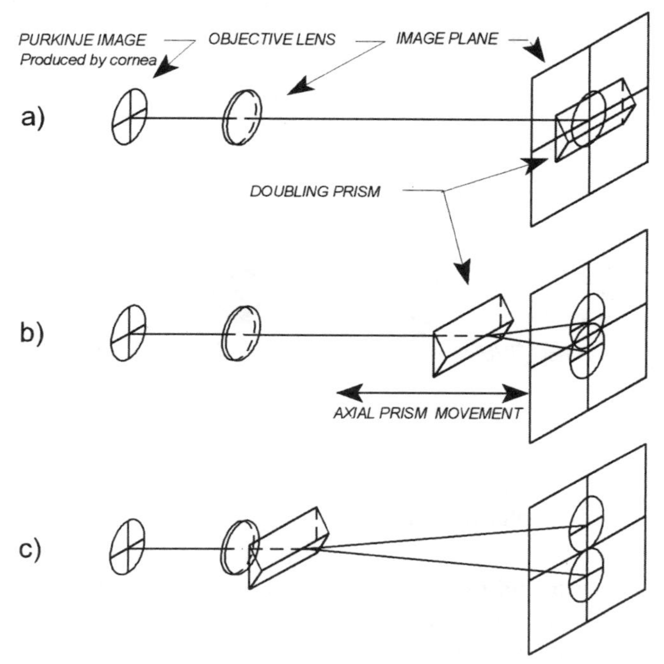

Figure 17.4 The doubling action of a biprism in a fixed mire ophthalmometer. (a) Biprism in the image plane produces nodoubling. (b) Biprism moved axially to produce some doubling. (c) Biprism position that produces tangency of images.

Example 1: Image doubling by positioning a biprism.

Assume that the diameter **D'** of the first Purkinje image is 1.8 mm, as calculated above, and the objective lens has a lateral magnification **Y = - 4**. Find the distance of a biprism of 10^{Δ} (each half) that produces tangency of the doubled images in the focal plane of the eyepiece.

The diameter of the mire image in the focal plane of the eyepiece is equal to Y(D') = - 7.2 mm. Therefore, the total biprism deviation of 20^{Δ} must separate the split images by 7.2 mm. Convert prism diopters to tan θ, and solve for distance **d**.

$$P = 100 \tan\theta, \therefore \tan\theta = \frac{20}{100} = 0.2 = \frac{7.2}{d}, \therefore d = 36\ mm$$

Each mm of translation of the biprism produces 0.2 mm of doubling.

Measurement of Toric Corneal Surfaces

Astigmatism due to toricity of the cornea is common. The variable image type of ophthalmometer is designed to measure the corneal radius in the two principal meridians by incorporating an aperture disk between the objective lens and the focal plane of the eyepiece. The disk contains four small openings arranged at 90° increments. See Figure 17.5. The mire and its Purkinje image are annuluses surrounded by four small crosses at 90° increments. The light from the Purkinje image is collected by the objective lens. Only the portion of the light transmitted through the four windows is converged to the plane that is conjugate to the Purkinje image. The rest of the light is vignetted. As explained below, the principal radii will be found when three appropriately aligned mire images are seen through the eyepiece.

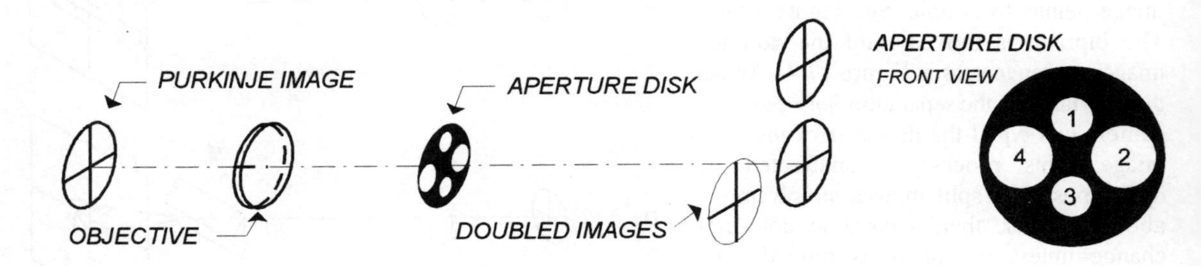

Figure 17.5 Aperture disk (also shown at the right) replaces the biprism. It doubles in two orthogonal meridians.

Focusing adjustment: Apertures 1 and 3 are clear openings. They produce the image used for focusing the microscope. The convergent light through these apertures will form a *single* image in the *conjugate* image plane. If this plane is *not coincident with the focal plane of the eyepiece*, a double image will be seen. This shows an out-of-focus condition (Figure 17.6a). The microscope is moved toward or away from the cornea until a single, in-focus condition is obtained. Figure 17.6b.

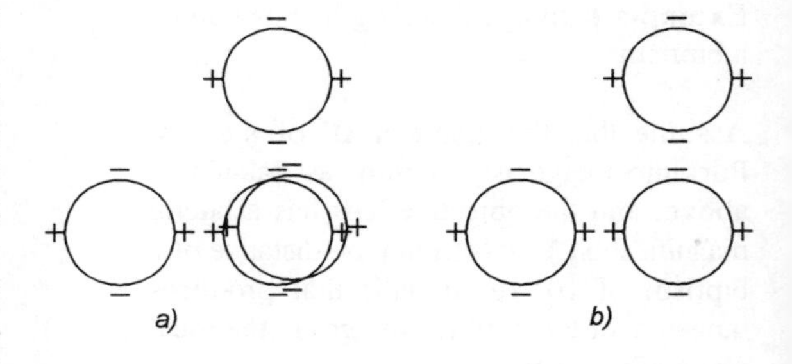

Figure 17.6 a) The relayed Purkinje images are out of focus, (b) the images are in focus.

Axis alignment: Light transmitted by apertures 2 and 4 go through a base-up and a base-out prism, respectively. They produce the upper and left images. An off-axis condition exists if the crosses are misaligned. This indicates misalignment between the principal meridians and the prism apertures. See Figure 17.7a. The aperture disk is rotated to obtain the on-axis condition, i.e., to produce doubling along the two principal meridians. See Figure 17.7b.

Radii or power measurement: This is done separately for the two principal meridians by moving the prisms toward or away from the objective lens until the crosses of the mire images along each meridian are superimposed. See Figure 17.8. The radii or refractive power is shown on the corresponding measuring drums. The axis is displayed on separate scale.

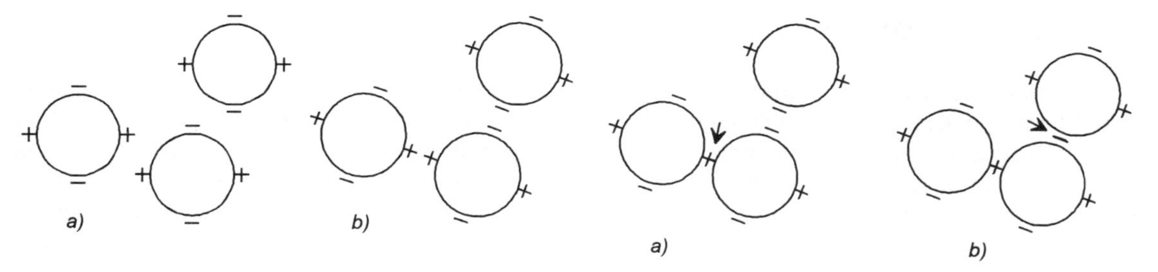

Figure 17.7 a. An astigmatic cornea is indicated when the crosses are misaligned; b. Principal meridians.

Figure 17.8 The power is measured when the crosses are superimposed in each principal meridian.

17.3.2 Adjustable Mire Size Ophthalmometer

Positioning the mires: In the Javal-Schiotz type ophthalmometer, the doubling device is not adjusted. It produces a constant separation in the images of the mires. The separation between the mires is varied until the rectangular and stepped mire images appropriately adjoin each other.

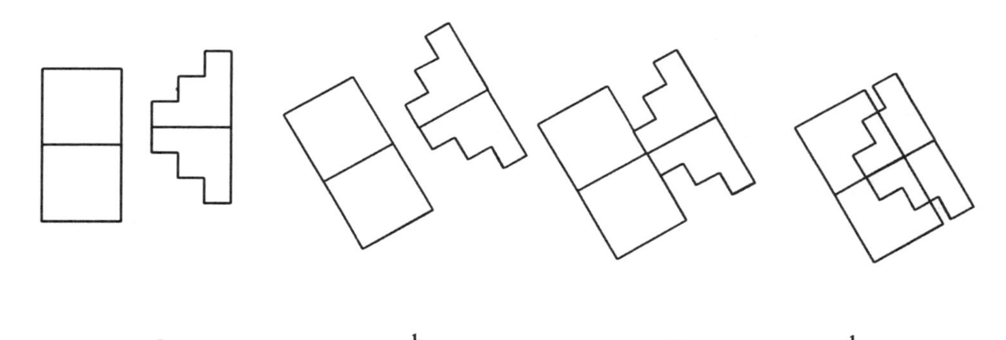

Figure 17.9 Javal-Schiotz or variable mire ophthalmometer. a. Principal meridians misaligned; b. Principal meridians aligned by rotating the arc; c. Tangency setting indicates power along this meridian; d. Each step of overlap corresponds to a diopter of power.

If the central lines in the two parts of the mire are not aligned, the surface is toric. See Figure 17.9a. The first adjustment is to find a principal meridian by rotating the arc until the broken lines are aligned, as in Figure 17.9b. At this point the mires may overlap or, as shown, not touch. The radius of the surface in this meridian is obtained by adjusting the separation of the mires until the images appear as in Figure 17.9c. The arc is rotated 90° and the mires are readjusted to measure the power in this meridian. Alternatively the previous mire setting may be retained and the astigmatic error indicated by the number of steps of overlap, as in Figure 17.9d. Each step corresponds to one diopter of astigmatism.

17.4 BADAL OPTOMETER

Optometers measure refractive errors and accommodation. The Badal optometer uses a doublet lens to do this. One way to measure the amplitude of accommodation is to take a 20/20 letter, which subtends 5' of arc at a designated distance, for example 20 ft, and bring it closer to the eye until it blurs. The demand on accommodation increases, but so does the subtense of the letter. At a distance of 2 ft it subtends 50' of arc, and is no longer useful for measuring high acuities or finding the near point of myopes. Moving an eye chart through a distance of 20 ft is not merely cumbersome physically, but positioning it at the far point of hyperopes is physically impossible. These disadvantages are avoided by the Badal optometer. The optometer measures the far and near points of the eye. It provides a target of constant angular subtense that may be positioned practically anywhere between $\pm\infty$. Yet the optometer is small enough to sit on a table.

This remarkable feat is accomplished with a positive lens and a target moved axially on an optical bench. Figure 17.10 illustrates the optometer. A +10 D Badal lens is mounted about 93 mm in front of the cornea so that its second focal point **F'** coincides with the nodal points of the eye (about 7 mm behind the cornea). The initial or zero position of the acuity chart target is at the first focal point $\mathbf{F_B}$ of the Badal lens. **MQ** = **y** is the target height. A reference ray parallel to the axis, through **Q** is refracted toward $\mathbf{F'_B}$. Its path through the eye is undeviated, and it strikes the retina at **Q'**. The Badal lens has collimated the light from the target, therefore, the eye views an image at infinity. The retinal image height **y'** is constant no matter the distance of the target from the Badal lens, but the image may be blurred. Only if the eye is emmetropic or accommodated for infinity will the image **y'** be sharp.

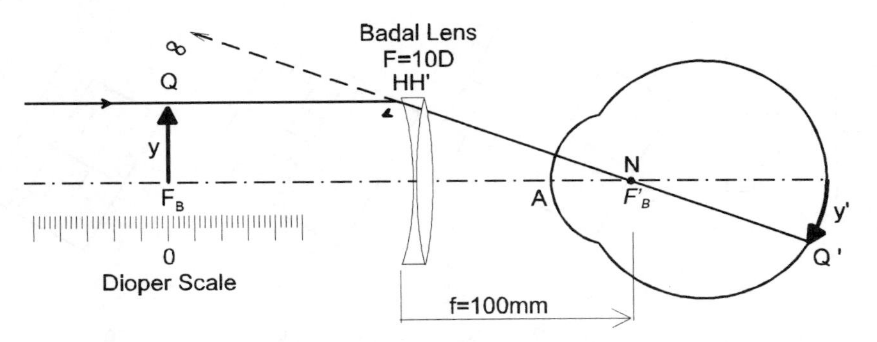

Figure 17.10 Badal optometer. Target at F (zero setting). The second focal point of the Badal lens coincides with the nodal points of the eye.

Virtual images will be produced as the target moves from the zero position toward the Badal lens. These images, which serve as objects for the eye, move from infinity toward the eye. That is, they cover a range of target positions corresponding to the loci of far points of myopes. When the target contacts the lens, the image coincides with it. This closest image distance represents the maximum degree of myopia that can be measured or the maximum stimulus to accommodation. It is nearly 11 D for a +10 D Badal lens.

As the target is moved away from the zero position and further from the lens, real images approach the eye from behind. These images are suitable for finding the far points of hyperopes. The maximum degree of hyperopia that is measurable depends on how far the target is moved. For example, if the target is positioned at -2f the image will be at +2f, or 100 mm behind the nodal points of the eye. This corresponds to about 10 D of hyperopia. Thus, an optometer length of less than 300 mm, from target to eye, will measure about ± 10 D of ametropia.

The retinal image size **y'** is constant for all settings of the target because it is determined by the same reference ray.

17.4.1 Badal Target Position and Ametropia

Figure 17.11. shows the Badal optometer with the target **M** moved toward the lens, through a distance **x** = $\mathbf{F_BM}$, so that the virtual image **M'** coincides with the far point of the eye. The distance from the nodal point of the eye to the far point is equal to the distance **x'** = $\mathbf{F'_BM'}$. The reciprocal of this distance is the *ametropia* **A**, that is, **A = 1/x'**. According to the Newtonian equation for a thin lens

$$xx' = \frac{x}{A} = -f^2$$

If distances are in meters, A and F will be in diopters, therefore, $x_{meters} = -\dfrac{A}{F^2}$

Solve for A, $$A = -xF^2$$

Convert **x** units (movement of the target) to centimeters by dividing meters by 100.

$$Ametropia:\ A = -\frac{xF^2}{100}\ diopters \qquad 17.6$$

Example 2: Given a +10 D Badal lens, what is the equation for Ametropia if x is measured in cm?

Ametropia = -x (Diopters)

In other words, the +10 D Badal lens measures one diopter of ametropia for each cm of target movement.

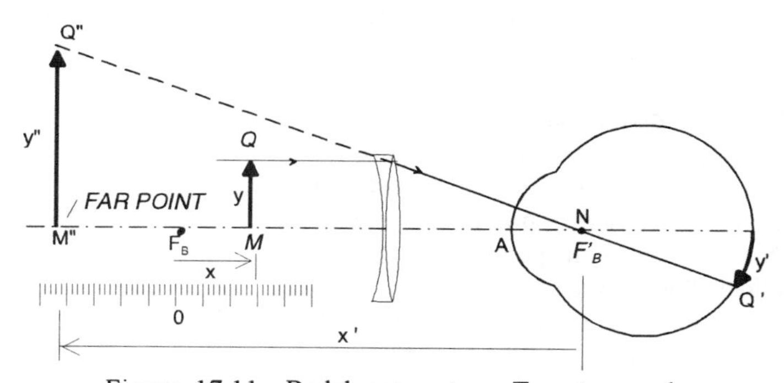

Figure 17.11 Badal optometer. Target moved toward lens. Far point at M", indicative of myopia.

Example 3: The target, initially at the zero setting of a +10 D Badal lens, appears blurred. It must be moved 2 cm toward the lens, before its image first appears sharp. This means that the image is at the far point of the eye. The ametropia = -2 D, or the eye is 2 D myopic.

The amount of ametropia per unit displacement of the target increases quadratically with Badal lens power. Therefore, the stronger the Badal lens the more sensitive it is to small positional errors. For example, if a 20-D Badal lens is used, and x is measured in cm, the ametropia = -4x. Each *cm* of target movement corresponds to 4D of ametropia. An error of one mm in the target position will produce an ametropia measurement error of 0.1 D for the +10 D Badal lens, and an error of 0.4 D for the +20 D lens.

17.4.2 ALTERNATE ARRANGEMENT OF BADAL OPTOMETER

A constant retinal image size also is obtained if the second focal point **F** of the Badal lens is made coincident with the first focal point $\mathbf{F_e}$ of the eye. The reference ray is collimated by the eye. The ray intersects the retina at **Q'** whatever the distance of the target. See Figure 17.12. This arrangement does not

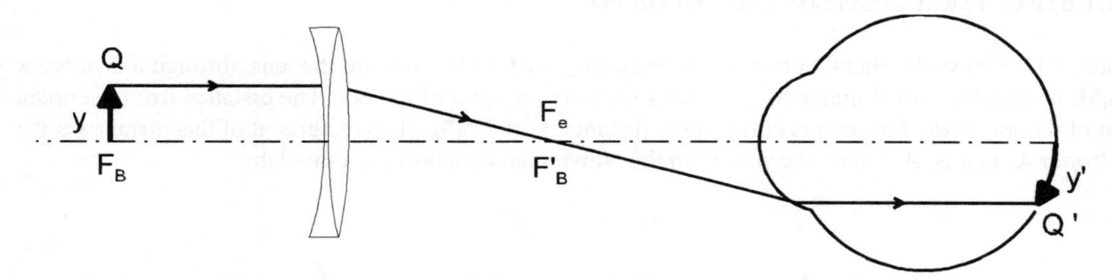

Figure 17.12 Alternative version of optometer. The second focal point of the Badal lens coincides with the first focal point of the eye.

yield the simple linear relationship between target movement and ametropia.

17.4.3 TELECENTRIC SYSTEMS

Telecentric optical systems have pupils at infinity. In the Badal system where the second focal point of the lens coincides with the nodal points and the iris diaphragm of the eye, the iris diaphragm is the aperture stop. This system is *telecentric* in object space because the entrance pupil is at infinity.

The alternative Badal system is essentially a telescope. The Badal lens and the eye correspond to the objective and ocular lenses. If a small stop is placed at **F'**, the entrance and exit pupils are formed at infinity. Such a system is telecentric in object and image space. The reference ray is, in fact, a *chief* ray. It goes through the center of the aperture stop. Chief rays delimit the 50% vignetted linear size of the object or image. Because chief rays originate at off-axis points, they ordinarily are inclined to the axis. Consequently, the separation between chief rays from opposite edges of an object changes with the axial position at which it is measured. This is illustrated in Figure 17.13.

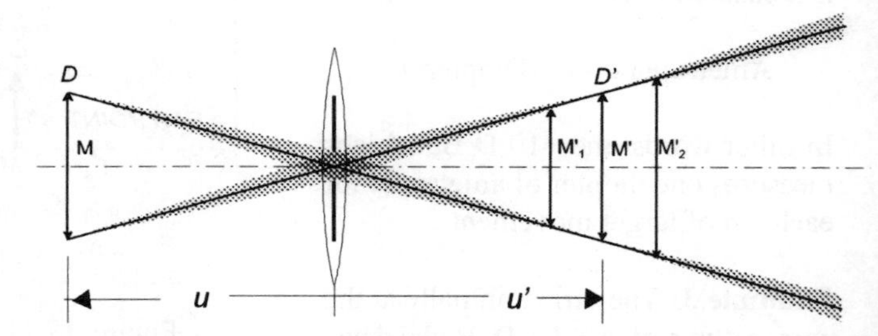

Figure 17.13 Non-telecentric system. The stop is at the lens. The image size varies with measurement plane because the chief rays diverge.

Let two points, separated by a distance D, be two focal lengths from a lens. Their separation in the image plane D' equals D. To measure D' accurately a scale must be precisely placed at M'. Otherwise, the measurement will be too small or large. In other words, the separation between the blur centers of out-of-focus images differs from the in focus image size. The special virtue of chief rays in telecentric systems is that their separation is constant because they travel parallel to the optical axis. The system in Figure.17.14

can be made telecentric simply by placing a small stop at the first focal point. Now the chief rays travel parallel to the optical axis. The separation between blur centers and the in focus images is constant. This property of telecentric systems makes them highly desirable for remote measurements. The contact lens Radius Scope is an optical comparator that uses telecentricity to measure and examine contact lenses.

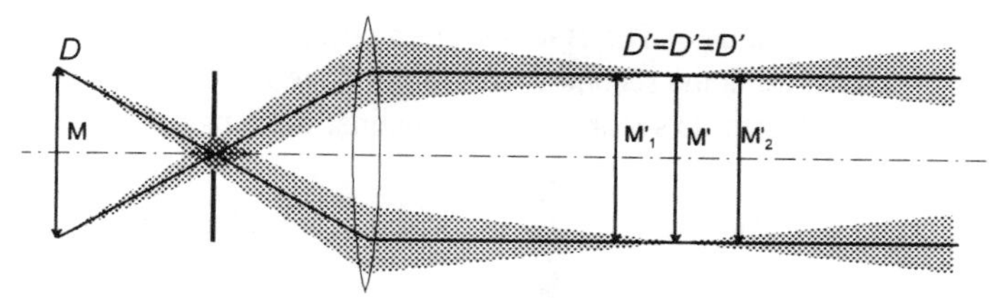

Figure 17.14 Telecentric system. The stop is at the first focal plane. Chief rays emerge parallel. The images measured from centers of blurs are all equal.

Another application of telecentric systems is to eliminate the parallax error that is common to the measurement of the interpupillary distance with a millimeter scale. In this method, the examiner places the scale about an inch or two from the patient's eyes. The examiner, sighting with his left eye, aligns the zero mark of the ruler with the subject's right eye. The examiner's right eye then reads the ruler mark aligned with the patient's left eye. See Figure 17.15. This procedure minimizes the gross parallax error that occurs should the examiner sight with only one eye, as indicated by the dashed line. Parallax error occurs because the ruler is not in contact with the patient's eyes. A telecentric arrangement to measure the interpupillary distance is shown in Figure 17.16. Although the scale is not in contact with the patient's eyes, no parallax error occurs.

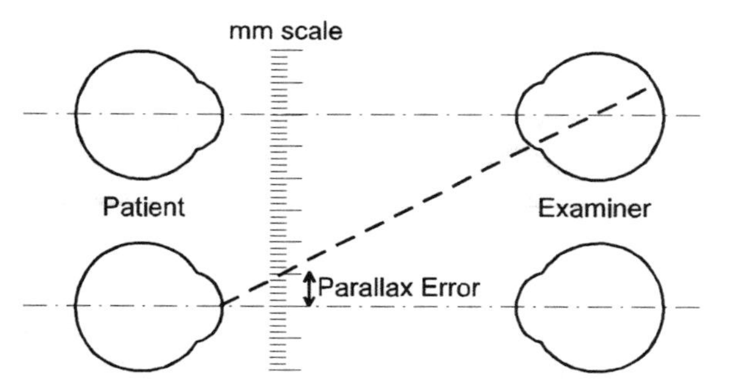

Figure 17.15 Parallax error in measuring the interpupillary distance with a ruler.

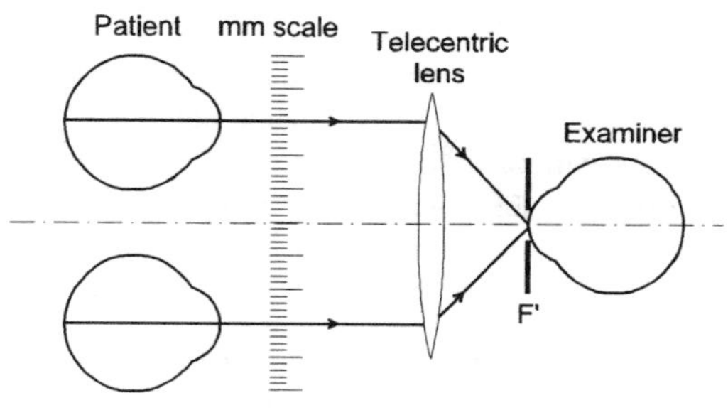

Figure 17.16 Elimination of parallax error with telecentric system.

17.5 THE LENSOMETER

The lensometer measures the vertex power of spherical lenses and the vertex powers along the principal meridians of toric lenses. The lensometer finds the cylinder axis in plus or minus cylinder form. It also finds the optical center and measures the prismatic power of these lenses. Other names for the lensometer are focimeter and vertometer. See Figure 17.17.

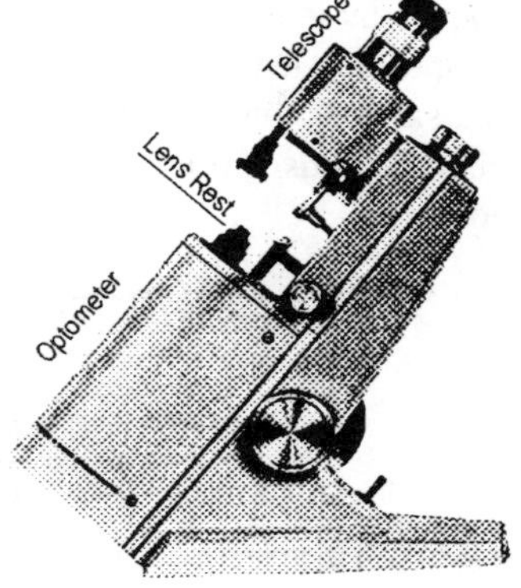

Figure 17.17 A Bausch & Lomb lensometer.

Lensometers contain a Badal system for generating an image at the second focal point of the spectacle lens under test, and a telescope to view the image. The Badal system comprises an axially adjustable target, a collimating or Badal lens and a lens rest at the second focal point of the Badal lens $\mathbf{F_B}'$, to hold the spectacle lens. The spectacle lens is placed with its second surface against the lens rest. See Figure 17.18.

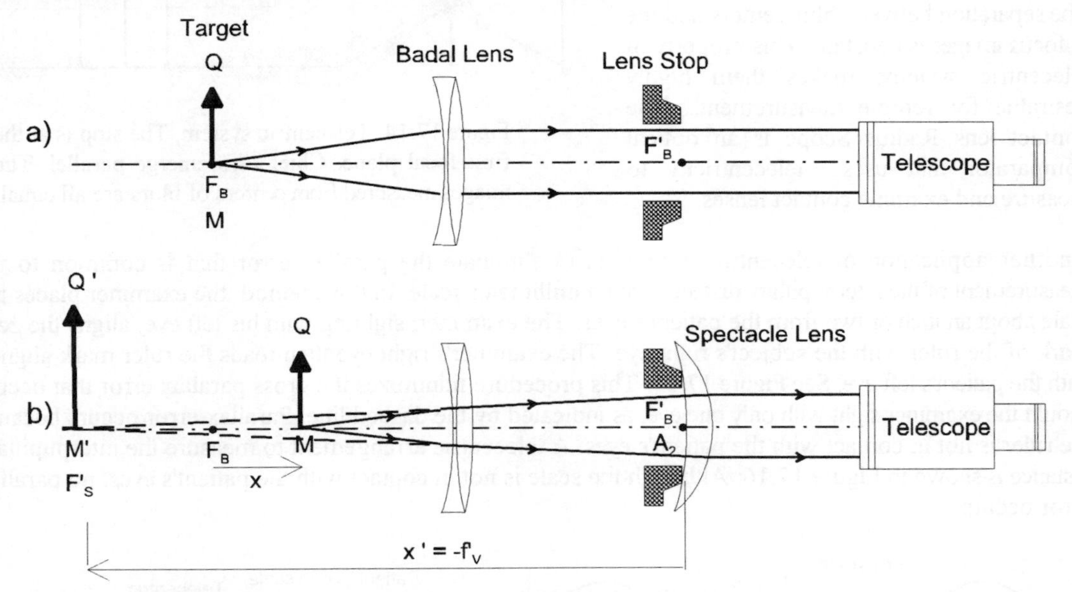

Figure 17.18 Optical schematic of optometer section and lens rest. a. Target at zero setting, no spectacle lens in the lens rest, collimated light entering the telescope. b. Spectacle lens introduced, target moved to position M so as to form an image at m', coincident with the second focal point of the reversed spectacle lens. Collimated light enters the telescope.

The Badal system: When the target is at zero position, i.e., at, the first focal point of the Badal lens $\mathbf{F_B}$, the emerging light is collimated. Absent a spectacle lens, a sharp image of the target is seen through the telescope. With the introduction of a spectacle lens, light to the telescope is decollimated and the image appears blurred. To restore a sharp image, the target **M** is moved from the zero position toward or away from the Badal lens until the target image **M'** coincides with the second focal point of the spectacle lens $\mathbf{F_s}'$. The spectacle lens collimates the light and restores the sharp image in the telescope. The distance $\mathbf{F_B M}$ = **x,** that satisfies this condition results in $\mathbf{x' = F_B'M' = f_v'}$. However, **x'** also equals $\mathbf{A_2F_s'}$ of the BFL of the spectacle lens. To compensate for the reversal of the spectacle lens it is necessary to set $\mathbf{x' = -f_v'}$. Consequently,

$$x\,x' = -x f_v' = -f_b^2$$

and

$$x = +\frac{f_b^2}{f_v'} = +\frac{F_v'}{F_b^2} \qquad 17.7$$

where F_b^2 is an instrument constant, and **x** is measured in meters. Measuring x in millimeters is preferable,

thus, $x_{mm} = +\dfrac{1000F_v'}{F_b^2}$

Since the denominator is a constant, the vertex power varies linearly with target displacement **x**. For example, if the power of the Badal lens equals 25 D, we find that

$$x = + \frac{1000 F_v^{/}}{625} = + 1.6 F_v^{/}$$

In other words, each 1.6 mm of target travel corresponds to one diopter of vertex power. To measure lens powers from +20 to -20 D the target must travel 40x1.6 = 64 mm.

17.5.1. LENSOMETER TARGET AND RETICLE

The illuminated lensometer target consists of thin focusing lines orthogonal to three broad lines. The target can be rotated to align it with any pair of normal oblique meridians. Concentric circles marked in prism diopters are engraved on a thin glass reticle. Before the lens is measured, the eyepiece is adjusted to focus the circles sharply. See Figure 17.19.

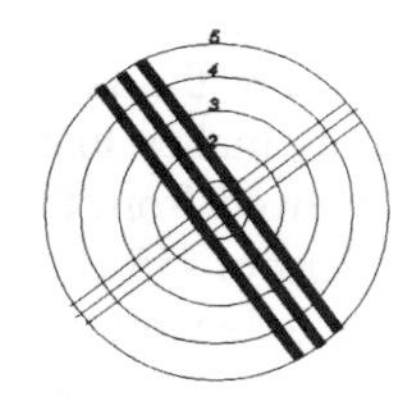

Figure17.19 Lensometer target and reticle.

The sphere and cylinder lens powers are measured consecutively. Both sets of target lines are blurred by moving the target to a position corresponding to greater plus power than the most plus meridional lens power. The target is then moved until the image of the thin lines is sharp. This is recorded as the power of the sphere of the lens. Continued movement of the target will then bring the three broad lines into focus. The dioptric difference between this reading and the sphere power is the cylinder power in minus form. Axis alignment is achieved when the lines appear straight and unbroken.

Decentration of the lens in the lens rest, or incorporation of a prism correction in the prescription will deviate the central intersection of the target lines from the center of the reticle rings. The prismatic effect corresponding to this eccentricity of the target lines is directly shown by the rings and scale in prism diopters.

17.6 SLIT LAMP or BIOMICROSCOPE

The slit lamp is a binocular microscope system for in vivo examination of the tissues and structures of the eye. It consists of a microscope and an illumination system. Each system rotates independently about the same vertical axis. A Bausch & Lomb slit lamp is shown in Figure 17.20. The Kohler illumination system comprises a lamp, condenser lenses, a slit of adjustable size and lenses to image the slit. The slit acts as a source. Its image is focused on the structures of interest and illuminates a slice of the ocular structure. In effect, it is an optical microtome. These

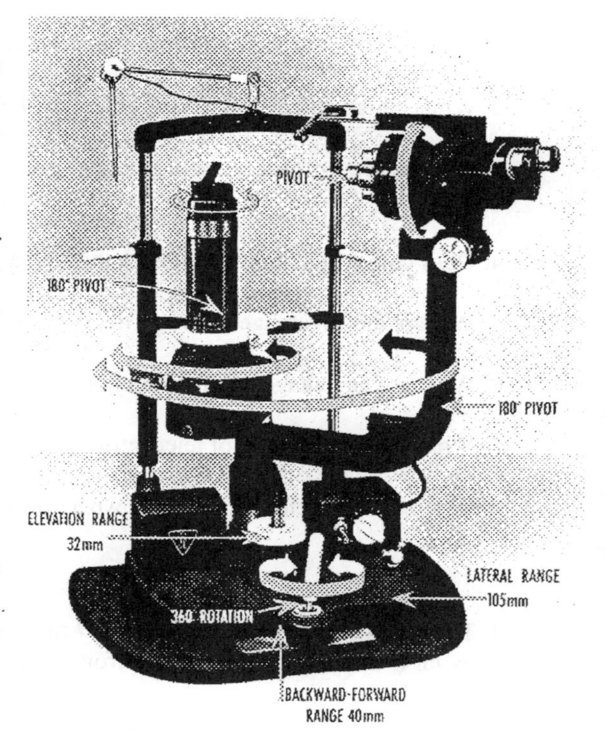

Figure 17.20 The B&L Thorpe Slit Lamp and various adjustments of system.

optical cross sections can then be examined under magnification by the microscope. The illumination and microscope systems can be pivoted ±90° in azimuth from the straight ahead direction to illuminate and view oblique sections. Once the binocular microscope is focused on the slit image, it remains in focus since it moves concomitantly with the slit image. Combinations of turret-mounted objective lenses and interchangeable eyepieces provide magnifications from about 10x to 60x.

The slit image can illuminate the layers of the cornea, the aqueous chamber, the crystalline lens, and a short distance into the vitreous chamber. Figure 17.21, from a B&L catalog, illustrates a bright slice through the cornea and lens, interrupted by the darker aqueous. Scattered light from these structures makes them visible against a dark background when viewed from the oblique angle of the microscope. A Hruby lens of -55 D is interposed to neutralize the power of the eye to enable the examination of the retina. can be examined by interposing. The slit lamp is also used to examine the eyelids and adnexa

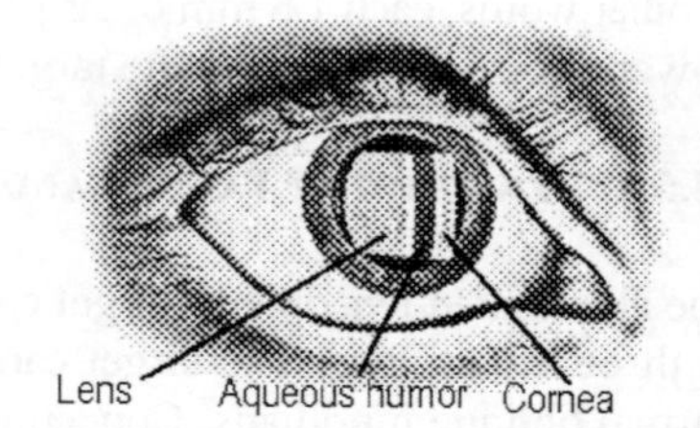

Figure 17.21 Light section through cornea and lens of eye.

REVIEW QUESTIONS

1. What image does the ophthalmometer measure?

2. Why is there an error in measurement by the ophthalmometer?

3. Why isn't the corneal refractive power, given by ophthalmometers, based on Gullstrand's index for the cornea?

4. Does convergent light that passes through a plate with several openings form more than one image?

5. What is the advantage of image doubling in an ophthalmometer?

6. What are advantages of the Badal optometer for measuring the range of accommodation?

7. What special property of telecentric systems is useful for measuring objects?

8. Why is a spectacle lens reversed when measured in a lensometer?

PROBLEMS

1. The radius of curvature of a cornea is 7 mm. A mire, 100 mm in diameter, is located 200 mm away. (a) How far is the first Purkinje image from the cornea? (b) How large is the P.I. No. 1?

Ans.: (a) 3.44 mm; (b) 1.720 mm.

2. A mire is 100 mm in diameter. Its image, formed by an 8 mm radius cornea, is 2.6 mm in diameter. (a) How far is the mire from the cornea? (b) How far is P. I. No 1 from the cornea?

Ans.: (a) 150 mm; (b) 3.9 mm

3. The mire of an ophthalmometer is 250 mm from a cornea, and 100 mm in diameter. The mire image (Purkinje No.1) is found to be 1.8 mm in diameter. Find (a) the exact and the (b) approximate radius of the corneal surface. (c) Find the % error.
Ans.: (a) 9.165 mm; (b) 9.0 mm; (c) -1.8%

4. The mire of an ophthalmometer is 100 mm from a cornea, and 100 mm in diameter. The mire image (Purkinje No.1) is found to be 4.5 mm in diameter. Find (a) the exact and the (b) approximate radius of the corneal surface. (c) Find the % error.
Ans.: (a) 9.424 mm; (b) 9.0 mm; (c) -4.5%

5. How far should the mire be from the cornea in order to measure a 9 mm radius with no more than 1% error?
Ans.: 445.5 mm.

6. The objective lens of an ophthalmometer has a focal length of 75 mm. P. I. No 1 is at a distance of 100 mm from the lens (working distance) and it is 2 mm in diameter. The objective re-images P.I. No. 1 at the first focal plane of the ocular. (a) Find the distance from the objective lens to this focal plane. (b) How large is the image produced by the objective lens? (c) How far must a biprism of 10^{Δ} (each half) be from this image plane in order produce tangency of the doubled images? (d) How far must a biprism of 5^{Δ} (each half) be from this image plane in order produce tangency of the doubled images?
Ans.: (a) 300 mm; (b) 6 mm; (c) 30 mm; (d) 60 mm

7. If the objective lens of an ophthalmometer has a focal length of 150 mm, and a working distance of 200 mm. How far must a biprism of 10^{Δ} (each half) be from the conjugate image plane of the lens in order attain tangency of the images? Assume the diameter of P. I. No. 1 is 1 mm.
Ans.: 15 mm.

8. A lensometer target, initially at the zero setting, is moved 8 mm away from the Badal lens. The Badal lens power is 25 D. Find (a) the corresponding power of the spectacle lens; (b) How many mm correspond to a diopter?
Ans.: (a) -5 D; (b) 1.6 mm/Diop.

9. A lensometer contains a +20 D Badal lens. (a) How far and in which direction must the target move from zero to measure a +6 D. lens? (b) How many mm per diopter?
Ans.: (a) 15 mm toward spectacle lens; (b) 2.5 mm/D.

10. You have just measured the power in a principal meridian of lens. It is +3.50 D. It is necessary to move the target an additional 3.75 mm toward the Badal lens (20 D) in order to measure the other principal meridian. (a) What is the power in this meridian? (b) What is the total distance x at this point?
Ans.: (a) +5.00 D; (b) +12.5 mm

11. A lensometer measures one diopter for every 10 mm of movement. What is the power of the Badal lens?
Ans.: +10 D.

12. A lensometer target is moved 15.625 mm to measure a +4 D lens. What is the power of the Badal lens?
Ans.: 16 D

13. You measure an Rx: -6.50 ⌔ - 2.50 x 180° in a lensometer with a +25 D. Badal lens. (a) Find the distances (u) of the target from the Badal lens to measure the power in the 90° and 180° meridians; (b) Find the travel of the target to measure the difference in powers at the principal meridians; (c) How many mm corresponds to one diopter?
Ans.: (a) u_{90} = -54.4 mm and u_{180} =-50.4 mm; ((b) 4 mm; (c) 1.6 mm/D.

NOTES:

CHAPTER 18

DISPERSION AND CHROMATIC ABERRATION

18.1 INTRODUCTION

Newton, in a famous experiment, passed a pencil of sunlight through a prism. The emergent light fanned out into a *spectrum* or rainbow of colors - red, orange, yellow, green, blue indigo and violet. In this continuous change in hues, red light was deviated least and violet most. This *dispersion* of sunlight proved that white light is composed of light of different colors. These spectral colors were not further dispersed by additional prisms. See Figure 18.1. According to the wave theory of light, the spectral colors vary in wavelength λ and frequency υ. Spectral reds have long wavelengths, about 700 nm; spectral blues have short wavelengths around 400 nm. All wavelengths travel at a constant velocity C in a vacuum where c is the product of wavelength and frequency.

$$C = \lambda \nu = 300{,}000 km/\sec$$

Refractive indices equal the ratio of the velocity of light in a vacuum **C** to its velocity **v** in the optical medium. Foucault proved that the velocity of light in any medium is less than in a vacuum, i.e., v<C. Therefore, the refractive indices of optical media are all greater than one.

$$n = \frac{C}{v}$$

Snell's law states that the ratio of the sines of the angles of incidence and refraction of a ray is equal to a constant. This constant is equal to the ratio of the refractive indices of the media.

$$\frac{\sin\alpha}{\sin\alpha'} = \frac{n'}{n}$$

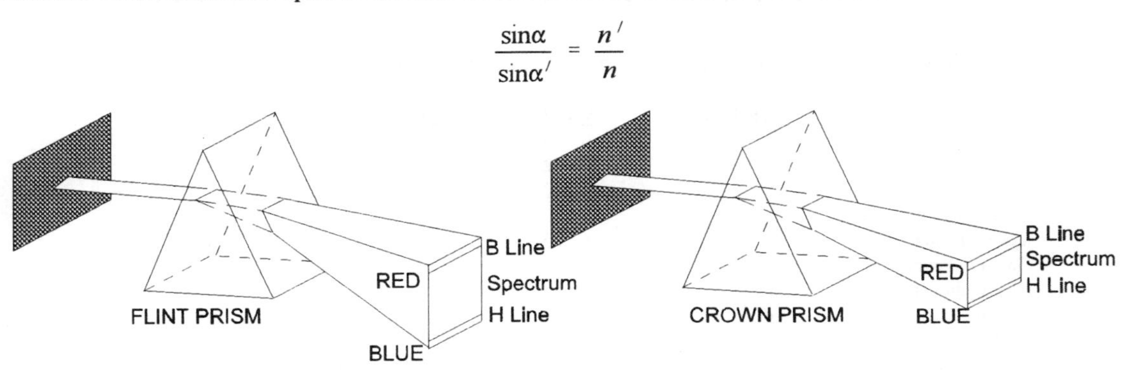

Figure 18.1 Dispersion spectra of glass prisms. The slit in the opaque screen admits a pencil of white light. The flint glass prism produces a larger spectrum than the crown glass prism.

18.2 WAVELENGTH and INDEX of REFRACTION

That light is dispersed by ordinary optical media, i.e., the angle of refraction varies with the wavelength of light, means that its velocity varies with wavelengths. Consequently, the index of refraction of the medium varies with wavelengths. The index corresponding to the least deviated long wavelength light is lower than the index for short wavelengths.

The wavelengths of light, commonly of interest in geometrical optics, are based on studies of the solar spectrum by Joseph von Fraunhofer. He found that the sun's spectrum was not completely continuous, as one would expect of an incandescent solid. It was interrupted by dark narrow lines distributed across the colors. The wavelengths of light corresponding to these lines were absorbed by gaseous elements in the sun's atmosphere. He identified the elements and absorption wavelengths for many of these lines. Certain of these Fraunhofer lines, due to their prominence, were assigned letters to identify them. These spectral lines correspond to monochromatic light. See Figure 18.2 and table below.

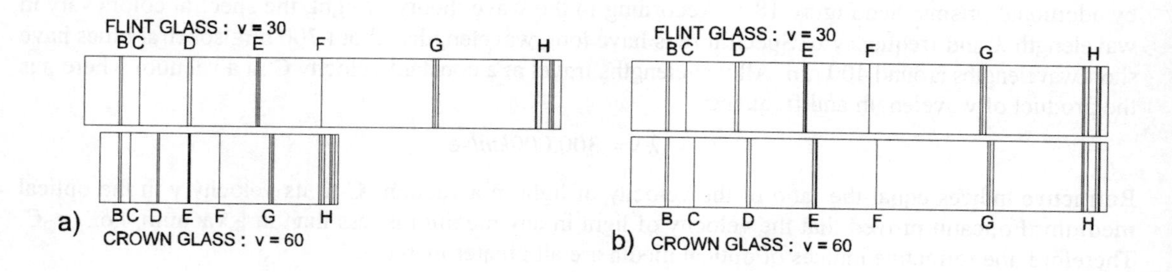

Figure 18.2 a) The length of the spectrum between lines B and H is twice as great for the flint glass compared to the crown glass; b) Irrationality of dispersion: when the crown glass spectrum is scaled up by 2x , the B and H lines are aligned for the two glasses, but the intermediate spectral lines are not. (Adapted from Southall).

Table 1. Wavelengths and sources of some prominent Fraunhofer lines.

COLOR	SPECTRAL LINE	WAVELENGTH (nm)	ELEMENT
Infrared	t	1013.98	Hg
Red helium	r	706.52	He
Red hydrogen	C	656.27	H
Yellow sodium	D	589.3	Na
Yellow helium	d	587.56	He
Green mercury	e	546.07	Hg
Blue hydrogen	F	486.13	H
Blue mercury	g	435.83	Hg
Dark blue	G'	434.05	H
Violet mercury	h	404.66	Hg
Ultraviolet	i	365.0	Hg

18.2.1 INDEX OF REFRACTION

The index of refraction of an optical glass depends on the wavelength of light. The index corresponding to a particular wavelength is identified with a subscript. Thus, n_d, n_F and n_C correspond to indices in yellow, blue and red light. For example, a common optical glass called BK7 has indices $n_d = 1.5168$, $n_F = 1.52238$ and $n_C = 1.51432$. When no subscript is used, the index in yellow helium light is understood. Thus, the index of BK7 is written simply as n = 1.5168. Indices of refraction of optical glasses vary from 1.44 to 2.04.

18.2.2 DISPERSION

The length of the spectra or the dispersion produced by prisms with equal refracting angles varies with the composition of the glass. Figure 18.2 shows the spectra of a flint and a crown glass with dispersions in the ratio of about 2:1. In Figure 18.2b, the spectra between lines B and H have been scaled to equal lengths. The non-alignment of intermediate lines is called the *irrationality of dispersion*. It frustrates the complete correction of chromatic aberration.

The dispersive power ω of a glass is $\omega = \dfrac{n_F - n_C}{n_d - 1}$

18.1

The dispersive power of BK7 glass is 0.015595. The *Abbe number* **V** is more commonly used to describe the dispersion of a glass. It is the reciprocal of the dispersive power:

$$V_d = \frac{n_d - 1}{n_F - n_C} \tag{18.2}$$

V_d is the Abbe number for d, F and C wavelengths. These are the usual wavelengths used for correcting the dispersion or chromatic aberration of visual optical systems. Other V numbers are used when correcting chromatic aberration in other parts of the visible, infrared or ultraviolet spectrum. In the visible spectrum, Abbe numbers of optical glasses range from 20 to 100. *The Abbe number is inversely proportional to the dispersive power of the glass.* BK7 glass has an Abbe number of

$$V_d = \frac{1.5168 - 1}{1.522378 - 1.514324} = 64.17$$

The quantity n_F - n_C is the *mean dispersion*. It is 0.00806 for BK7 glass.

Figure 18.3 illustrates the variation in refractive indexes with the wavelength of BK7 (crown), F4 (flint) and SF8 (dense flint) glasses. The flint glasses have greater indices than the crown glass. Further, the flint glasses are more dispersive, i.e., the rate of change in their slopes increases more rapidly compared with the crown glasses, especially, at the short wavelengths.

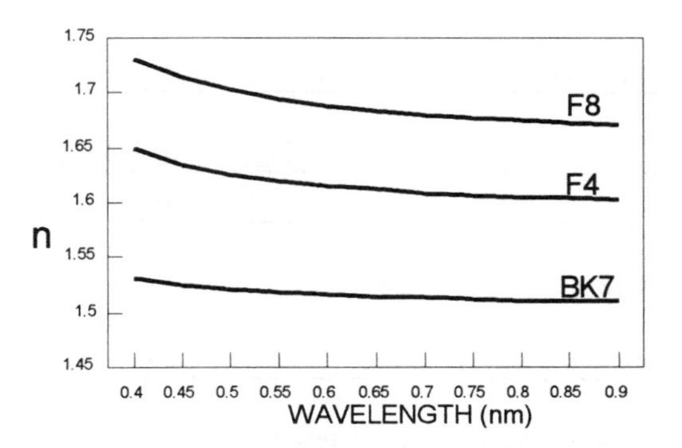

Figure 18.3 Variation in refractive index with glass type.

A six-digit code number is used to designate optical glasses. For example the number 517642 identifies BK7 glass. The first three digits, 517, are the rounded off first three digits following the decimal point in n_d. The last three digits, 642, are the first three digits of the Abbe number obtained by omitting the decimal point. A flint glass, identified as F4 has a code number of 617366. Its index $n_d = 1.61659$ and $V_d = 36.63$.

18.3 OPTICAL GLASS

Glass is an inorganic non-crystalline amorphous substance. It is a super-cooled liquid. Glasses are made of the oxides of silicon, boron and phosphorus. Other oxides are added as fluxes to vary specific properties, such as melting temperatures, index of refraction, dispersion, and the color of the glass. Optical glasses are precisely formulated to control their optical, chemical and mechanical properties. More than 250 different optical glasses are manufactured.

The two most important optical properties of a glass are the indexes of refraction and the dispersion of the glass. Chemical properties that affect the surface quality of optical elements will also affect the quality of the images they produce. The quality of polished glass surfaces is affected by water vapor, acids and gases. These factors may cause surface haze and staining. The mechanical properties of interest are hardness and abrasion resistance. In addition, optical glasses must possess a high degree of homogeneity, i.e., a uniform refractive index. Incomplete stirring of the molten glass can result in threadlike streaks of glass of different composition, density and index of refraction called *striae*. They refract light differently. Other properties that are rigorously controlled are bubbles and stress birefringence.

18.3.1 CROWN AND FLINT GLASSES

Optical glass manufacturers plot the variety of glasses they make on a n_d/V_d diagram or glass map such as in Figure 18.4 The glasses are divided into two broad categories, known as *crowns* and *flints*. Crown glasses largely have lower refractive indices and dispersive power than flints. Crown glasses have $n_d > 1.60$, and $V_d > 50$, or $n_d < 1.60$ and $V_d > 55$. All other glasses are flints. Glasses on the n_d/V_d diagram are also grouped according to their main chemical components as in Table 3. The BK7 glass, referred to earlier, belongs to the borosilicate crown group. BK7 is the abbreviation used by the Schott Optical Glass Co. to identify borosilicate crown glass No 7. Other companies use other abbreviations for the equivalent glass type. Some optical data for BK7 and flint glass - F4 are shown in Table 2. Besides refractive indices and dispersion properties, manufacturers provide mechanical, physical and chemical properties, and light transmittances for wavelengths between 280 and 700 nm.

Table 2. Indices of refraction of a crown and a flint glass.

Glass code	Refractive indices for λ (nm)				
	n_r=706.5	n_c=656.3	n_d=587.6	n_F=486.1	n_g=435.8
BK7 - 517642	1.51289	1.51432	1.51680	1.52238	1.52669
F4 - 617366	1.60891	1.61164	1.61659	1.62848	1.63828

Some of the more common glass types are crown (K), borosilicate crown (BK), barium crown (BaK) and lanthanum crown (LaK). Common flint glasses include flint (F), dense flint (SF) and lanthanum flint (LaF).

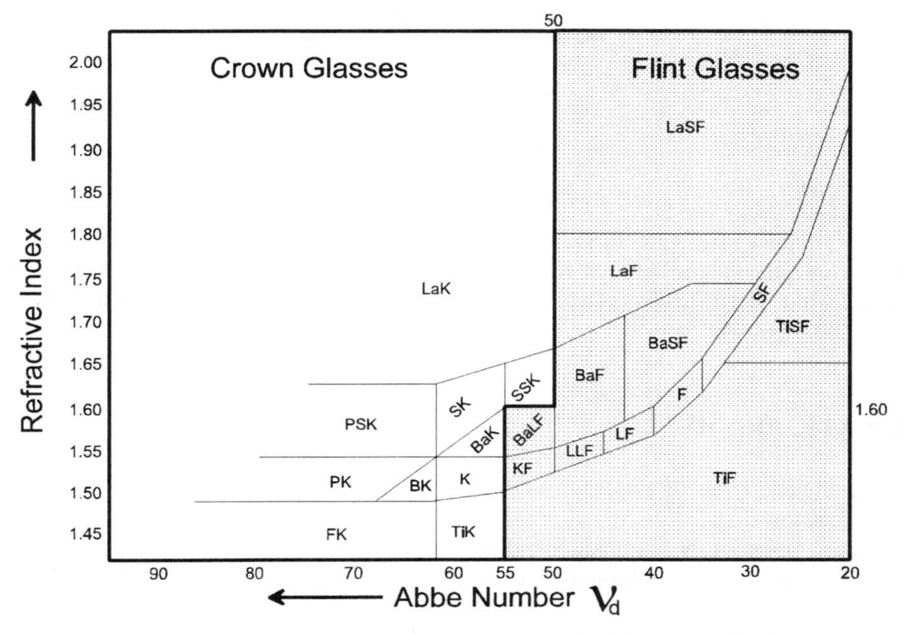

Figure 18.4 A glass map showing the indices and dispersions of optical glasses manufactured by the Schott Optical Glass Co.

18.4 CHROMATIC ABERRATION OF THIN PRISMS

The dispersion of d, F and C light by a thin prism of refracting angle β^Δ is shown in Figure 18.5. The deviation of these rays in prism diopters is P_d^Δ, P_F^Δ and P_C^Δ. Where,

$$P_d^\Delta = (n_d - 1)\beta^\Delta \qquad 18.3a$$

$$P_F^\Delta = (n_F - 1)\beta^\Delta \qquad 18.3b$$

$$P_C^\Delta = (n_C - 1)\beta^\Delta \qquad 18.3c$$

The chromatic aberration of the prism is the deviation in blue minus the deviation in red light.

$$\text{Chr. Ab.} = P_F^\Delta - P_C^\Delta. \qquad 18.4$$

$$= (n_F - 1)\beta^\Delta - (n_C - 1)\beta^\Delta$$

$$= (n_F - n_C)\beta^\Delta$$

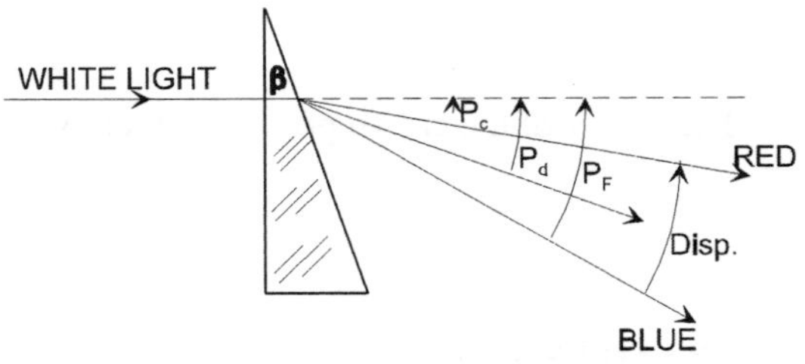

Figure 18.5 Chromatic aberration of a prism in prism diopters.

Thus, the chromatic aberration of the prism is equal to the mean dispersion times the refracting angle. According to Eq. 18.3a

$$\beta = \frac{P_d}{n_d - 1}$$

Substitution into Eq. 18.4 results in $$Chr.Ab. = \frac{n_F - n_C}{n_d - 1}P = \omega P$$

thus,

$$\boxed{Chr.Ab. = \frac{P^{\Delta}}{V}} \qquad 18.5$$

A prism of P = 10^{Δ}, made of BK7 glass (517642), will have a chromatic aberration of 10/64.2 = 0.156^{Δ}. The angle between the red and blue light will be 0.156^{Δ} or 5.35 minutes of arc. This chromatic blur is about the size of a 20/20 letter. A 10^{Δ} prism made of F4 glass (617366) will produce a blur of 9.39 minutes of arc.

18.5 ACHROMATIC PRISMS

Newton, based on the few types of glasses available to him, thought that the dispersions of all glasses were similar. He erroneously concluded that chromatic aberration could not be eliminated with refractive elements. However, as the previous calculations show, a weaker F4 prism can produce a chromatic aberration equal to that of a BK7 prism. This opens the possibility of correcting the chromatic aberration of prisms by combining, *with bases opposite*, a crown prism $\mathbf{P_1}$ with a weaker flint prism $\mathbf{P_2}$. The combination will have a net power **P**.

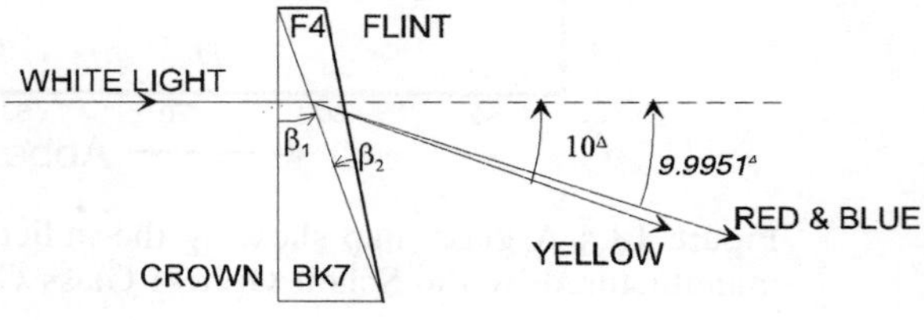

Figure 18.6 An achromatic prism.

$$\mathbf{P = P_1 + P_2} \qquad 18.6$$

The condition for achromatism is that the chromatic aberrations of the two prisms are equal. Because their bases are opposite, the aberrations will cancel. This can be stated as follows:

$$\frac{P_1}{V_1} + \frac{P_2}{V_2} = 0 \qquad 18.7$$

The power of the crown prism is found by solving Eq. 18.6 for P_2 and substituting it into Eq. 18.7.

$$P_1 = \frac{PV_1}{V_1 - V_2} \qquad 18.8$$

EXAMPLE 1: Design a 10^{Δ} achromatic prism with BK7 and F4 glasses. See Figure 18.6.

The glass data are BK7 (517642) $n_d = 1.51680$, $V_d = 64.2$
F4 (617366) $n_d = 1.61659$, $V_d = 36.6$

Step 1. Substitute: P = 10^{Δ}, $V_1 = 64.2$, and $V_2 = 36.6$ into Eq. 18.8.

$$P_1 = \frac{10(64.2)}{64.2 - 36.6} = 23.289^{\Delta}$$

Solve Eq. 18.6 for the power of the flint prism, $P_2 = 10 - 23.289 = -13.289^{\Delta}$ The negative sign means that the base of the flint prism is opposite that of the crown.

Step 2. Find the refracting angles of the prisms with Eq. 18.3a.

Refracting angle of crown prism = $\beta_1 = 23.289/0.5168 = 45.064^{\Delta}$
Refracting angle of flint prism = $\beta_2 = -13.289/0.61659 = -21.553^{\Delta}$

The refracting angles have been computed for index n_d. To check on the achromatism provided by these prisms, calculate the deviation in F and C light with Eqs. 18.3b and 18.3c.

Deviation in F light:
Crown prism: $P_1 = (0.522378)\ 45.064^{\Delta} = 23.5405^{\Delta}$
Flint prism: $P_2 = (0.62848)(-21.5526^{\Delta}) = -13.5454^{\Delta}$
Achromatic prism: $P = 23.5405 - 13.5454 = 9.9951^{\Delta}$

Deviation in C light:
Crown prism: $P_1 = (0.514324)\ 45.064^{\Delta} = 23.1776^{\Delta}$
Flint prism: $P_2 = (0.61164)(-21.5526^{\Delta}) = -13.1825^{\Delta}$
Achromatic prism: $P = 23.1776^{\Delta} - 13.1825^{\Delta} = 9.9951^{\Delta}$

The achromatic prism deviates red and blue light identically (9.9951^{Δ}) but different from yellow light (10^{Δ}). This slight error of 0.005Δ, due to the irrationality of dispersion, is called *secondary color*.

18.6 THE LONGITUDINAL CHROMATIC ABERRATION OF LENSES

18.9

The focal length of a thin lens in air, according to the lensmaker's equation, is

$$\frac{1}{f_d} = (n_d - 1)\left(\frac{1}{r_1} - \frac{1}{r_2}\right)$$

We will find, when we use the appropriate indices, that the respective focal lengths in blue (f_F) and red light (f_C) will be shorter and longer than in yellow light for positive and negative lenses. See Figure 18.7. Longitudinal chromatic aberration **LCA** results in a chromatically blurred image. Consider the positive lens. The image seen on a screen placed at the d light focal plane would have a purplish halo due to the overlapping red and blue blurs. If the screen is moved toward the blue focus, a reddish halo will be seen. The reverse occurs at the red focus. Longitudinal chromatic aberration is a very objectionable aberration.

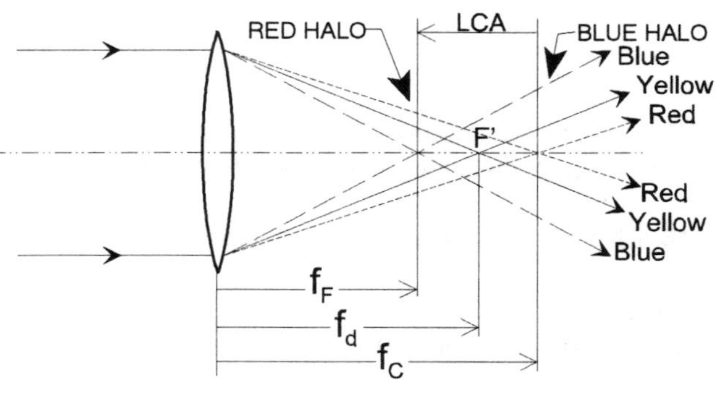

Figure 18.7 Longitudinal chromatic aberration of a positive lens.

18.6.1 LONGITUDINAL CHROMATIC ABERRATION IN DIOPTERS

Distant Object: The lensmaker's equation may be solved for dioptric power by substituting $F = 1/f$. Furthermore, the radii do not vary with the wavelength, thus, $(1/r_1 - 1/r_2)$ is a constant. The lensmaker's equations in d, F and C light are

$$F_d = (n_d - 1)k \qquad 18.10a$$

$$F_F = (n_F - 1)k \qquad 18.10b$$

$$F_C = (n_C - 1)k \qquad 18.10c$$

Longitudinal chromatic aberration is a difference in power with wavelength, ΔF:

$$LCA = \Delta F = F_F - F_C \qquad 18.11$$

Substitute Eqs. 18.10b and c into the above equation. $LCA = (n_F - n_C)k$. Solve Eq. 18.10a for **k** and substitute into the above equation. $$LCA = \frac{n_F - n_C}{n_d - 1} F_d \qquad 18.12$$

The longitudinal chromatic aberration in diopters is $$\boxed{LCA = \frac{F}{V}} \qquad 18.13$$

Compare this equation with Eq.18.5 for prisms.

EXAMPLE 2: A thin equiconvex lens with radii of 100 mm is made of BK7 glass. It has a focal length $f_d = 96.749$ mm and its power is +10.336 D. Find the longitudinal chromatic aberration in diopters.
From Eq. 18.13 we find, **LCA** = 10.336/64.2 = **0.161 D**.

Near Object: The longitudinal chromatic aberration for a near object is $\mathbf{LCA = \Delta U' = U'_F - U'_C}$. Substitution of $U_F + F_F$ and $U_C + F_C$ into this equation, where $U_F = U_C$, results in

$$\mathbf{LCA = F_F - F_C = F/V} \qquad 18.14$$

The dioptric value of longitudinal chromatic aberration is F/V independent of the position of the object. Linear longitudinal chromatic aberration is not.

18.6.2 LINEAR LONGITUDINAL CHROMATIC ABERRATION

Longitudinal chromatic aberration was expressed as a difference in refractive power $\Delta F = F_F - F_C = F/V$. Since power = 1/f, this equation can be written in terms of focal length, and LCA as a linear aberration may be found.

$$\frac{1}{f_F} - \frac{1}{f_C} = \frac{1}{fV}$$

Solve for f_C - f_f.

$$f_C - f_F = \frac{f_C f_F}{fV} \approx \frac{f^2}{fV}$$

or, approximately

$$LCA = f_C - f_F = \frac{f}{V} \qquad 18.15$$

In this form of the equation longitudinal chromatic aberration is a linear distance between the red and blue foci, for an object at infinity.

The chromatic focal lengths found with the lensmaker's equation depend only on the dispersion of the glass medium, not on the radii of curvature or bending of the lens. Consequently, longitudinal chromatic aberration is a *paraxial* aberration. Positive LCA values mean *undercorrected* LCA, or that the red focus is longer than the blue.

EXAMPLE 3: We can find the linear LCA of the thin BK7 equiconvex lens in Example 2 by finding the focal lengths f_d = 96.749 mm, f_F = 95.716 mm, and f_C = 97.215 mm and solving for **LCA = f_C - f_F = 97. 215 - 95.716 = 1.499 mm**. Alternatively, we can use Eq. 18.15, where, **LCA = f/ V = 96.749/64.2 = 1.507 mm**.

The linear longitudinal chromatic aberration for a *near* object is derived similarly. From Eq. 18.14.

$$LCA = U_F' - U_C' = \frac{1}{u_F'} - \frac{1}{u_C'} = \frac{F}{V}$$

From this equation we find that the approximate linear difference (u_C' - u_F') in the red and blue image distances is

$$LCA = u_C' - u_F' = \frac{F u'^2}{V} \qquad 18.16$$

EXAMPLE 4: If the object for the BK7 lens in the previous example is two focal lengths away, i.e., u_d = -2(96.949) = -193.498 mm, we can find the longitudinal chromatic aberration by solving the thin lens equation for u_C' with the red focal length and u_F' with the blue focal length.

$$\frac{1}{u_C'} = -\frac{1}{193.498} + \frac{1}{97.215} \qquad \therefore +195.371$$

$$\frac{1}{u_F'} = -\frac{1}{193.498} + \frac{1}{95.716} \qquad \therefore = +189.410$$

Thus, **LCA** will equal u_C' - u_F' = **5.961 mm.**

A more direct solution is to solve Eq. 18.16. Find the image distance in yellow light from the +10.336 D lens (V = 64.2). **U' = U+*F*** = -5.168 +10.336 =5.168 D, ∴the image distance **u'** = 1/5.168 m = 0.1935 m.

According to Eq. 18.16. **LCA** = $10.336(0.1935^2)/64.2$ = .00603 m = **6.028** mm. This differs negligibly from the previous result.

The above method for designing achromatic doublets is based on thin lenses. It results in a very close but inexact correction when actual lenses with non-negligible thicknesses must be fabricated. Trigonometric ray tracing is employed to adjust the powers of the thick lenses to optimize the longitudinal chromatic aberration.

18.7 ACHROMATIC LENSES IN CONTACT

Positive and negative lenses have excessive power in blue light and deficient power in red light, with respect to yellow light. The blue focus is closest to the lens in both cases. The excessive positive vergence of blue light can be neutralized by the negative vergence of the minus lens. If the two lens elements have the same dispersion their refracting powers will also be neutralized, as Newton concluded. However, if the dispersion of the negative lens is greater than that of the positive lens, the combination of a crown and flint element will have a net positive power. See Figure 18.8.

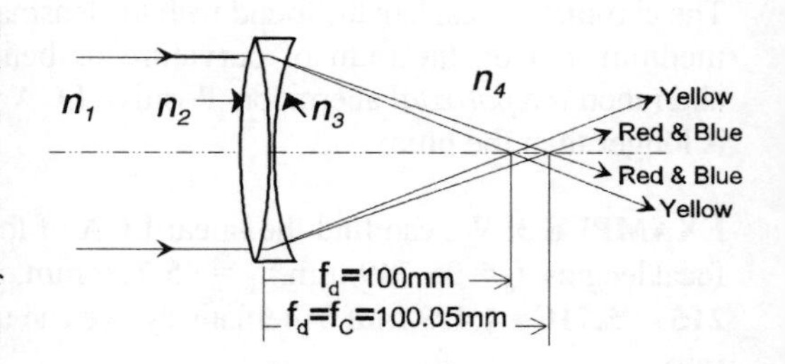

Figure 18.8 An achromatic doublet.

The conditions for designing an achromatic doublet of two thin lenses in contact are:

1. Net power. $F = F_1 + F_2$. 18.17

2. Zero longitudinal chromatic aberration. $$\frac{F_1}{V_1} + \frac{F_2}{V_2} = 0 \qquad 18.18$$

Given F and the V-numbers, the power of the Crown lens is equal to $$F_1 = \frac{FV_1}{V_1 - V_2} \qquad 18.19$$

EXAMPLE 5: Design an achromatic doublet with a power of 10 Diopters. Use BK7 and F4 glass. Given: F =10 D, V_1 = 64.1669, and V_2 = 36.6146. The power of the crown lens, from Eq. 18.18 is F_1 = +23.2891 D. Therefore, F_2 = -13.2891.

Let the lenses be plano convex and plano concave. Their plano faces will be cemented together.
Step 1: Find the radii of the convex r_1 and concave r_3 surfaces.

$$r_1 = \frac{1.5168 - 1}{23.2891} = 22.191\,mm \quad and \quad r_3 = \frac{1 - 1.61659}{-13.2891} = +46.398\,mm$$

Step 2: Calculate the powers of the crown and flint lenses in blue light.

$$\text{Crown: } F_{1F} = \frac{1.522378 - 1}{0.02219} = +23.5405\ D. \qquad \text{Flint: } F_{2F} = \frac{1 - 1.62848}{0.046398} = -13.5454\ D$$

Step 3: Add the powers to find the power of the doublet in blue light. $F_F = +23.5405 - 13.5454 = +9.9951$ D.

Repeat steps 2 and 3 for red light. Accordingly, $F_C = +9.9951$ D.

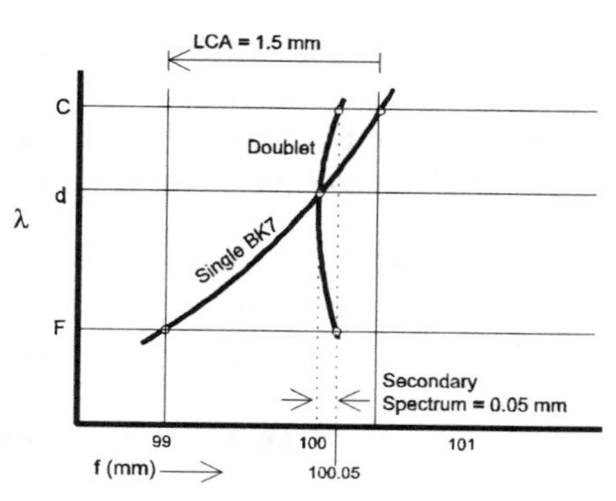

Figure 18.9 Longitudinal chromatic aberration of a singlet and an achromatic doublet.

The red and blue light comes to a common focal plane at 100.0489 mm from the lens. Yellow is focused at 100 mm. See Figure 18.8. All the intermediate wavelengths focus between these extremes. The slight difference in focus is called the *secondary spectrum*. Figure 18.9 shows the longitudinal chromatic aberrations of the BK7 singlet and the achromatic doublet.

18.8 TRANSVERSE CHROMATIC ABERRATION

18.8.1 LINEAR TRANSVERSE CHROMATIC ABERRATION

The principal planes coincide at the optical center of a thin lens. A white light chief ray from object point Q, with slope θ, that goes through the optical center is undeviated. In the presence of longitudinal chromatic aberration, extended images will be formed in the paraxial image planes that correspond to the d, F and C wavelengths. The ends of the images (Q_F' and Q_C') will lie on the chief ray. The images vary in size, as shown in Figure 18.10. They produce a fringe of color called transverse chromatic aberration. **Transverse chromatic aberration** is the difference in size of the red and blue images.

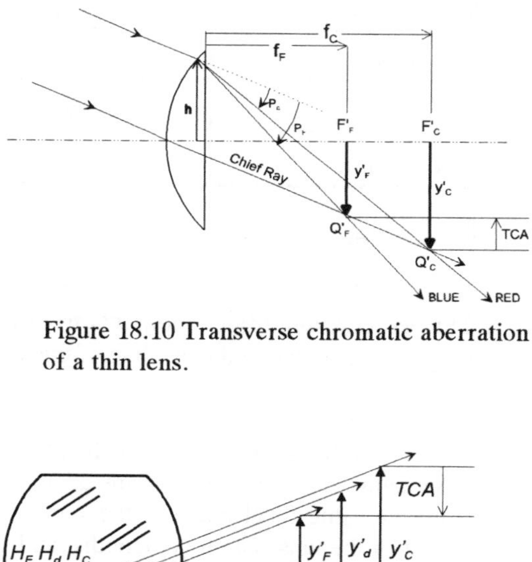

Figure 18.10 Transverse chromatic aberration of a thin lens.

$$TCA = y_C' - y_F'. \qquad 18.20$$

However, $y_C' = f_C \tan\theta$, and $y_F' = f_F \tan\theta$.

Therefore, $\quad TCA = (f_C - f_F)\tan\theta.$

According to Eq. 18.16, $f_C - f_F = LCA$, thus,

$$TCA = LCA\tan\theta = \frac{f}{V}\tan\theta \qquad 18.21$$

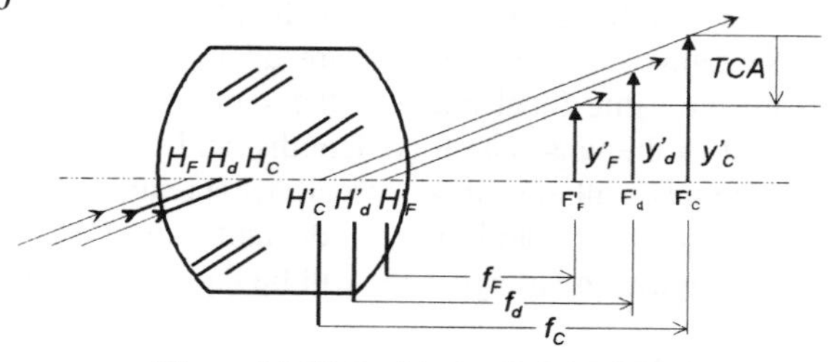

Figure 18.11 Positions of the principal planes and transverse chromatic aberration of a thick lens.

EXAMPLE 6: Find the TCA produced by the thin, 96.749 mm focal length lens referred to in the Example in Sect. 18.7, if an infinitely distant object subtends 5.7106°.

$$TCA = 1.507(\tan 5.7106°) = 0.151 \text{ mm}$$

When the thickness of lenses cannot be neglected, the positions of their principal planes depend on the lens thickness, bending (i.e., the surface powers) and index of refraction. This last condition means that the positions of the principal planes vary with the wavelength. See Figure 18.11. As shown in the following example of an equiconvex lens, the principal planes in F light are closest to the vertices of the lens.

EXAMPLE 7: A positive equiconvex lens, made of BK7 glass, has radii of 100 mm and a thickness of 20 mm. Find the Transverse chromatic aberration in mm for a star at an angle of 5.7106° to the optical axis. The equivalent focal length of the lens is 100.1619 mm. The V number is 64.167. According to Eq. 18.21

$$\mathbf{TCA} = (100.1619/64.167) \tan 5.7106° = \mathbf{0.1561} \text{ mm.}$$

The transverse chromatic aberration also can be found by finding f_C and f_F. However, this requires solving the Gullstrand equations using, in turn, $n_d = 1.5168$, $n_F = 1.522378$, and $n_C = 1.514324$.

Table 3. Variation in optical properties with wavelength of a thick lens.

	d (λ = 587.6 nm)	F (λ = 486.1 nm)	C (λ = 656.3 nm)
F (diopters)	9.9838	10.0891	9.9371
f (mm)	100.1619	99.1172	100.6328
A_2H' (mm)	-6.8254	-6.8021	-6.8358
A_2F' (mm)	93.3365	92.3151	93.7971
$y' = f \tan\theta$	10.0162	9.9117	10.0633

From the above table we find that **LCA** = $f_C - f_F$ = 100.6328 - 99.1172 = **1.5156** mm. **TCA** = **0.1516.** mm. This exact result differs about 3% from the result obtained with the Equation 18.22. The actual image heights are shown in the last line of this table.

An achromatic doublet, if treated as a *thin* lens, will *not* exhibit transverse chromatic aberration. Essentially, the focal lengths are all measured from the same focal plane to the same optical center of the lens. The focal lengths and chief-ray angles will be equal for all wavelengths. Since, $y' = f \tan\theta$ is a constant for all wavelengths, the image sizes will be equal.

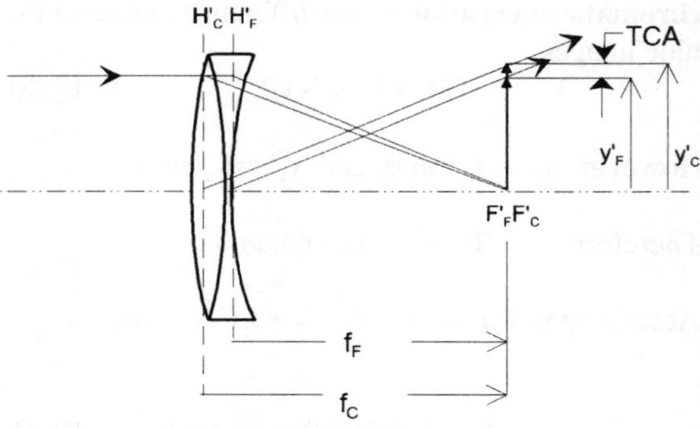

Figure 18.12 Transverse chromatic aberration in a thick achromatic doublet.

A *thick* achromatic doublet focuses the red and blue light at the same focal plane, however, the red focal length is longer and the emergent red chief ray is displaced more than the blue. Consequently, the red image is larger than the blue. See Figure 18.12. Although achromatic doublets are free of longitudinal chromatic aberration, they are not corrected for transverse chromatic aberration.

18.8.2 TRANSVERSE CHROMATIC ABERRATION IN PRISM DIOPTERS

We saw that the chief ray from an off-axis object point Q is undeviated and undispersed in going through the optical center of a thin lens. However, a ray from Q that strikes the lens at some height **h** is chromatically aberrated. The difference in slopes can be expressed as a prismatic effect by the lens. According to Prentice's rule: $\mathbf{P_F^\Delta = h}F_F$; **and** $\mathbf{P_C^\Delta = h}F_C$

Transverse chromatic aberration in prism diopters is equal to the difference in the blue and red deviations.

$$TCA = hF_F - hF_C = h(F_F - F_C)$$

But, $F_F - F_C = F/V$. Thus,

$$\boxed{TCA = \frac{hF}{V}} \qquad 18.22$$

This equation is simply Prentice's rule divided by V-number. It says that the angular subtense of color fringes seen through a lens increases linearly with decentration of the lens. To find the subtense of the color fringe in prism diopters, the ray height **h** must be in cm.

EXAMPLE 8: A white light ray from a star at 5.711° strikes the thin lens of the previous example at a height h = 20 mm. Find the Transverse chromatic aberration in prism diopters. Note: V = 64.2, f = 96.749, thus, F = 10.336 D.

$$\mathbf{TCA} = 2(10.336)/64.2 = \mathbf{0.322^\Delta}$$

On looking through the lens at a height of 20 mm the image of the star will be a spectrum that subtends 0.322^Δ, or about 11 minutes of arc.

18.9 ACHROMATIC COMBINATION OF TWO AIR-SPACED LENSES

The angular fields of view of the objective lenses of microscopes and telescopes are small, compared with the oculars. Uncorrected longitudinal chromatic aberration is the more serious of the chromatic aberrations of the objectives because it results in a blurred image of a small axial object. In visual instruments, the axial displacements of the chromatic images form an object with depth for the ocular. Consequently, a difference in the vergences at the ocular of the blue and red wavefronts result in a difference in accommodative demand.

EXAMPLE 9: A thin BK7 objective lens has a focal length of 250 mm. LCA = 250/64.17 ≈ 4 mm. The ocular has a focal length of 25 mm (a 10x telescope). Assume that the blue focus falls 25 mm from the ocular. U_F = 1000/-25 = -40 D. U_F' = -40 +40 = 0. The blue image is at infinity and the vergence at the eye is zero. The red image will be 21 mm from the ocular. U_C = -47.62 D. Therefore, U_C' = -7.62 D. The red image is formed 131.25 mm away. The apparent thickness of the image is huge. Objective lenses are routinely made as cemented doublets to correct longitudinal chromatic aberration. It should be noted that the longitudinal chromatic aberration of the ocular was neglected in this example.

Transverse chromatic aberration is the more important chromatic aberration to correct in oculars, because they cover wider fields than objectives. For example,a 50x telescope with a 1° real field of view requires an ocular with a 50° apparent field of view. If the ocular is not corrected, color fringes surrounding the image of an extended object will be visible. Because transverse chromatic aberration is a chromatic variation in magnification due to a chromatic variation in focal length, it can only be corrected by a lens system with a *constant* focal length or power in all wavelengths.

18.9.1 THE HUYGENS EYEPIECE

Huygens discovered that two air-spaced lenses can be designed to have a constant focal length at all wavelengths near, for example, the **d** line. Furthermore, they could be made of the *same* glass type. The power of two separated lenses is,

$$F = F_1 + F_2 - dF_1F_2$$

where,

$$F_1 = (n-1)\left(\frac{1}{r_1} - \frac{1}{r_2}\right) = (n-1)k_1$$

and

$$F_2 = (n-1)\left(\frac{1}{r_3} - \frac{1}{r_4}\right) = (n-1)k_2$$

Thus, the thick lens equation may be written as follows:

$$F = nk_1 + nk_2 - k_1 - k_2 - dn^2k_1k_2 + 2dnk_1k_2 - dk_1k_2 \qquad 18.23$$

In order for the power to be constant with wavelength, the power must be independent of the variation in index with wavelength. That is,

$$\frac{dF}{dn} \quad \textit{must equal zero}$$

Equation 18.23 is differentiated with respect to index **n**, and set equal to zero.

$$\frac{dF}{dn} = k_1 + k_2 - 2dk_1k_2(n-1) = 0$$

Multiplying this equation by (n-1) converts the terms to refractive power.

$$F_1 + F_2 - 2d_1F_1F_2 = 0$$

Solve for d

$$d = \frac{F_1 + F_2}{2F_1F_2} \qquad 18.24a$$

Thus, two lenses of the same glass will have a constant focal length for a range of wavelengths near that used for calculating f_1 and f_2, if their separation is equal to the mean of these focal lengths.

$$\boxed{d = \frac{f_1 + f_2}{2}} \qquad 18.24b$$

The Huygens eyepiece is corrected for transverse chromatic aberration in this way. It consists of two plano-convex lenses, with convex surfaces facing the incident light. The first lens is called the field lens; the second is the ocular. Empirically, a ratio in focal length of **k** = f_1/f_2 between 2 and 2.5 is the best choice for telescopes. See Figure 18.13.

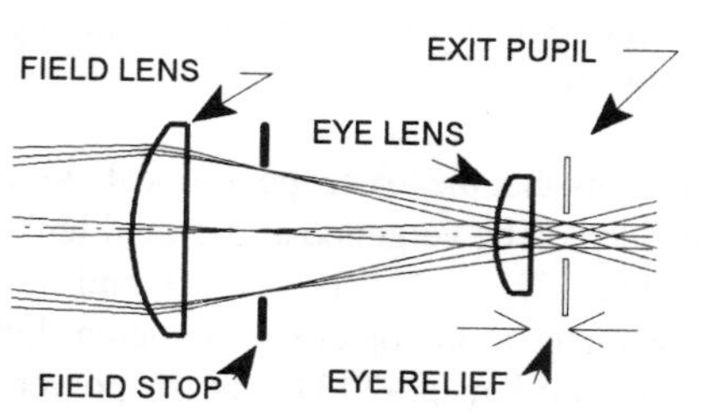

Figure 18.13 Huygens' eyepiece has an internal field stop.

EXAMPLE 10: Design a Huygens eyepiece with a focal length of 66.667 mm, and a k = 2. Given: $k = 2 = f_1/f_2$, or $f_1 = 2f_2$; therefore, $d = 1.5f_2$.

Solve the thick lens equation, $f = \dfrac{f_1 \cdot f_2}{f_1 + f_2 - d}$

$$66.667 = \frac{2f_2^2}{1.5f_2} = \frac{2f_2}{1.5}$$

We will obtai n $f_2 = 50$ mm, $f_1 = 100$ mm, and d = 75 mm for this Huygens eyepiece design.

There is an alternate solution using power: $F =$ 15 D; $F_2 = 2F_1$. According to Eq. 18.24a

$$d = \frac{3F_1}{4F_1^2} = \frac{0.75}{F_1}$$

Substitute these quantities into the thick lens equation and solve for F_1,

$$15 = 3F_1 - 1.5F_1 = 1.5F_1$$

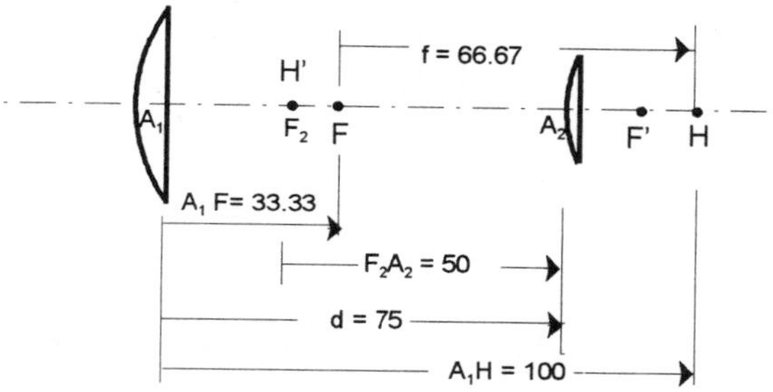

Thus, $F_1 = 10$ D, $F_2 = 20$, and d = 75 mm.

Figure 18.14 The focal points F and F', and the principal points H and H' of a Huygens eyepiece. The objective lens converges rays toward F. After refraction by the field lens, the rays converge on F_2 (the first focal point of the eyelens). They emerge collimated.

The Huygens eyepiece, as part of a telescope, must be positioned with its first focal point F_e coincident with the second focal point of the objective F_o'. Finding Fe to determine the separation between the objective and eyepiece is necessary.

EXAMPLE 11: Given the data for the above eyepiece: $F_e = 15$D; $F_1 = 10$ D; $F_2 = 20$ D; and d = 75 mm, find A_1H and A_1F, where A_1 is the vertex of the field lens. See Figure 18.14

$$A_1H = cF_2/F_e = 0.075(20)/15 = 0.10 \text{ m} = 100 \text{ mm}$$

$$A_1F = A_1H + HF = 100 - 66.667 = 33.333 \text{ mm.}$$

The first focal point of the eyepiece lies 33.333 mm to the right of the field lens. If the eyepiece is part of an afocal 2x Keplerian telescope, what is the distance between the objective and the field lens?

$$M = -2 = -f_o/66.667.$$

Therefore, f_o = 133.333 mm, and the separation between the objective and the field lens = 133.333-33.333 = 100 mm.

To show that the telescope is afocal, we will find the position of the target image after it is refracted by the field lens. The vergence at the field lens is U_1 = 1000/33.333 = +30 D. The power of the field lens is +10 D. Thus, U_1 ' = 30 +10 = 40 D, and u_1 ' = 25 mm.
This image is an object for the ocular lens: $u_2 = u_1' - d$ = 25 - 75 = -50 mm. Since the object vergence for the ocular lens is -20 D, and the power of the ocular is +20 D, the telescope is afocal.

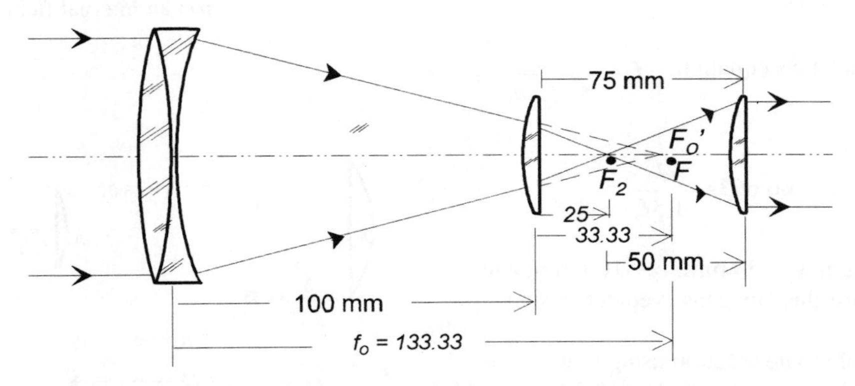

Figure 18.15 The positions of the objective lens and Huygens' eyepiece in an afocal telescope. The second focal point F'_o of the objective lens coincides with the first focal point F of the ocular. A real internal image is formed at F_2, the first focal plane of the eyelens.

The Huygens eyepiece covers a small FOV of ±15° and has short eye relief. For example a 10X telescope, with f_o = 100 mm and f_e = 10 mm, will have a 3.5 mm eye relief. It is low-cost and used in low-priced telescopes and microscopes.

Ordinarily, having a cross-hair or other type reticle in the focal plane of the objective lens is desirable for aiming and making measurements. The target image and the reticle are then viewed through the eyepiece. The Huygens eyepiece design is not suitable for reticles because the target image falls between the field and ocular lenses. Although the target image is acted on by the entire eyepiece and so is free of transverse chromatic aberration, the reticle between the lenses of the eyepiece is observed through the ocular lens alone, and is not corrected for transverse chromatic aberration.

18.9.2 THE RAMSDEN EYEPIECE

The Ramsden eyepiece, designed to be used with a reticle, also consists of a field and ocular lens. Its first focal plane is in front of the field lens. However, the spacing between the eyepiece lenses is less than the mean of the sum of their focal lengths. Transverse chromatic aberration is not as well corrected compared with the Huygens eyepiece. Better correction will be

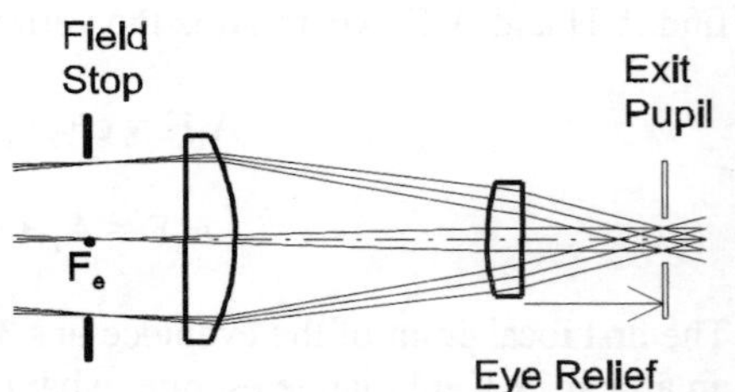

Figure 18.16 The Ramsden eyepiece.

obtained by designing the eyepiece to place the first focal plane close to the field lens. However, the reticle must not be too close to the field lens because, to see clearly, uncorrected myopes push the eyepiece toward the objective. The reticle will obstruct such adjustments. The optimum separation of the lenses of a Ramsden eyepiece is $\mathbf{d = 0.7f_1}$. Values of $\mathbf{k} = f_1/f_2$ between 1 - 1.4 are best. See Figure 18.16.

Although the Ramsden design is inferior in correcting transverse chromatic aberration, it is superior to the Huygens eyepiece in correcting all other aberrations,. Like the Huygens eyepiece, it covers a FOV of ±15°, but is preferred by spectacle wearers for its longer eye relief.

EXAMPLE 12: Design a Ramsden eyepiece with a focal length of 50 mm using $\mathbf{k} = 1$.

Given: $d = 0.7f_1$, and $f_1 = f_2$.
$$f = 50 = \frac{f_1 f_2}{f_1 + f_2 - d} = \frac{f_1^2}{2f_1 - 0.7f_1} = \frac{f_1}{1.3}$$

And so $f_1 = +65\ \text{mm} = f_2$; $d = +45.5\ \text{mm}$

18.9.3 THE KELLNER AND OTHER EYEPIECES

As stated above, the Huygens and Ramsden eyepieces cover small field angles. The Kellner eyepiece, shown in Figure 18.17, is a Ramsden design with an achromatized ocular. It reduces the longitudinal and transverse chromatic aberration as well as other aberrations. Consequently, the field coverage is increased to ±20°.

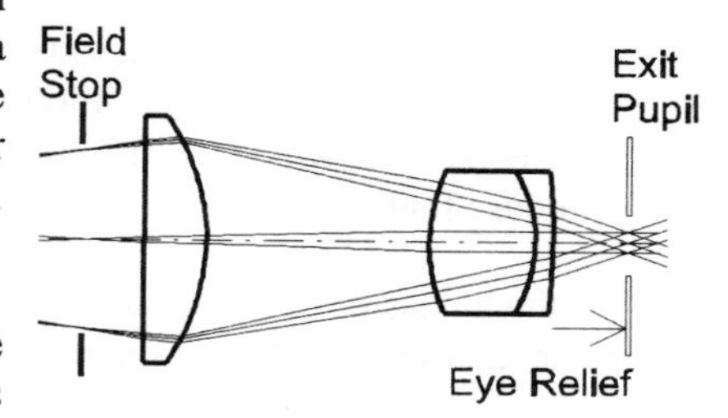

Figure 18.17 The Kellner eyepiece.

Interchangeable eyepieces are often used to change the magnification of a telescope or microscope. Magnification is increased with shorter focal length eyepieces, but the field coverage of the eyepiece must also increase to view the full field of the objective. Wider field coverage requires more complex and costly designs. The designs of several eyepieces may be compared in Figure 18.17. Four element designs such as the symmetrical and Berthele eyepieces cover as much as ±25°. The five element Erfle is a wide-angle eyepiece that covers ±30°.

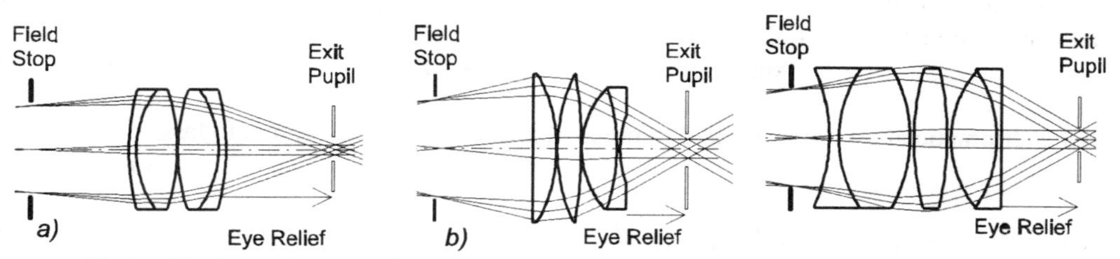

Figure 18.18 Eyepieces: a) Symmetrical; b) Berthele; and c) Erfle eyepieces. From MIL HDBK-141.

QUESTIONS

1. Is the velocity in an optical medium greater for red or blue light?
2. Is the index of the medium greater for red or blue light?
3. Is the dispersion by a medium directly proportional to the V number?
4. Do crown glasses disperse light more than flints?
5. Will an achromatic doublet bring all visible light to the same focal point?
6. What is secondary color?
7. Is it more important to correct the LCA or TCA of eyepieces?
8. Why doesn't correcting the LCA of a thick achromat also correct TCA?

PROBLEMS

1. Calculate the reciprocal dispersions (V-numbers) and mean dispersions of the following substances:

Substance	n_C	n_d	n_F	Answers: V	Mean Disp
cornea	1.3751	1.3771	1.3818	56.28	0.0067
vitreous	1.3341	1.336	1.3404	53.33	0.0063
quartz	1.4565	1.4585	1.4632	68.43	0.0067
spectacle crown	1.5204	1.5231	1.5293	58.77	0.0089
dense flint	1.7103	1.7174	1.7346	29.52	0.0243

2. Find the angles of deviation in d, F and C wavelengths by a 15^Δ prism, made of BK7 glass, and the chromatic aberration of the prism in minutes of arc.
Ans.: $P_d\Delta = 15^\Delta$, $P_F\Delta = 15.162^\Delta$, $P_C\Delta = 14.928^\Delta$, Chr. Ab. $= 0.234^\Delta = 8.04'$.

3. Find the chromatic aberration (in $^\Delta$ and minutes of arc) of a 15^Δ prism made of glass 517642.
Ans.: 0.234^Δ, 8.04'

4. Find the angles of deviation in d, F and C wavelengths by a 9^Δ prism, made of F4 glass, and the chromatic aberration of the prism in minutes of arc.
Ans.: $P_d\Delta = 9^\Delta$, $P_F\Delta = 9.174^\Delta$, $P_C\Delta = 8.928^\Delta$, Chr. Ab. $= 0.246^\Delta = 8.45'$.

5. Find the chromatic aberration (in $^\Delta$ and minutes of arc) of a 9^Δ prism made of glass 617366.
Ans.: 0.246^Δ, 8.45'

6. Design an achromatic prism with a power of 10^Δ using glasses: crown (500600) and flint (600200).
Ans.: crown = 15^Δ; flint = 5^Δ.

7. An achromatic prism is made with a 16^Δ prism made of glass 550500 and an 8^Δ prism made of flint glass with a mean dispersion of 0.026. Find the power of the achromat; and the index of refraction and V number of the flint glass?
Ans.: 8^Δ; 1.650; 25.

8. A thin lens has a focal length in d wavelength of 100 mm. The lens indices in d, F and C wavelengths are: 1.51, 1.52 and 1.505. Find its focal length in F and C wavelengths, and LCA.
Ans.: f'_F =98.077 mm; f'_C=100.990 mm; LCA=+2.913 mm.

9. A 100 mm focal length lens is made of glass with a V number of 34. Find LCA for an object at infinity. What is the V number of the glass in Problem 8? Ans.: 2.941 mm; 34.

10. An object is located two focal lengths in front of the lens in Prob. 8. Find the image distances in d, F and C wavelengths and the LCA in the image.
Ans.: u'_d = 200 mm; u'_F = 192.4528 mm; u'_C = 204 mm; LCA=+11.547 mm.

11. Use V = 34 to find the LCA in Prob. 10. Ans.: LCA = +11.765 mm

12. Find the TCA, by the lens Prob. 8, of an object at infinity that subtends 15°. Ans.: 0.78 mm.

13. Find TCA, for Problem 12, in prism diopters, if ray height is 20 mm. Ans.: 0.588^{Δ}

14. Design a Huygens eyepiece with field and ocular lenses of focal lengths 37.5 and 18.75 mm, respectively. Find the equivalent focal length of the eyepiece, and its magnifying power.
Ans.: EFL = 25 mm; 10x.

15. Let the vertices of the eyepiece in the above problem be numbered A_2 and A_3. Find the positions of the focal and principal points: A_2F, A_2H, A_3F', A_3H'.
Ans.: A_2F = 12.5 mm; A_2H = 37.5 mm: A_3F'= 6.25 mm; A_3H'=-18.75 mm.

16. Find the focal lengths of the field and ocular lenses of a Huygens eyepiece with a focal length of 10 mm and **k** = 2.5. What is the magnifying power?
Ans.: f_{field} = 17.5 mm; f_{ocular} = 7 mm; Mag = 25x.

17. What power objective lens is required to make a 2x telescope with the above eyepiece?
Ans.: +50 D.

18. Design a Ramsden eyepiece with a 50 mm focal length field lens, and **k** = 1.25.
Ans.: f_{ocular} = 40 mm; $f_{eyepiece}$ = 36.36 mm; d = 35 mm; Mag = 6.88x.

19. Let the vertices of the eyepiece in the above problem be numbered A_2 and A_3. Find the positions of the focal and principal points: A_2F, A_2H, A_3F', A_3H'.
Ans.: A_2F = -4.545 mm; A_2H = 31.818 mm: A_3F'= 10.90 mm; A_3H'=-25.454 mm.

NOTES:

CHAPTER 19

TRIGONOMETRIC RAY TRACING

19.1 INTRODUCTION

Paraxial theory dictates perfect imagery by optical systems. All rays from each point in the object meet at conjugate image points. The image is an exact duplicate of the object except, perhaps, in scale. However, paraxial rays comprise only a minuscule fraction of all rays refracted and reflected by most real optical systems so, in fact, the paraxial image is not a true representation of the object. The rays from an axial object point that strike the aperture of an optical system at increasing distances from the axis, or that originate from object points at increasing angles from the axis will be aberrated. They will not intersect the image plane at the same points as the paraxial rays. A blurred or distorted image will be produced. In Figure 19.1a an axial and off-axis *tangential* pencil of rays is incident across the full aperture of a lens. Tangential rays lie in the *meridional* plane that includes the optical axis. The ray at the edge of the aperture of any pencil of rays is called a *marginal* ray. It is one of a ring of rays at the perimeter of the aperture. Rays that strike the lens at heights between the margin and the axis are *zonal* rays. Zonal rays are specified by their fractional height at the lens aperture, for example, 0.7 or 0.5 of the full aperture. A chief ray that goes through the center of the aperture stop is contained by the pencils of rays from each object point. The *sagittal* plane, shown in Figure 19.1b is perpendicular to the meridional plane. Sagittal pencils also contain zonal and marginal rays. Rays that do not lie in a plane that contains the optical axis are called skew rays.

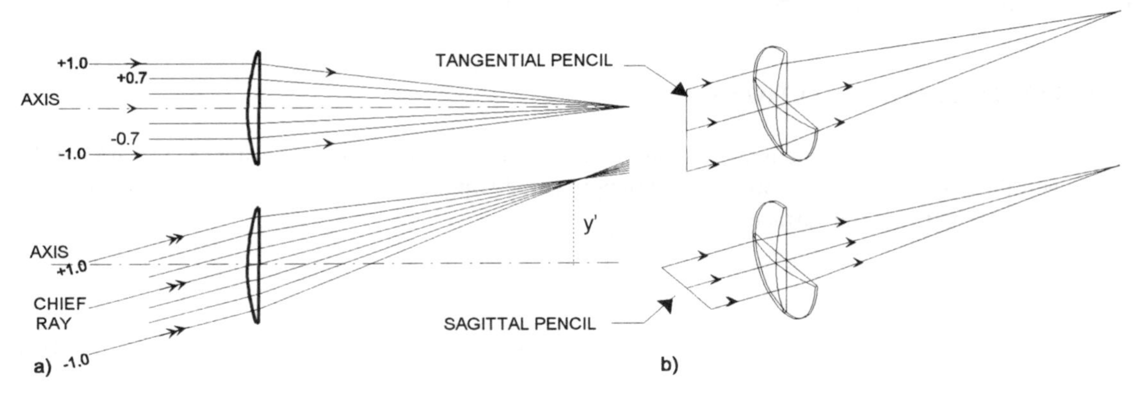

Figure 19.1 (a) Meridional pencils of rays from axial and off-axis points. Each pencil contains upper and lower marginal (± 1) and zonal rays (± 0.7) rays; (b) meridional and sagittal ray pencils.

19.2 RAY TRACING SIGN CONVENTIONS AND NOMENCLATURE

The paraxial image equations were derived by replacing the sines of angles in Snell's law with the angles themselves because the angles were assumed to be small. In trigonometric ray tracing Snell's law is rigorously applied. Ray tracing reveals the aberrations or defects in an image.

The sign convention for refracting a ray through a single spherical refracting surface is unchanged from prior usage. In Figure 19.2, **C** is the center of curvature of a convex spherical surface. **C** lies on the optical axis that intersects the surface at the vertex **A**. An object **M** lies on the axis and sends a ray to the point **B** on the surface. **BC** is a radius to the surface and, therefore, the normal. Its slope is the central angle ϕ. After refraction, the ray crosses the axis at the image point, **M'**. The point **A** is the origin of a coordinate system in the meridional section, where the tangent to the surface at **A** is the **y** axis. A perpendicular dropped from point **B** to the optical axis at point **D** is the height of incidence **h** of the ray. Angles that are measured counter clockwise (CCW) are positive, clockwise CW angles are negative. M and M' are no longer paraxial points. Object and image distances L and L' represent trigonometrically traced or non-paraxial distances.

Angles	Symbol	
Incidence:	NBR	$= \alpha$
Refraction:	N'BR'	$= \alpha'$
Slope of incident ray:	AMB	$= \theta$
Slope of refracted ray	AM'B	$= \theta'$
Central angle:	BCA	$= \phi$

Distances:		
Radius of curvature:	AC	= r
Object distance:	AM	= L
Image distance:	AM'	= L'
Object height:	MQ	= y
Image height	M'Q'	= y'

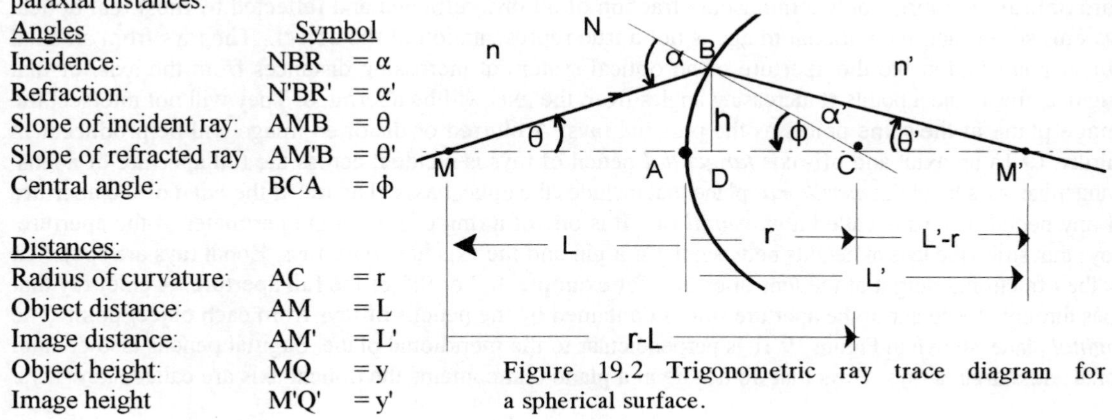

Figure 19.2 Trigonometric ray trace diagram for a spherical surface.

19.3 RAY TRACING EQUATIONS

Four equations are needed to trace rays trigonometrically. It is assumed that the distance **L** = AM of the object and the slope θ of the ray are given. Triangle CBM supplies the equation for finding the angle of incidence α of the ray at the surface.

$$\frac{\sin\alpha}{r-L} = \frac{\sin\theta}{r}$$

thus

$$\sin\alpha = \frac{r-L}{r}\sin\theta \tag{19.1}$$

The second equation, Snell's law, yields the angle of refraction.

$$\sin\alpha' = \frac{n}{n'}\sin\alpha \tag{19.2}$$

From Figure 19.2 we find that $\phi = \alpha - \theta = \alpha' - \theta'$

Therefore, the slope of the refracted ray is found with the third equation.

$$\theta' = \alpha' - \alpha + \theta \tag{19.3}$$

Finally, triangle CBM', provides the equation needed to find the distance **L' = AM'** at which the ray crosses the axis.

$$\frac{\sin\alpha'}{L'-r} = \frac{\sin\theta'}{-r}$$

Thus,

$$L' = r\left(1 - \frac{\sin\alpha'}{\sin\theta'}\right) \tag{19.4}$$

EXAMPLE 1: A SSRS has a radius of curvature of +50 mm and separates air from glass of index 1.5. Trace a ray with a slope of 5.739° from an object 200 mm in front of the surface.
Given: L = -200 mm, r = +50 mm, θ = 5.739°, n = 1, n' = 1.5.

Solve Eq. 19.1:

$$\sin\alpha = \frac{+50+200}{+50}\sin(+5.739°) = +0.5000$$

$$\alpha = 30°$$

Solve Eq. 19.2:

$$\sin\alpha' = 0.5/1.5 = 0.3333$$

$$\alpha' = +19.471°$$

Solve Eq. 19.3:

$$\theta' = +5.739 - 30 + 19.471 = -4.790°$$

$$\sin\theta' = -0.08350$$

Solve Eq. 19.4:

$$L' = +50\left(1 - \frac{0.33333}{-0.08350}\right) = +249.609\,mm$$

The ray crosses the axis 249.609 mm to the right of the vertex. See Figure 19.3. The paraxial image distance u' = +300mm.

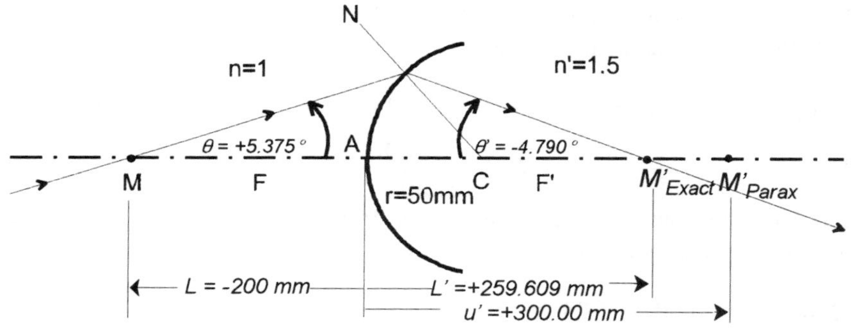

Figure 19.3 Illustration for the Example 1 ray trace.

Special case: Axial objects at infinity. Equation 19.1 fails for objects at infinity, since the quotient (r-L)/r = (r-∞)/r = -∞, and the ray slopes all equal zero. Figure 19.4 applies to this special case. The angle of incidence of a ray from a distant axial object depends on its height **h**. Evidently, the angle of incidence α equals the angle ϕ of the normal. Eq. 19.1 is replaced with

$$\sin \alpha = h/r \qquad 19.1a$$

EXAMPLE 2: Repeat example 1, except place the object at infinity and set the ray height h = 20 mm. Note, $\theta = 0°$.

Solve Eq. 19.1a: $\sin \alpha = 20/50 = +0.4$, $\therefore \alpha = +23.57817°$

Solve Eq. 19.2 $\sin \alpha' = +0.266667$, $\therefore \alpha' = +15.46600°$

Solve Eq. 19.3 $\theta' = -8.112168°$ $\therefore \sin \theta' = -0.141111$

Solve Eq. 19.4 $L' = +144.4879$ mm

Compare this with the paraxial image distance u' = 150 mm.

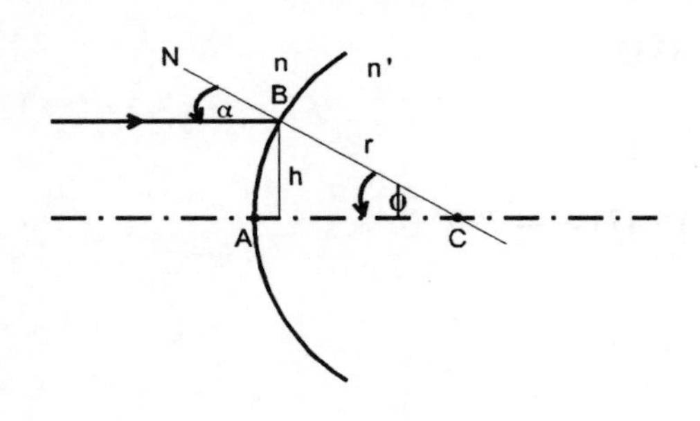

Fig. 19.4 Incidence angle for axial ray from infinity.

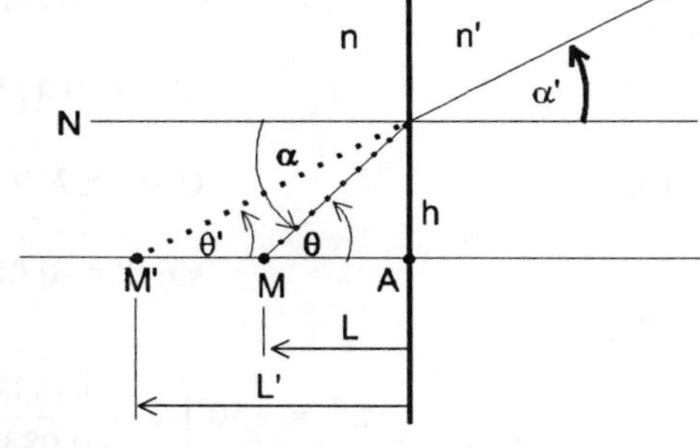

Figure 19.5 Ray trace through a plano refracting surface.

Special case: Plano surfaces. Another special case occurs when the surface is plano, i.e., r = ∞. The quotient in Eq. 19.1 must be interpreted as follows: (∞-L)/∞ = 1. Therefore, as shown in Figure 19.5, $\alpha = \theta$, and $\alpha' = \theta'$. Snell's law can be used to find θ'

$$n \sin \theta = n' \sin \theta' \qquad 19.2b$$

Eq. 19.4 fails for plano surfaces because r = ∞. To find L', note that $\tan \theta = h/\text{-}L$ and $\tan \theta' = h/\text{-}L'$.

Consequently,

$$L' = \frac{L \tan\theta}{\tan\theta'} \qquad 19.4a$$

EXAMPLE 3: A plano refracting surface separates air from glass of index 1.5. Trace a ray with a slope of 5.74° from an object 200 mm in front of the surface.

Given: L = -200 mm, r = ∞, $\theta = +5.74°$, n = 1, n' = 1.5.

Angle α equals angle θ, thus $\sin \theta = 0.100014$

.Solve Eq. 19.2b $\sin \theta' = 0.100014/1.5 = 0.066676$. Therefore, $\theta' = 3.823115°$

Find tangents θ and θ' and solve Eq. 19.4a:

$$L' = -200(0.10052)/0.066825 = -300.840 \text{ mm}$$

Special case: spherical mirrors. Trigonometric ray tracing Equations Nos. 19.1-19.4 apply to mirrors simply by substituting index n' = -n.

EXAMPLE 4: Suppose the SSRS in the earlier example is replaced with a spherical mirror. All else remains the same. Find the distance at which the ray crosses the axis. See Figure 19.6.
Given: L = -200 mm; r = +50 mm; θ = 5.74°; n = 1; n' = -1. Solve Eq. 19.1:

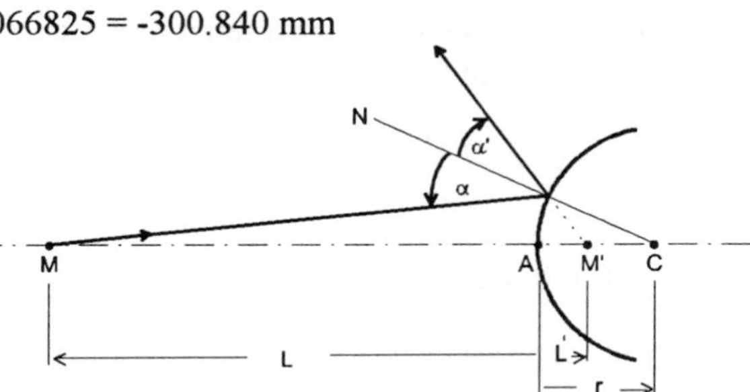

Figure 19.6 Ray trace at a spherical mirror.

$$\sin\alpha = \frac{50+200}{+50}\sin(5.74°) = +0.500$$

$$\alpha = 30°$$

Solve Eq. 19.2: $\sin\alpha' = 0.5/-1 = -0.500, \quad \therefore \alpha' = -30°$

Solve Eq. 19.3: $\theta' = 5.74 - 30 - 30 = -54.2500° \;\therefore \sin\theta' = -0.8117$

Solve Eq. 19.4:

$$L' = +50\left(1 - \frac{0.5000}{0.8117}\right) = +19.1995mm$$

The ray virtually crosses the axis at 19.1995 mm from the vertex of the mirror. The paraxial image distance would have been u' = +22.22 mm.

19.3.1 Transfer Equations

To trace rays through additional surfaces transferring ray data to the subsequent surfaces is necessary. You will recall that in paraxial calculations the image produced by one surface became the object for the next surface, that is, $u_2 = u'_1 - d$, where d was the axial separation between the surfaces. The corresponding transfer equation for the exact ray is

$$L_2 = L_1' - d_1. \qquad 19.5$$

The second transfer equation simply recognizes that the refracted ray continues to the next surface, therefore, the slope angle θ_1' of the refracted ray becomes the slope of the ray at the next surface.

$$\theta_2 = \theta_1' \qquad 19.6$$

19.4 RAY TRACING THROUGH A SERIES OF SURFACES

In general, to trace rays through a series of surfaces Equations 19.1 - 19.6 are used iteratively, and carry subscripts to denote the current surface j.

$$\sin\alpha_j = \frac{r_j - L_j}{r_j}\sin\theta_j \qquad 19.1, \qquad \sin\alpha_j' = \frac{n_j}{n_j'}\sin\alpha_j \qquad 19.2$$

$$\theta_j' = \theta_j - \alpha_j + \alpha_j' \qquad 19.3 \qquad L_j' = r_j\left(1 - \frac{\sin\alpha_j'}{\sin\theta_j'}\right) \qquad 19.4$$

$$L_{j+1} = L_j' - d_j \qquad 19.5 \qquad \theta_{j+1} = \theta_j' \qquad 19.6$$

EXAMPLE 5: The ray trace started in Example 1 will be continued by making the SSRS into an equiconvex lens with a thickness of 15 mm. The previous data are, $r_1 = 50$ mm and $L_1 = -200$ mm. Note: the equivalent focal length of the lens is 52.6316 mm. See Figure 19.7. The additional given data are, $r_2 = -50$ mm, $d = 15$ mm, $n_1' = n_2 = 1.5$, $n_2' = 1$.

The transfer data are, $\quad L_2 = L_1' - d_1 = 249.609 - 15 = 234.609$ mm

and, $\quad \theta_2 = \theta_1' = -4.790°$

From Eq.19.1: $\quad \sin\alpha_2 = \frac{(-50-234.609)}{-50}\sin -4.790° = -0.475$

$\therefore \alpha_2 = -28.377°$

Solve Eq. 19.2. $\quad \sin\alpha_2' = \frac{1.5}{1}\sin(-28.377) = -0.713$

$\therefore \alpha_2' = -45.472°$

Solve Eq. 19.3 $\quad \theta_2' = -4.790 + 28.377 - 45.472 = -21.884°$

$\therefore \sin\theta_2' = -0.373$

Solve Eq. 19.4 $\quad L_2' = -50\left(1 - \frac{-0.713}{-0.373}\right) = 45.632\,mm = A_2M_2'$

The ray crosses the axis +45.632 mm from the second surface of the lens.

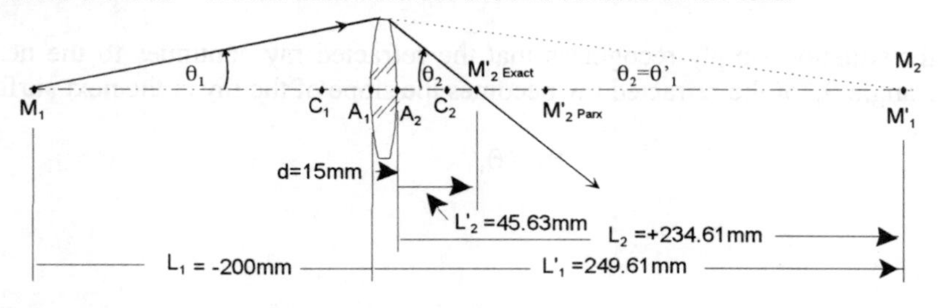

Figure 19.7 Trigonometrically traced ray through a thick lens.

19.5 LONGITUDINAL AND TRANSVERSE SPHERICAL ABERRATION

Our purpose in tracing the above ray through the equiconvex lens is to find its aberration. By that we mean how much its path differs from the paraxial ray. We will calculate the position of the paraxial image formed by the equiconvex lens by applying the refraction equation at each surface.

$$\frac{1.5}{u_1'} = \frac{1}{-200} + \frac{1.5-1}{50}$$

thus, $u_1' = 300$ mm.

Transfer to second surface, $u_2 = 300 - 15 = 285$ mm

$$\frac{1}{u_2'} = \frac{1.5}{285} + \frac{1-1.5}{-50}$$

the paraxial image distance is $u_2' = 65.5172$ mm

Longitudinal spherical aberration is the distance from the paraxial image plane to the real ray intersection with the optical axis. See Figure 19.8.

$$\mathbf{LSA = L_2' - u_2'} \qquad 19.7$$

The LSA of the lens for the marginal ray with slope 5.74° is, LSA = 45.632 - 65.517 = -19.885 mm
A negative LSA, or *undercorrected* spherical aberration, shows that the marginal ray crossed the axis closer to the lens than the paraxial ray.

LSA is not a good indicator of image quality. Lenses of long focal length and small aperture (high f/No) may exhibit large LSA yet not as severely blur the image as short focal length fast lenses. The blur of the axial image is a *transverse spherical aberration*. It is a circle with a semi-diameter equal to the intersection height of the real marginal ray in the paraxial image plane.

$$\mathbf{TSA = LSA \tan \theta_k'} \qquad 19.8$$

where subscript k refers to the last surface. According to Figure 19.8, we see that TSA = -19.885 tan 21.8904° = -7.990 mm. *The equiconvex lens produces a 16 mm diameter blur!*

Figure 19.8 also shows that there is a minimal blur for spherical aberration at the *disk of least confusion* or *best focus* (BF), i.e., where the upper marginal ray intersects the -0.7 zonal ray. This occurs at about 0.75xLSA, or at a defocus of -14.9 mm.

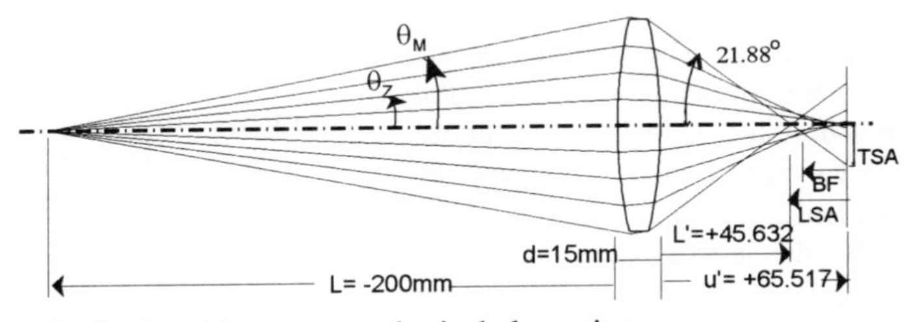

Figure 19.8 Longitudinal and transverse spherical aberration.

19.6 RAY TRACE CALCULATIONS FOR A PENCIL OF RAYS

The ray chosen in the above example struck the lens at a height of about 20 mm. Let this be the full aperture height of the lens. We will investigate the way spherical aberration varies with ray height by tracing rays at the following zones: full aperture, 0.75 aperture, 0.5 aperture and 0.25 aperture. To organize the calculations we will use the following record sheet to tabulate our results. The ray heights **h**, at the top of columns 2-5, are 20, 15, 10 and 5 mm. They were computed to provide convenient slopes (tangents), then converted into degrees to initiate the calculations. The sixth column is a paraxial ray trace. The *paraxial equations* are

$$\alpha = \frac{r-u}{r}\theta. \qquad Eq.\ 19.1p \qquad\qquad \alpha' = \frac{n}{n'}\alpha \qquad Eq.\ 19.2p$$

$$\theta' = \theta - \alpha + \alpha' \qquad Eq,\ 19.3p \qquad\qquad u' = r\left(1-\frac{\alpha'}{\theta'}\right). \qquad Eq.\ 19.4p.$$

19.7 A FORMAT FOR TRACING RAYS THROUGH A LENS

Table 1 illustrates a format for calculating and recording the essential data for tracing a pencil of rays from a point through an equiconvex thick lens of 102.564 mm focal length. **Given:** r_1 =100 mm; r_2 = -100 mm; d = 15 mm; $n_1 = 1 = n_2'$; $n_1' = 1.5 = n_2$; L_1 = -400 mm. Four rays striking the first surface (entrance pupil) at heights of 40, 30, 20 and 10 mm, and one paraxial ray are traced.

19.8 IMAGE EVALUATION

The purpose for ray tracing is to evaluate the quality of the image. Spherical aberration can be depicted as a longitudinal or transverse aberration. Ray intercept curves can, additionally, indicate the type and amount of aberrations present in off-axis points. Additional methods for evaluating the quality of images include spot diagrams, radial energy and knife edge distributions.

19.8.1 SPHERICAL ABERRATION CURVES

The paths of four rays from an infinitely distant axial object are shown in Figure 19.9. The ray trace data is used to plot longitudinal spherical aberration as a function of aperture height to evaluate the lens. The ray trace results suggest that LSA increases approximately as $\mathbf{h^2}$, i.e., the curve is parabolic. As **h** is doubled, LSA increases about four times. LSA does not vary exactly with h^2 for this ray trace example because the lens is very fast, i.e., the f/no is very small. Consequently, the lens exhibits significant amounts of higher order spherical aberration.

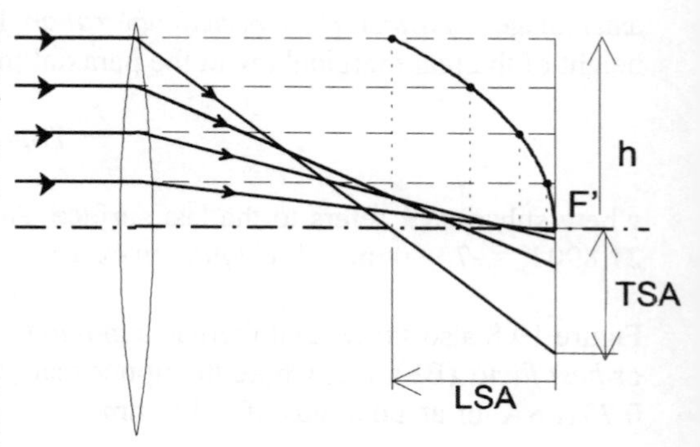

Figure 19.9 Plot of Longitudinal Spherical Aberration. The height of the ray at the lens (h) plotted vs. LSA.

TABLE 1. TRIGONOMETRIC RAY TRACES THROUGH A THICK LENS.

FIRST SURFACE OF LENS					
	Trigonometric				Paraxial
h_1	40.000	30.000	20.000	10.000	10.000
r_1	100.0000	100.0000	100.0000	100.0000	100.0000
L_1	-400.000	-400.000	-400.000	-400.000	-400.000
n_1	1.0000	1.0000	1.0000	1.0000	1.0000
n_1'	1.5000	1.5000	1.5000	1.5000	1.5000
d_1	15.0000	15.0000	15.0000	15.0000	15.0000
θ_1	5.7106	4.2829	2.8553	1.4276	0.0250
$\sin\theta_1$	0.0995	0.0747	0.0498	0.0249	
$\sin\alpha_1$	0.4975	0.3734	0.2491	0.1246	0.1250
α_1	29.8360	21.9260	14.4224	7.1561	
$\sin\alpha_1'$	0.3317	0.2489	0.1660	0.0830	0.0833
α'_1	19.3707	14.4148	9.5580	4.7638	
θ'_1	-4.7547	-3.2283	-2.0091	-0.9646	-0.0167
$\sin\theta_1'$	-0.0829	-0.0563	-0.0351	-0.0168	
L_1'	500.1477	542.0482	573.6244	593.3077	600.0000
LSA	-99.8523	-57.9518	-26.3756	-6.6923	0.0
TSA	-8.3053	-3.2687	-0.9253	-0.1127	0.0
SECOND SURFACE OF LENS.					
r_2	-100.0000	-100.0000	-100.0000	-100.0000	-100.0000
L_2	485.1477	527.0482	558.6244	578.3077	585.0000
n_2	1.5000	1.5000	1.5000	1.5000	1.5000
n_2'	1.0000	1.0000	1.0000	1.0000	1.0000
θ_2	-4.7547	-3.2283	-2.0091	-0.9646	-0.0167
$\sin\theta_2$	-0.0829	-0.0563	-0.0351	-0.0168	
$\sin\alpha_2$	-0.4850	-0.3531	-0.2309	-0.1142	-0.1142
α_2	-29.0140	-20.6784	-13.3503	-6.5571	
$\sin\alpha_2'$	-0.7275	-0.5297	-0.3464	-0.1713	-0.1713
α_2'	-46.6802	-31.9841	-20.2646	-9.8628	
θ_2'	-22.4208	-14.5340	-8.9234	-4.2704	-0.0738
$\sin\theta_2'$	-0.3814	-0.2510	-0.1551	-0.0745	
L_2'	90.7508	111.0678	123.2912	130.0345	132.2034
LSA	-41.4526	-21.1356	-8.9122	-2.1689	
TSA	-17.1032	-5.4794	-1.3993	-0.1619	

19.8.2 RAY INTERCEPT CURVES

Perhaps a better way to depict the effect of aberrations on an image is to plot TSA. This represents the blur in the paraxial image plane. This latter diagram is called a *ray intercept curve*. Figure 19.10a is such a plot of the rays from the axial object point. The height of the ray in the paraxial image plane (δy') is plotted on the ordinate and the height of the ray in the entrance pupil is plotted on the abscissa. Both the LSA and ray intercept curves illustrate the rapid increase in spherical aberration with aperture height of simple lenses. Because TSA = LSA tan θ_k', LSA varies as h^2 and tan $\theta'_k = h/L'_k$ TSA evidently increases as h^3. The ray intercept curve is a cubic function.

The ray intercept curve is also used to evaluate the aberration of rays from off-axis points. Here we plot the height of the ray in the paraxial image plane (δy') with respect to the height of the chief ray height. Figure 19.10b is a plot of the rays from the point at 15° off axis. It shows a comatic blur elongated along the y-axis.

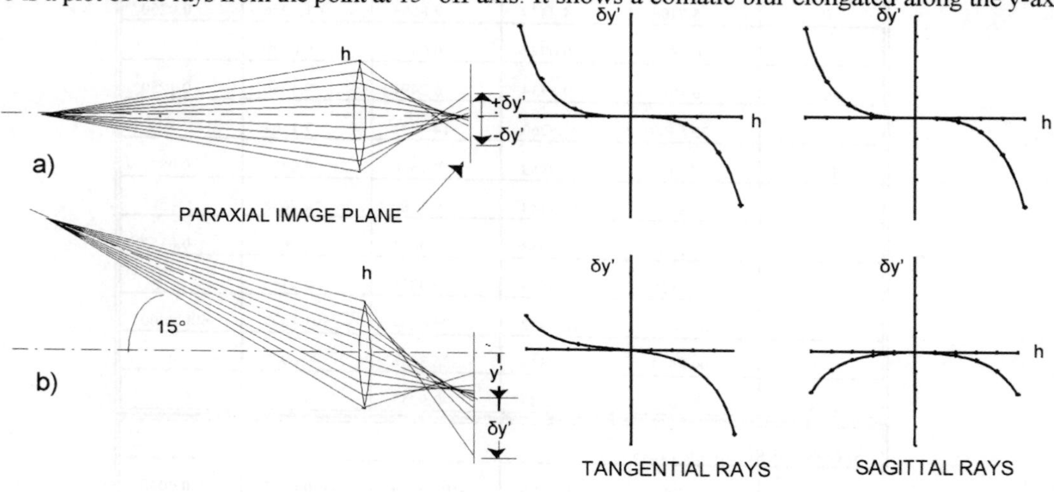

Figure 19.10 Ray intercept curves for a singlet lens: (a) axial object point; (b) 15° off-axis.

19.8.3 SPOT DIAGRAMS

A powerful depiction of aberrations is provided by spot diagrams. They are used to calculate the geometrical modulation transfer function of an optical system. A square array of rays across the entrance pupil from the object point is traced through the lens. Figure 19.11a illustrates the array on the entrance pupil of the lens in Table 1. Each ray is considered to contribute a proportionate amount of light to the image. The intersections of the rays in or near the image plane is called a *spot diagram*. Figure 19.11b is the spot diagram, in the paraxial image plane, of the spherically aberrated axial object point. Compare the scale with TSA in Table 1. The spots are concentrated in the center but surrounded by a large flare that will reduce contrast in adjacent image points. Clearly, this is not the plane of best-focus. The smallest spot diagram is obtained by defocussing to the disk of least confusion. As seen in Figure 19.11c, the ray intersections are more closely packed and the flare is eliminated. An off-axis object point at y = 100mm produces the asymmetrical spot diagram shown in Figure 19.11d. Note the scale is 2.5 times larger than for the axial object point.

When an optical system is very highly corrected, the size of the spot diagram may be smaller than the diameter of the Airy disc.

$$Diam\ of\ Airy\ disk = \frac{2.44\lambda f}{D} = 2.44\lambda f/No.$$

Diffraction does not allow for such a distribution of rays in the image plane. The spot diagram no longer represents the light distribution in the image. Such highly corrected systems produce diffraction patterns and are said to be *diffraction limited.*

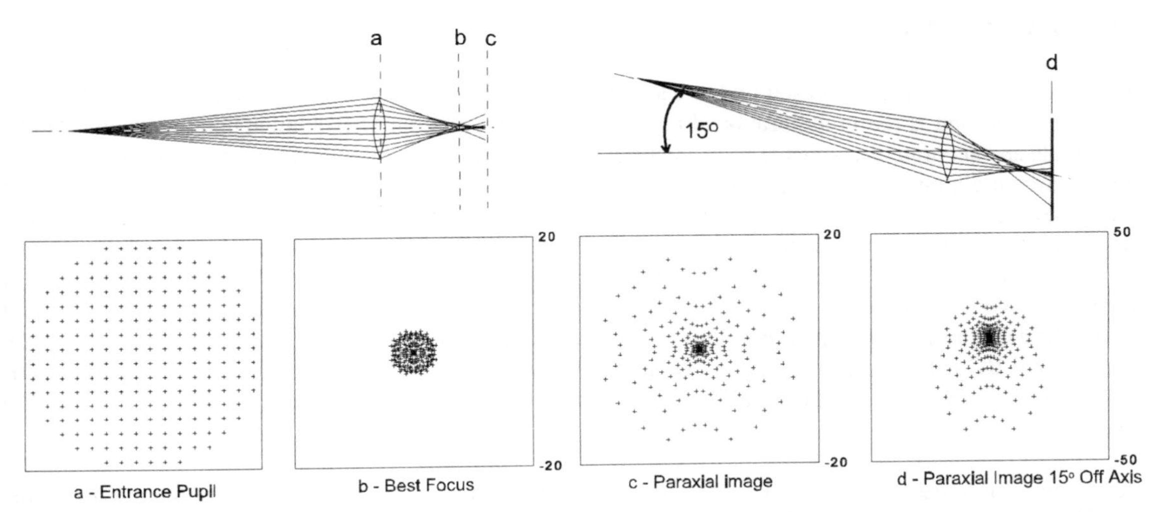

Figure 19.11 Spot diagrams of axial object rays at (a) the entrance pupil, (b) the plane of best focus, (c) the paraxial image plane. The spot diagram in the paraxial image plane (d) of an off-axis object is shown at the right. Note scale.

19.8.4 Radial Energy Diagrams

One way to evaluate spot diagrams is to plot the energy enclosed in a series of circles of increasing diameters, centered for example on the chief ray intersection point. Each ray intersection point in the spot diagram represents an equal amount of energy. The number of rays within each expanding circle is summed. See Figure 19.12a. The plot, called the *radial energy distribution*, is shown in Figure 19.12b. This distribution is most meaningful for axial images.

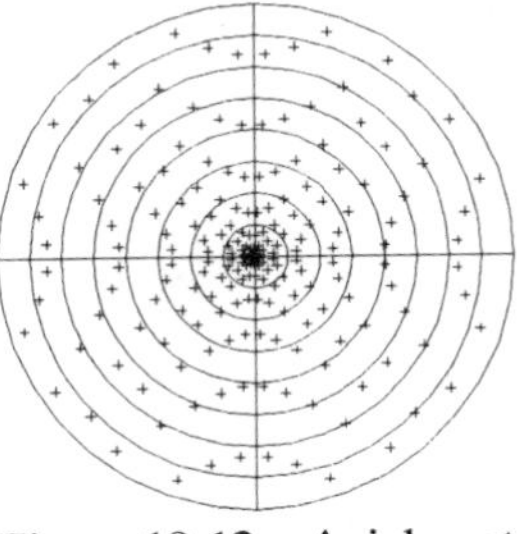

Figure 19.12a. Axial spot diagram with radial energy circles superimposed.

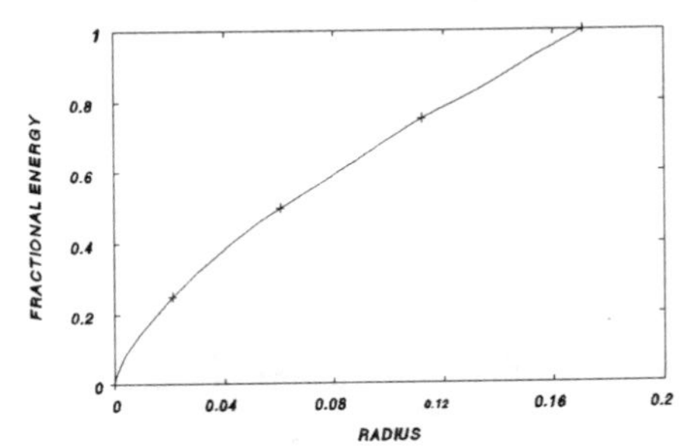

Figure 19.12b. Radial energy diagram for an axial image at the paraxial focal plane.

19.8.5 Knife Edge Distributions

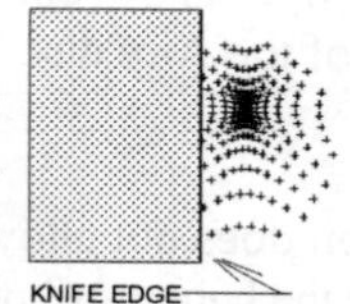

Figure 19.12c. Knife edge scan

The knife edge distribution is analogous to physically sliding a straight sharp edge across the spot diagram and summing the number of spots under the edge as a function of its position. The scan can be done across and/or up and down the spot diagram. This is particularly useful for off-axis images which are not circularly rotational.

19.9 GRAPHICAL RAY TRACE

Trigonometric ray tracing is based on Snell's law. It establishes the path of refracted and reflected rays to whatever precision is required. Graphical ray tracing also is based on Snell's law. High precision requires large scale and accurate drawings. The procedure, you will recall, involves drawing arcs in proportion to n and n', drawing a ray to the surface and translating the normal to the point of incidence to the appropriate arc. If several rays are to be traced, it is desirable to avoid cluttering the drawing with separate arcs for each ray. Instead one set of arcs is constructed at a convenient location. Triangles are used to translate normals and rays back and forth. In Figure 19.13a, Ray 1 has been drawn to point B on the refracting surface. A parallel axis with a point **O** on it is drawn below. Two arcs centered on **O**, with radii proportional to n and n', are drawn. Ray 1 is translated to point **O**. It intersects the arc corresponding to object space at point **D**. The normal **BC** is translated to the point **D** and its intersection with the image space arc **E** is marked. **OE** is the refracted ray direction. This direction is translated to the point **B** on the ray diagram, and crosses the axis at **M'**.

A second ray is traced in Figure 19.13b. This trace uses the same construction diagram as the previous one. It is drawn separately only for clarity. Additional rays will make the construction diagram more complex but the ray diagram will be free of extraneous construction lines.

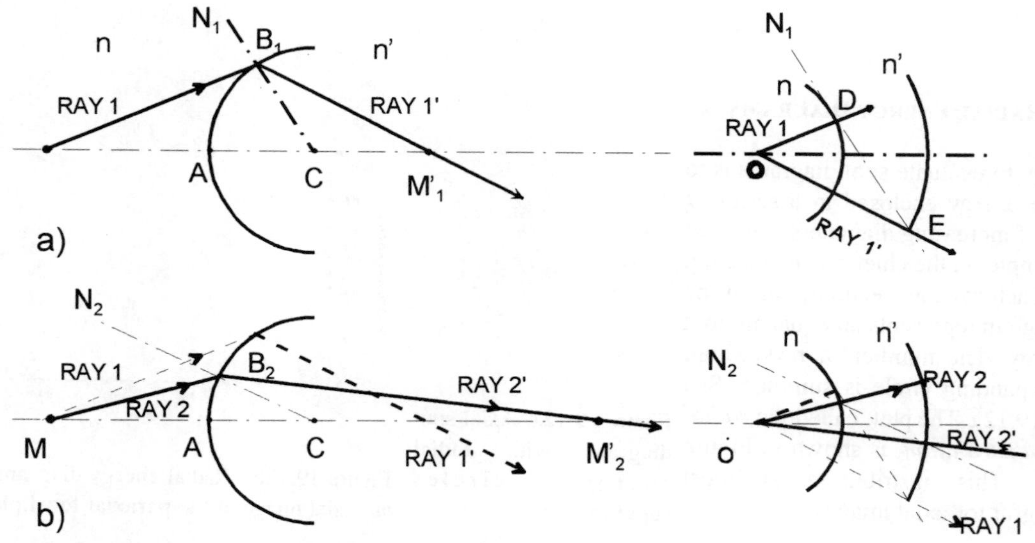

Figure 19.13 Graphical method for exact ray tracing.

PROBLEMS

1. Given the following data for a SSRS: r = +20 mm; n = 1; n' = 1.5. Trace a ray from an axial object point at infinity, with a height of 15 mm. Find (a) u', (b) L', (c) LSA, (d) TSA. Draw the ray path for this ray and the paraxial ray. Ans.: u'= +60 mm; L'= +51.368 mm; LSA = -8.632 mm; TSA = 2.904 mm.

2. Repeat Problem 1 for an object distance of -100 mm and a height of 15 mm at the *tangent plane*.
Ans.: L'= +58.503 mm; LSA = -41.497 mm; TSA = 13.445 mm.

3. Given the following data for a SSRS: r = -50 mm; n = 1; n' = 1.5. Trace a ray from an axial object point at infinity, with a height of 20 mm. Find: u', L', LSA, TSA. Draw the ray path for this ray and the paraxial ray.
Ans.: u' = -150 mm; L'= -144.487 mm; LSA = +5.512 mm; TSA = 0.786 mm.

4. A spherical mirror has a radius of curvature of -100 mm. Trace a ray from an axial object point at infinity, with a height of 25 mm. Find: u', L', LSA, TSA. Draw the ray path for this ray and the paraxial ray; (d) What is the f/No of the mirror?; (e) What is the semi-diameter s of the blur (Airy disk) due to diffraction (λ=589 nm)? Compare TSA with Airy disk blur.
Ans.: u' =-50 mm; L'= -48.360 mm; LSA = +1.640 mm; TSA = 0.907 mm; f/1 ; s = 0.000718 mm.

5. Repeat Problem 4 for a ray height of 5.18 mm. Compare TSA with Airy disk blur.
Ans.: L' = -49.933 mm; LSA = +0.0672 mm; TSA = 0.00699 mm; f/4.826 ; 2s = 0.00694 mm.

6. An axial object is located 50 mm in front of a parallel plate, 10 mm thick, made of index 1.5. A ray strikes the first surface of the plate at a height of 30 mm. Find: u_2', L_2', LSA, TSA. Draw the ray path for this ray and the paraxial ray.
Ans.: u_2'= -56.667 L_2'= -56.0858 mm; LSA = 0.581 mm; (c) TSA = 0.349 mm.

7. The following four lenses of index 1.5 and thickness d = 15 mm, have approximately the same equivalent power. Trace one ray from infinity, at a height of 10 mm through each lens. Find LSA. Note: R_1 is the curvature of the first surface.

Lens	r_1 (mm)	r_2 (mm)	R_1 (m^{-1})
1	+15	+24	+66.7
2	+25	+500	+40
3	+50	-50	+20
4	-500	-25	- 2

Draw a diagram of each lens to scale. Plot LSA (ordinate) of each lens against R_1. What do you estimate to be the curvature that will minimize LSA? Ans.: 1) -8.647; 2) -2.373; 3) -2.872; 4) -10.301 mm

8. Given the lens in Example 5, trace rays from an axial point at infinity at heights of 20,15,10 and 5 mm. Find L_2', LSA, and TSA. Plot LSA.
Ans.: h_1 = 20 mm: L_2'= +33.322 mm; LSA = -14.047 mm; TSA = 7.197 mm.
h_1 = 15 mm: L_2'= +40.420 mm; LSA = - 6.948 mm; TSA = 2.276 mm.
h_1 = 10 mm: L_2'= +44.497 mm; LSA = - 2.872 mm; TSA = 0.577 mm.
h_1 = 5 mm: L_2'= +46.678 mm; LSA = - 0.691 mm; TSA = 0.067 mm.

NOTES:

CHAPTER 20

MONOCHROMATIC ABERRATIONS

20.1 INTRODUCTION

In paraxial optics pencils of rays from each point in the object plane converge on a conjugate point in the paraxial image plane. The intersections of the chief ray of each pencil at the paraxial image plane determine the positions of the ideal image points. In real systems, all the rays from an object point do not converge on the ideal image point, the chief ray does not intersect the ideal image point, nor is the image formed on a plane surface. These departures from paraxial perfection are transverse aberrations. The aberrated image exhibits defects in sharpness and shape. Aberrations increase dramatically with increases in entrance pupil diameter and field angle. The small pupil diameters required to limit the aberrations of simple optical systems also limits these systems to low numerical apertures or large f/numbers. These systems exhibit large diffraction blurs and dim images. Aberrations must be corrected to obtain bright, sharp and flat images across the full field.

Although aberrations can be precisely calculated by trigonometric ray tracing, this is a very tedious procedure without a high speed computer, especially, if the system contains many lens elements. Seidel systematically analyzed the defects in images. He defined five monochromatic aberrations: spherical aberration, coma, astigmatism, curvature of field and distortion, and he showed how the refractive index, aperture diameter and field angle affect these aberrations.

20.2 THIRD ORDER THEORY

The sine of an angle can be expressed as the following infinite series, where α is in radians.

$$\sin\alpha = \alpha - \frac{\alpha^3}{3!} + \frac{\alpha^5}{5!} - \frac{\alpha^7}{7!} + \ldots \qquad 20.1$$

The paraxial equations replace the sines of angles in Snell's law with the angles themselves in radians. That is, the series is truncated after the first term, and $n \sin\alpha = n' \sin\alpha'$, becomes $n\alpha = n'\alpha'$. Paraxial optics is called *first order theory* because paraxial equations use only the first term of the expansion. The first *two* terms are used in *third order theory*. This better approximation to Snell's law holds for larger apertures and field angles than the first order theory. The more orders used the better the approximation to Snell's law and the greater the accuracy in computations for fast, wide angle systems. Very complex systems of aberration equations are based on fifth and seventh order approximations. The sines and approximations for angles to 60° are tabulated below.

Table I. Comparison of the sine of an angle with its approximations to 9th order.

α	sin α	1st Order	3st Order	5th Order	7th Order	9th Order
1	0.0174524	0.0174533	0.0174524	0.0174524	0.0174524	0.0174524
5	0.0871557	0.0872665	0.0871557	0.0871557	0.0871557	0.0871557
10	0.1736482	0.1745329	0.1736468	0.1736482	0.1736482	0.1736482
30	0.5000000	0.5235988	0.4996742	0.5000021	0.4999999	0.5000000
60	0.8660254	1.047198	0.8558008	0.8662953	0.8660213	0.8660254

This shaded cells show how many terms must be included to compute the sin of an angle to seven significant figures. For example, we must include all terms to the ninth order to compute the sin 60°

20.3 THIRD ORDER ABERRATION POLYNOMIAL

The aberration polynomials include third, fifth, seventh and higher order terms to describe the five monochromatic aberrations. For optical systems with small numerical apertures and small fields of view, only the *third order* terms are needed to obtain a good estimate of the aberrations. As the numerical aperture and field of view increase, higher order terms are needed to obtain accurate measures of the ray aberrations. The aberration polynomials containing only the third order aberration terms are

$$\delta_y = k_1 h^3 \cos\psi + k_2 h^2 y(2+\cos 2\psi) + (3k_3+k_4)hy^2\cos\psi + k_5 y^3 \quad 20.2a$$

$$\delta_z = k_1 h^3 \sin\psi + k_2 h^2 y \sin 2\psi + (k_3+k_4)hy^2\sin\psi \quad 20.2b$$

δ_y and δ_z *denote the transverse aberrations of any ray with respect to the position of the ideal image point*, where the *y* and *z* subscripts correspond to the Cartesian coordinates in the image surface. Each term corresponds to a Seidel aberration. Spherical aberration is the term containing the k_1 constants; the k_2 term is coma; astigmatism and curvature of field are given by the k_3 and k_4 terms; and distortion by the k_5 term. The values of the **k** coefficients are found by paraxially tracing an axial and chief ray through the optical system being studied. It is beyond the scope of this chapter to compute these coefficients.

It is assumed that the optical system is rotationally symmetrical, i.e., it contains spherical surfaces whose centers of curvature all lie on the optical (x) axis, and bilaterally symmetrical about the meridional plane.

Given the symmetry of the optical system, it suffices to specify any particular ray by its height in the meridional plane of the object, and its coordinates in the entrance pupil. However, rays are not confined to the meridional plane. Such rays are called *skew* rays. The ray height in the object plane is $\mathbf{y_o}$. The Cartesian coordinates of a ray in the entrance pupil are $(\mathbf{z_1}, \mathbf{y_1})$. In polar coordinates, as shown in Figure 20.1, they are $(\mathbf{h}, \boldsymbol{\psi})$, thus,

$$\mathbf{z_1} = \text{h} \sin\psi \quad 20.3a$$

and

$$\mathbf{y_1} = \text{h} \cos\psi \quad 20.3b$$

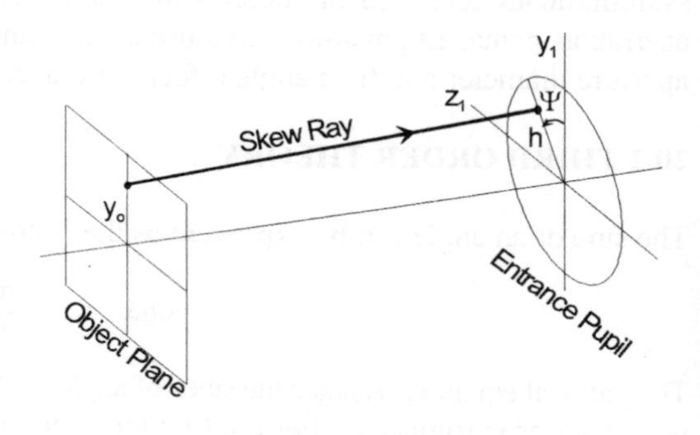

Figure 20.1 Polar coordinates of a ray in the entrance pupil of an optical system. Adapted from Sinclair Optics OSLO manual.

20.3.1 THE SPHERICAL ABERRATION TERMS OF THE POLYNOMIAL

Third order spherical aberration is given by the first terms in Eqs. 20.2a and b. Consider a meridional pencil of rays from an axial object point. Because the rays are confined to the meridional section of the system, $\psi = 0$ and $\sin\psi = 0$. Therefore, $\delta_z = 0$ and the ray intersections for the pencil are in the yz section of the image surface. The transverse spherical aberration of these rays is given by $\delta_y = k_1 h^3 \cos\psi$, where $\cos 0° = 1$. Note that third order transverse spherical aberration (TSA) is a cubic function of the height of the ray in the entrance pupil, as indicated in Chapter 19. A plot of δ_y (TSA or ray height in image plane) against the height (h) of the ray in the entrance pupil is called a *ray-intercept curve*.

Table II compares the TSA calculated from $\delta_y = k_1 h^3 \cos\psi$ with the trigonometrically ray-traced TSA of the lens of Table I in Chapter 19. Coefficient $k_1 = -0.00016$ for this lens. These data form the ray-intercept curves of Figure 20.2. At a ray height in the pupil >20 mm the trigonometrically traced TSA increases more rapidly than the 3rd order TSA. The ray-traced aberrations are worse because the lens is very fast (effective f/No ≈ f/1.6) and exhibits significant amounts of higher order (5th and 7th) aberrations.

Table II. Comparison of the Third Order and the Trigonometrically Calculated Transverse Spherical Aberration of an Equiconvex Lens (f=102.564 mm).

Ray height in Entrance Pupil	TSA: 3rd Order. (k_1 =-0.00016)	TSA (mm). Ray Trace. Table I
0	0	0
10	-0.160	-0.162
20	-1.280	-1.399
30	-4.320	-5.479
40	-10.240	-17.103

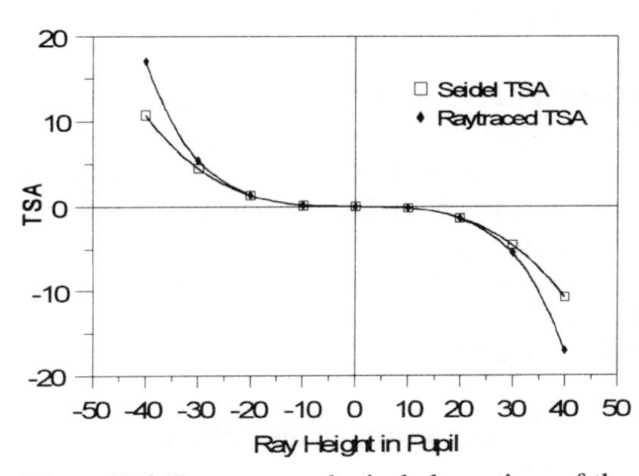

Figure 20.2 Transverse spherical aberration of the lens in Table II. Ray trace (+); 3rd order (□).

20.3.2 Spherical Aberration of a Thin Lens

We can use an alternative equation for spherical aberration that avoids the need to find k_1, but is applicable to thin lenses only. The equation calculates the longitudinal spherical aberration (LSA). **LSA** is the longitudinal difference between the intersection with the optical axis (distance L') of a ray incident at some height **h** in the pupil and the paraxial image distance **u'**. That is, **LSA = L' - u'.** Spherical aberration is *undercorrected* when **L'** is less than **u'.** LSA = TSA/tan θ'. *LSA increases: 1) as the square of the height of the ray in the pupil, and depends on; 2) the bending or shape factor of the lens and 3) the position of the object,* **L`**. For a thin lens, it is given by

$$L' = \frac{u'}{Gu' + 1} \quad 20.4a$$

$$where, \quad G = \frac{1}{L'} - \frac{1}{u'} = \frac{h^2}{f^3}(As^2 + Bsp + Cp^2 + D) \quad 20.4b$$

$$\therefore \; LSA = L' - u' = \frac{-Gu'^2}{Gu' + 1} \quad 20.4c$$

The $\mathbf{h^2}$ term in Eq. 20.4b shows the dependance of LSA on the square of the ray height. The dependence of LSA on lens bending is given by *shape factor* **s,** where

$$s = \frac{r_2 + r_1}{r_2 - r_1} \quad 20.5$$

The position of the object is denoted by *position factor* **p**, where $p = \frac{u' + u}{u' - u}$ 20.6

The other constants are functions of the lens index **n.**

$$A = \frac{n+2}{8n(n-1)^2} \;;\quad B = \frac{n+1}{2n(n-1)} \;;\quad C = \frac{3n+2}{8n} \;;\quad D = \frac{n^2}{8(n-1)^2} \qquad 20.7$$

EXAMPLE 1: Given a thin equiconvex lens of 100 mm focal length where, $r_1 = +100 = -r_2$, n = 1.5, and an object 400 mm in front of the lens. Find **L'** and **LSA** for a ray from the axial object that is incident on the lens at a height of 40 mm. Eqs. 20.5-20.7 given:
s = 0; **p** = -0.5, **A** = 1.1667, **B** = 1.6667, **C** = 0.5417, and **D** = 1.125.
Solve Eq. 20.4b

$$G = \frac{40^2}{100^3}[1.1667(0) + 1.6667(0)(-0.5) + 0.5417(-0.5^2) + 1.125] = 0.0020$$

Solve the thin lens equation to obtain u' = 133.333 mm.
Solve Eq. 20.4a

$$L' = \frac{133.333}{133.333(0.0020) + 1} = 105.263$$

Table III Comparison of LSA for 3rd order and ray traced lenses of 100 mm focal lengths.

Ray height h (mm)in Entrance Pupil	Third Order LSA: From Eq. 20.4	LSA (mm). From Ray Trace. Chap. 19: Table I
0	0	0
1	-0.0224	-0.0216
5	-0.558	-0.542
10	-2.204	-2.169
20	-8.398	-8.912
40	-28.070	-41.453

Therefore, **LSA = L'-u' = 105.263-133.333 = -28.070 mm.** (Note: Eq. 20.4c yields LSA directly). Compare -28.255 mm with the trigonometrically traced LSA of -41.453 mm of the thick 102.564 mm focal length equiconvex lens in Chapter 19, Table I.

20.3.3 Spherical Aberration and Lens Bending

Figure 20.3 shows the LSA of a 50-mm focal length thin lens as a function of bending based on the data in Table IV.

Evidently LSA is minimized for a shape factor between +0.5 and +1. To find this *best-form* lens, differentiate Eq. 20.4b with respect to **s.** Note: $\mathbf{h^2/f^3}$, Cp^2 and D are constants. The shape factor of the best-form lens is

$$s = -\frac{Bp}{2A} \qquad 20.8$$

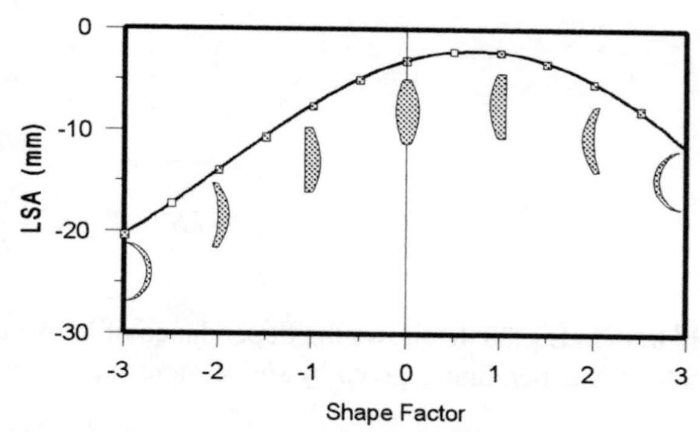

Figure 20.3 The longitudinal spherical aberration as a function of lens shape based on Table IV. Note: simple thin lenses of index 1.5 cannot be corrected for LSA; they may be bent to minimize it.

Table IV. Longitudinal Spherical Aberration of a Thin Lens vs. Shape Factor.

f = 50 mm; p = -1; h = 10 mm; and n = 1.5.

SHAPE	SHAPE FACTOR	RADII: r_1/r_2 (mm)	LSA (mm)
Concave/convex meniscus	-3.0	-25/-12.50	-20.36
Concave/convex meniscus	-2.5	-33.33/-14.29	-17.21
Concave/convex meniscus	-2.0	-50.00/-16.67	-13.94
Concave/convex meniscus	-1.50	-100.00/-20.00	-10.68
Plano/convex	-1.00	plano/-25.00	-7.63
Biconvex	-0.50	+100.00/-33.33	-5.02
Equiconvex	0	50.00/-50.00	-3.13
Biconvex	+0.50	33.33/-100.00	-2.15,
Convex/plano	+1.00	+25.00/plano	-2.23
Convex/concave meniscus	+1.50	+20.00/+100.00	-3.34
Convex/concave meniscus	+2.00	+16.67/+50.00	-5.36
Convex/concave meniscus	+2.50	+14.29/+33.33	-8.04
Convex/concave meniscus	+3.00	+12.50/+25.00	-11.14

EXAMPLE 2: Find the shape factor for minimum LSA of the 50-mm focal length lens, where, B = 1.667; p = -1; and A = 1.1667.

$$s = -\frac{(1.667)(-1)}{(2)(1.1667)} = +0.7143$$

The radii corresponding to a shape factor are obtained with the lensmakers equation.

$$r_1 = \frac{2f(n-1)}{s+1}; \quad r_2 = \frac{2f(n-1)}{s-1} \qquad 20.9$$

EXAMPLE 3: Find the radii of this best form 50 mm focal length lens.

$$r_1 = \frac{2(50)(0.5)}{(+0.7143+1)} = +29.167\ mm; \quad r_2 = \frac{2(50)(0.5)}{(+0.7143-1)} = -175.0\ mm$$

The lens is biconvex and the ratio of the radii is roughly 1 to 6. The more steeply curved surface faces the collimated light.

Figure 20.4 shows that the LSA of this best form lens increases quadratically with aperture diameter. For a ray height of 10 mm the LSA is -2.055 mm, the distance L' is 47.945 mm, and the ray slope is tan θ'= 0.2086. Consequently, TSA = 0.429 mm, or the blur diameter is **0.858 mm**. See Figure 20.5. A 50-mm focal length lens with a 20-mm diameter pupil corresponds to an f/2.5 lens. The Airy disk diameter due to diffraction is $2.44\lambda f/No$ = $2.44(0.555x10^{-3})2.5$ = **0.0034 mm**. Spherical aberration by the best form lens produces a blur diameter that is 253 times larger than the Airy disc

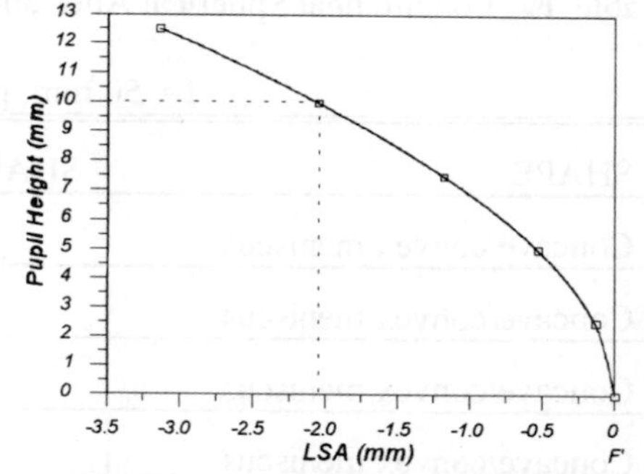

Figure 20.4 The longitudinal spherical aberration of the best-form f=500 mm, f/2.5 lens is -2.05 mm

20.3.4 Correction of Spherical Aberration

No single lens bending provides zero LSA. However, negative lenses produce an opposite LSA. Spherical aberration can be corrected by combining a strong, best form positive lens with a weaker negative lens that is bent to have an opposing spherical aberration. Two such lenses form a doublet. They may be closely spaced, or cemented together if their contacting surfaces have the same radius of curvature. Generally, the correction obtained with a doublet is to focus the marginal rays at the paraxial focus. See Figure 20.6. The bulge in the curve is called *zonal spherical aberration.* It is caused by fifth and higher order aberrations.

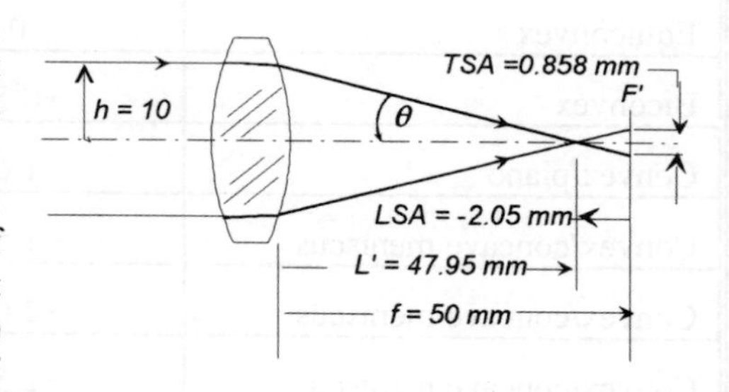

Figure 20.5 The blur due to transverse spherical aberration of the best-form 50 mm, f/2.5 lens.

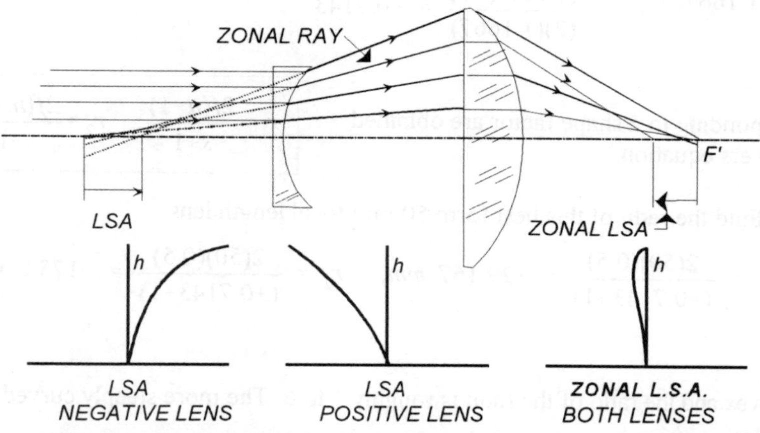

Figure 20.6 A combination of positive and negative lenses, in the form of a doublet, will correct marginal spherical aberration, but there will be a residual zonal spherical aberration.

20.4 THE COMA TERMS OF THE ABERRATION POLYNOMIAL

Coma is an unsymmetrical blur that gets larger as an object point is moved off-axis. It is a variation in lateral magnification with the lens zone. As in spherical aberration, coma increases as the square of the ray height in the pupil **h**. In addition, coma increases directly with the height of the object or the field angle, as indicated by Eq. 20.10. The lens can be thought of as having increasing power from center to margin. In pure coma, the chief ray from an off-axis point intersects the image plane at a height y'_{ideal}. Rays entering the various zones of the lens aperture, as shown in Figure 20.7a undergo a complicated transformation. Figure 20.7a is a view looking into the lens. The corresponding meridional section is shown in Figure 20.7b.

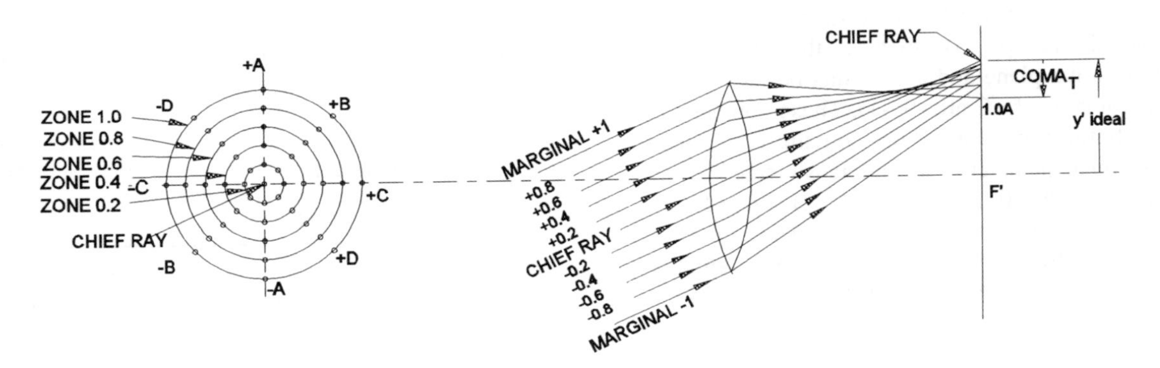

Figure 20.7 a) Distribution of rays in various zones of the entrance pupil. Zone 1 is the meridional zone. A-D locate diametrical pairs of rays in each zone. b) The medidional (AA) section. Collimated rays from an off-axis point are incident at each zone and exhibit coma.

Rays at the top and bottom of the *meridional section*, (±A) are refracted by the marginal zone of the lens. This zone has the shortest focal length, therefore, the rays meet at a lower height than y'_{ideal}. Rays that are incident at the top and bottom of successively smaller aperture zones meet in the meridional section of the image plane at shorter distances from y'_{ideal}. Marginal rays ±C, on either side of the aperture, in the *sagittal section*, will also intersect at the meridional plane as shown. Rays ±B are skew rays. All opposing pairs of skew rays such as ±B and ±D will meet at points in the image plane to form a ring shaped image. See Figure 20.8. The entire ring is displaced down from y'_{ideal}. This process is repeated for rays entering at successively smaller zones. They form a series of rings that get progressively smaller and approach y'_{ideal}. Their envelope is comet like. *Tangential coma* or $Coma_T$ is the distance from y'_{ideal} to the bottom edge of the largest ring.. *Sagittal coma* or $Coma_S$ is equal to the semi-diameter of the marginal ring image. Because coma produces an asymmetrical blur, it is a very objectional aberration.

The ring shaped ray intersections illustrated in Figure 20.8 were found with the third order comatic terms of the aberration polynomial.

$$Coma_T = \delta_y = k_2 h^2 y(2+\cos 2\psi); \quad Coma_S = \delta_z = k_2 h^2 y \sin 2\psi \qquad 20.10$$

We will assume some simplifying hypothetical values to illustrate the results of these equations: object height y=1; the entrance pupil diameter is equal to 2, therefore, h = 1 at the edge of the pupil; we will also calculate the image coordinates at zones h_z = 0.8, 0.6, 0.4, and 0.2. Finally, we assume $k_2 = -1$, then $k_2h^2y = -1$ for the edge or marginal zone. Note: the chief ray is at h= 0, therefore, the chief ray coordinates in the image plane (the off-axis image point Q') is the origin of a coordinate system (0,0) and δ_x and δ_y are measured with respect to this origin.

According to Table IV, the rays in Zone 1 at ±A coincide at (0, -3), that is, they lie in the meridional plane and are 3 units below Q'. The coordinates of ±C are (0, -1). These points lie at the bottom and top of the comatic circle. The horizontal diameter of this circle is located by the rays through ±B and ±D at (-1, -2) and (+1, -2). The center of the ring is at (0, -2). The centers of the comatic rings for the successively smaller pupil zones move toward Q' and shrink until the chief ray forms point Q'.

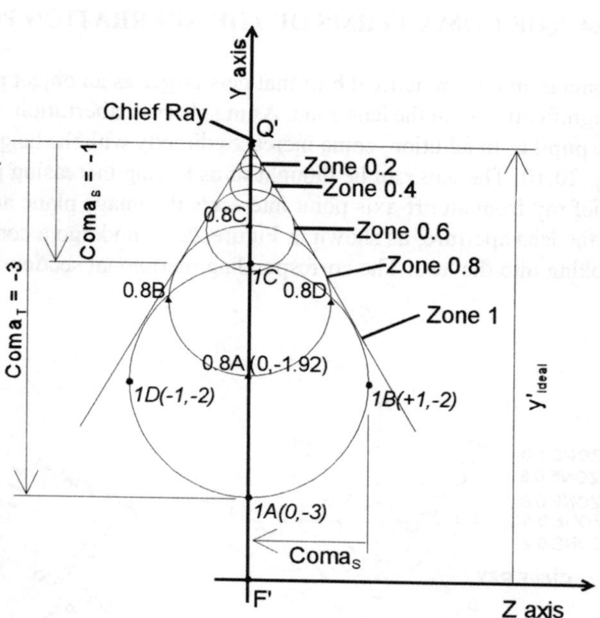

Figure 20.8 Comatic distribution of rays in the paraxial image plane.

According to Eqs. 20.10, the maximum values of $Coma_T$ and $Coma_S$ occur when cos 2 ψ = 1, or ψ = 0° and sinψ = 1, or ψ = 90°. As a result, $\mathbf{Coma_T = k_2h^2y(3)}$ **and** $\mathbf{Coma_s = k_2h^2y(1)}$. Thus, third order theory shows that

$$\mathbf{Coma_T = 3\ Coma_S} \qquad 20.11$$

This relationship is evident in Table V.

Table V. Comatic ray intersection coordinates in image plane.

		ZONE: h = 1		ZONE: h = 0.8		ZONE: h = 0.6		ZONE: h = 0.4		ZONE: h = 0.2	
ψ°	RAY	$\delta_{z'}$	$\delta_{y'}$	$\delta_{z'}$	$\delta_{y'}$	$\delta_{z'}$	$\delta_{y'}$	$\delta_{z'}$	$\delta_{y'}$	$\delta_{z'}$	$\delta_{y'}$
0	+A	0.0	-3.0	0.0	-1.92	0.0	-1.08	0.0	-0.48	0.0	-0.12
45	-D	-1.0	-2.0	-0.64	-1.28	-0.36	-0.72	-0.16	-0.32	-0.04	-0.08
90	-C	0.0	-1.0	0.0	-0.64	0.0	-0.36	0.0	-0.16	0.0	-0.04
135	-B	+1.0	-2.0	+0.64	-1.28	+0.36	-0.72	+0.16	-0.32	+0.04	-0.08
180	-A	0.0	-3.0	0.0	-1.92	0.0	-1.08	0.0	-0.48	0.0	-0.12
225	+D	-1.0	-2.0	-0.64	-1.28	-0.36	-0.72	-0.16	-0.32	-0.04	-0.08
270	+C	0.0	-1.0	0.0	-0.64	0.0	-0.36	0.0	-0.16	-0.0	-0.04
315	+B	+1.0	-2.0	+0.64	-1.28	+0.36	-0.72	+0.16	-0.32	+0.04	-0.08

20.4.1 Coma of a Thin Lens

The length of the comatic blur, $Coma_T$, of a thin lens is given by

$$Coma_T = \frac{3h^2 y'_{ideal}}{f^3}(Jp+Ks) \qquad 20.12$$

where; y' is the ideal image height and

$$J = \frac{3(2n+1)}{4n}; \; K = \frac{3(n+1)}{4n(n-1)} \qquad 20.13$$

Like spherical aberration, the coma of a lens depends on the shape **s** and position **p** factors. However, the coma of a thin lens can be made zero by bending it. The best form lens for coma is obtained by setting the (Jp+Ks) term in Eq. 20.12 to zero, and solving for the shape factor.

$$s = \frac{-Jp}{K} \qquad 20.14$$

EXAMPLE 4: Given: an object at infinity (p=-1) and a lens of index 1.5. From Eq. 20.13 we obtain:

$$J = \frac{3(3+1)}{6} = 2; \qquad K = \frac{3(2.5)}{6(.5)} = 2.5$$

Find $Coma_T$ for an equiconvex lens (s=0) of 100 mm focal length for which y'_{Ideal} =10 mm. From Eq. 20.12,

$$Coma_T = \frac{3(10)^2(10)}{100^3}[2(-1)+2.5(0)] = -0.006$$

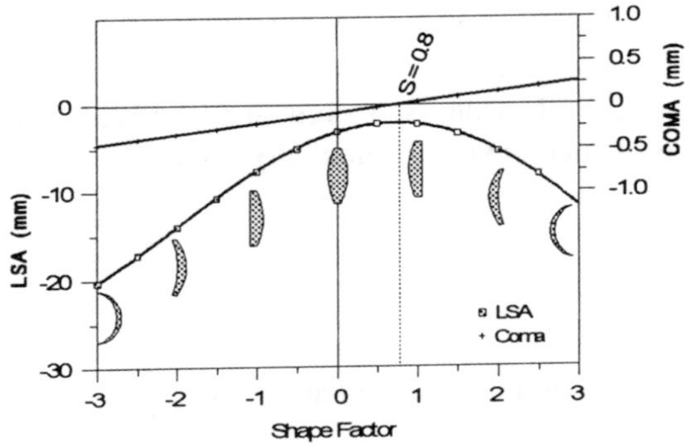

Figure 20.9 Coma varies linearly with shape factor and is zero for the 50-mm lens at s = 0.8.

Find the shape factor of a thin lens which corrects coma. Substitution in Eq. 20.14, s = -2(-1)/2.5 = 0.80, shows that coma can be made zero by bending the lens to a shape factor of 0.80.

Figure 20.9 illustrates the coma and spherical aberration for a thin lens as a function of the shape factor. Coma varies linearly with shape factor.

20.4.2 The Sine Condition

The sine condition requires that the lateral magnification of all zones of a lens be equal to theparaxial lateral magnification.

$$Y = \frac{y'}{y} = \frac{nu'}{n'u} = \frac{nL'}{n'L} \qquad 20.15$$

Systems that satisfy the sine condition are called *aplanatic*. They are free of spherical aberration and coma. A spherical refracting surface will be aplanatic for the following

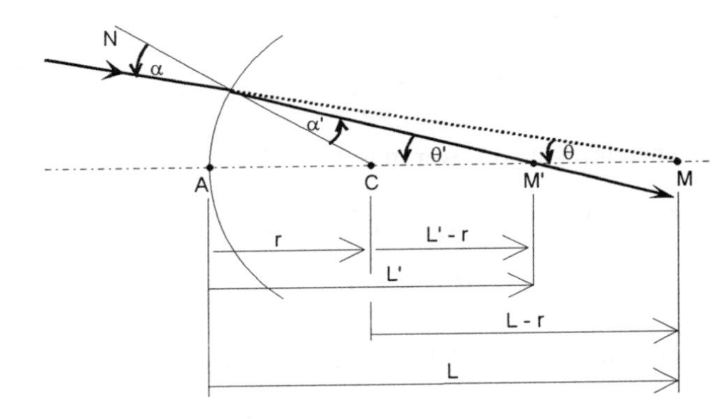

Figure 20.10 The aplanatic points of a spherical refracting surface. .

unique pair of object/image conjugates:

$$L = r\left(\frac{n'+n}{n}\right); \quad L' = r\left(\frac{n'+n}{n'}\right) \tag{20.16}$$

As shown in Figure 20.10, these conjugates will result in $\alpha = \theta'$, and $\alpha' = \theta$. Substitute these angles into Snell's law.

$$\frac{n}{n'} = \frac{\sin\theta}{\sin\theta'} \tag{20.17}$$

Substitute Eqs 20.16 and 17 into 20.15 to obtain: $Y = \frac{y'}{y} = \frac{n\sin\theta}{n'\sin\theta'}$

In the paraxial condition, $\theta = \mathbf{h/u}$, and $\theta' = \mathbf{h/u'}$. Therefore, $\mathbf{u'/u} = \theta/\theta'$. Substitute this into Eq. 20.15 for paraxial lateral magnification: $\mathbf{y'/y} = \mathbf{n\theta/n'\theta'}$. Thus,

$$Y = \frac{y'}{y} = \frac{n\theta}{n'\theta'} = \frac{n\sin\theta}{n'\sin\theta'}$$

The sine condition for small angles is

$$\frac{\theta}{\theta'} = \frac{\sin\theta}{\sin\theta'} = constant \tag{20.18}$$

This equation says that the lens will have a constant magnification for a near object for all ray slopes. For a spherical refracting surface, both object and image must lie on the same side of the surface. Thus, obtaining a real aplanatic image of a real object is not possible.

EXAMPLE 5: The first element of an oil immersion microscope objective lens is illustrated in Chapter 14. This hemispherical element and the layer of oil form an aplanatic surface for the microscope slide object.

Given: radius of curvature of the hemisphere $r = -1$ mm
index of lens and oil of object space. $n = 1.5$
index of image space. $n' = 1.0$

Step 1. Find L (object distance AM) with Eq. 20.16

$$L = r\frac{n'+n}{n} = -1\frac{1+1.5}{1.5} = -1.667 \ mm$$

Therefore, the oil layer must be 0.67 mm thick.

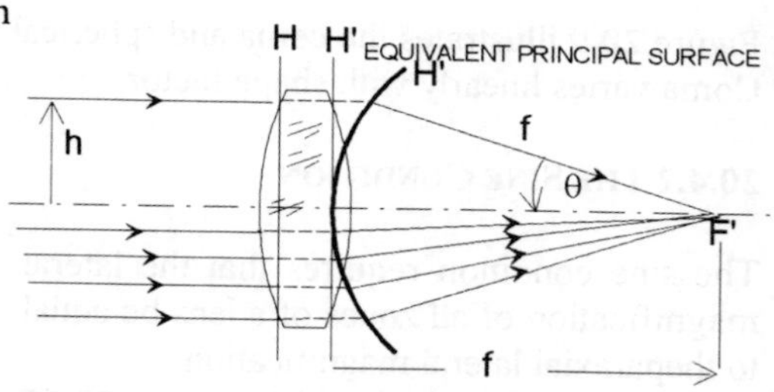

Step 2. Find L' (image distance AM')

$$L' = r\frac{n'+n}{n'} = -1\frac{1+1.5}{1} = -2.5 \ mm$$

The image is virtual, free of spherical aberration and coma, but laterally magnified.

$$Y = \frac{n}{n'}\frac{u'}{u} = \frac{(-1.5)(-2.5)}{(-1)(-1.67)} = 2.25$$

This initial stage of magnification means that the successive lens elements of the objective have less work to do to form a real image at the ocular.

Figure 20.11 An aplamatic lens has a spherical principal surface centered on the image point.

A lens will satisfy the aplanatic condition for collimated rays when **h/sinθ = kr = a constant = f.** This requires that the second principal plane be spherical. This equivalent refracting surface produces a spherical wavefront. All rays will have equal optical path lengths and will converge on F'. See Figure 20.11.

An aplanatic lens may have a maximum aperture equal to the diameter of the second principal surface, or **D = 2f**. Consequently, the limit on f/No = f/2f = **f/0.5**

A doublet may be designed to be *aplanatic*. It will be corrected for spherical aberration and coma. If it is made of a positive crown and a negative flint glass, the doublet can also correct longitudinal chromatic aberration. Such a lens is suitable for use as an objective for a telescope. Objectives need only be corrected for small field angles. When lenses are to cover large field angles, considering the remaining Seidel aberrations is necessary.

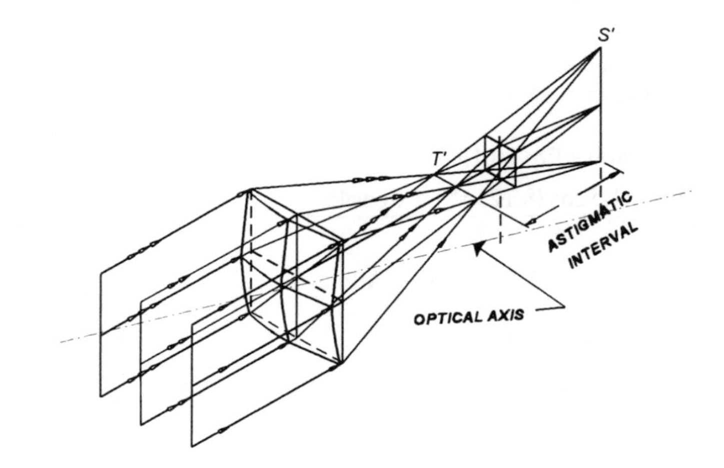

Figure 20.12 An astigmatic image formed by a spherical lens of an off-axis point.

20.5 ASTIGMATISM

20.5.1 ASTIGMATISM BY A SPHERICAL REFRACTING SURFACE

Lenses with spherical surfaces produce astigmatic images of off-axis object points. As seen in Figure 20.12, point Q sends tangential and sagittal rays to a lens. The tangential and sagittal pencils intersect the chief ray at different distances from the lens, forming focal lines T' and S', respectively. In undercorrected astigmatism, the focal lines fall in front of the paraxial image plane. The distance between the image plane and S', and the separation between T' and S' increase with the height of the object. Astigmatism at any point in the field is given by the separation between T' and S', measured along the chief ray. The tangential and sagittal astigmatic image distances along the chief rays are plotted in Figure 20.13. The Coddington equations for calculating the sagittal **s'** and tangential **t'** image distances for a spherical refracting surface follow:

$$\frac{n'}{s'} - \frac{n}{s} = \frac{n'\cos\alpha' - n\cos\alpha}{r} \quad 20.19$$

$$\frac{n'\cos^2\alpha'}{t'} - \frac{n\cos^2\alpha}{t} = \frac{n'\cos\alpha' - n\cos\alpha}{r} \quad 20.20$$

The object distances s and **t**, and the conjugate image distances **s'** and **t'** are measured along the chief ray.

EXAMPLE 6: Find the tangential and sagittal foci of a spherical refracting surface (r = 100 mm) for an infinitely distant object at 45°. The aperture stop is at the surface. Given: s = t = ∞; n = 1; n' = 1.5; r = 100 mm; α = 45°.

Step 1. Find the second focal plane. This is the reference plane with which to compare the astigmatic foci.

$$-\frac{1.5}{f'} = \frac{1.5 - 1}{100}; \quad \textit{thus} \quad AF' = +300 \ mm$$

Step 2. Find α'. From Snell's law, sin α' = sin 45°/1.5, therefore, α' = 28.1255°

Step 3. Solve Eq. 20.19.

$$\frac{1.5}{s'} = \frac{1.5\cos 28.1255° - 1.0\cos 45°}{100}$$

s' = +243.5978 mm = AS'

Step 4. Solve Eq. 20.20.

$$\frac{1.5\cos^2 28.1255°}{t'} = \frac{1.5\cos 28.1255° - 1.0\cos 45°}{100}$$

t' = +189.4650 mm = AT'

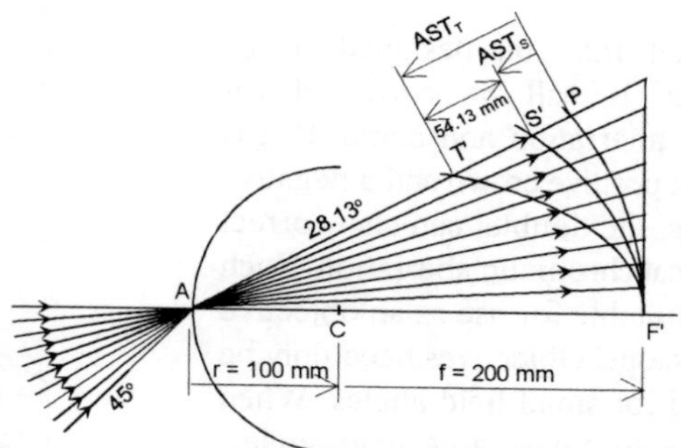

Figure 20.13 Tangential (T') and Sagittal (S') images by a refracting suface: r - 100 mm. P is the Petzval surface.

The astigmatic interval, **t' - s'** = -54.1328 mm.

Figure 20.13 illustrates the positions of the S' and T' foci for object field angles from 0 to 45°. Revolving the curves drawn through the foci about the optical axis results in curved tangential and sagittal image surfaces that are tangent to the paraxial image plane at the optical axis. See Figure 20.14a. The tangential surface is the more deeply curved of the two image surfaces of a spherical refracting surface.

A wheel, centered on the optical axis, is an object with a ring shaped rim and radial spokes. On the sagittal image surface all the points in the wheel are imaged as short lines that radiate out from the optical axis. Thus, the image of the rim is blurred but the spokes are sharp. On the tangential image surface, however, the spoke images are blurred because each point on a spoke is imaged as a short line perpendicular to the spoke. The rim will be sharply imaged because the short line images of each point will follow the contour of the rim. See Figure 20.14b.

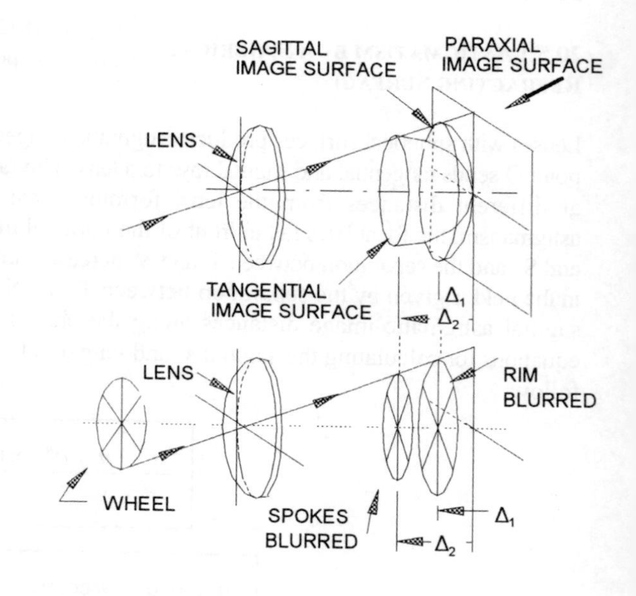

Figure 20.14 a) Tangential and Sagittal image surfaces; b) image of wheel at defocussed positions which sharpen either the rim or the spokes.

20.5.2 ASTIGMATISM BY A THIN LENS

The Lensmaker's equation was obtained by combining the equations for two spherical refracting surfaces. In an analogous way the Coddington equations for a spherical refracting surface can be combined to yield an equation for the astigmatism of a thin lens. The results of such a procedure are:

$$\frac{1}{s_2'} - \frac{1}{s_1} = \cos\alpha\left(\frac{n\cos\alpha'}{\cos\alpha} - 1\right)\left(\frac{1}{r_1} - \frac{1}{r_2}\right) \qquad 20.21$$

$$\frac{1}{t_2'} - \frac{1}{t_1} = \frac{1}{\cos\alpha}\left(\frac{n\cos\alpha'}{\cos\alpha} - 1\right)\left(\frac{1}{r_1} - \frac{1}{r_2}\right) \qquad 20.22$$

where, α is the angle of incidence and emergence, and α' is the angle of refraction in the lens.

Astigmatism is corrected when the tangential and sagittal image surfaces coincide with the Petzval surface. In uncorrected astigmatism, the distances **PS'** and **PT'** from the Petzval surface are the sagittal **(Ast_S,)** and the tangential astigmatism **(Ast_T)**. Their third order relationship is

$$\boxed{Ast_T = 3Ast_S} \qquad 20.23$$

Astigmatism can be corrected by a combination of bending the lens and choosing an appropriate position for an aperture stop. Correctly bending the lens causes the tangential and sagittal surfaces to become flatter. The tangential surface flattens three times faster than the sagittal surface. At the best bending, the surfaces coincide at the Petzval surface. Although this results in zero astigmatism, a blurred image will be produced on the flat paraxial image surface.

20.6 PETZVAL CURVATURE

20.6.1 PETZVAL CURVATURE OF A REFRACTING SURFACE

Ideally, images produced by cameras and projectors are formed on flat surfaces. Optical systems ordinarily do not form flat images. The simplest example is the image formed by a spherical refracting surface. No single optical axis exists when a small aperture stop is placed at the center of curvature of a spherical refracting surface. All chief rays can be considered to coincide with an optical axis. A distant wide-field object will be imaged on a spherical surface with its vertex at F' and its center at **C**. See Figure 20.15a. This image surface is called the Petzval surface. The radius of curvature this Petzval surface is $\mathbf{r_{ptz}} = \mathbf{F'C} = \mathbf{-f}$.

Since, $F = \frac{n}{f} = \frac{n' - n}{r} = \frac{-n}{r_{ptz}}$. Therefore, $r_{ptz} = -\frac{nr}{n' - n}$. 20.24

The Petzval curvature is $$R_{ptz} = -\frac{1}{f} = \frac{n - n'}{nr} = -\frac{F}{n}$$ 20.25

EXAMPLE 7: Find the Petzval radius of a spherical refracting surface for which r = +100 mm, n = 1, n' = 1.5.

$$r_{ptz} = -1(100)/(1.5-1) = -200 \text{ mm.}$$

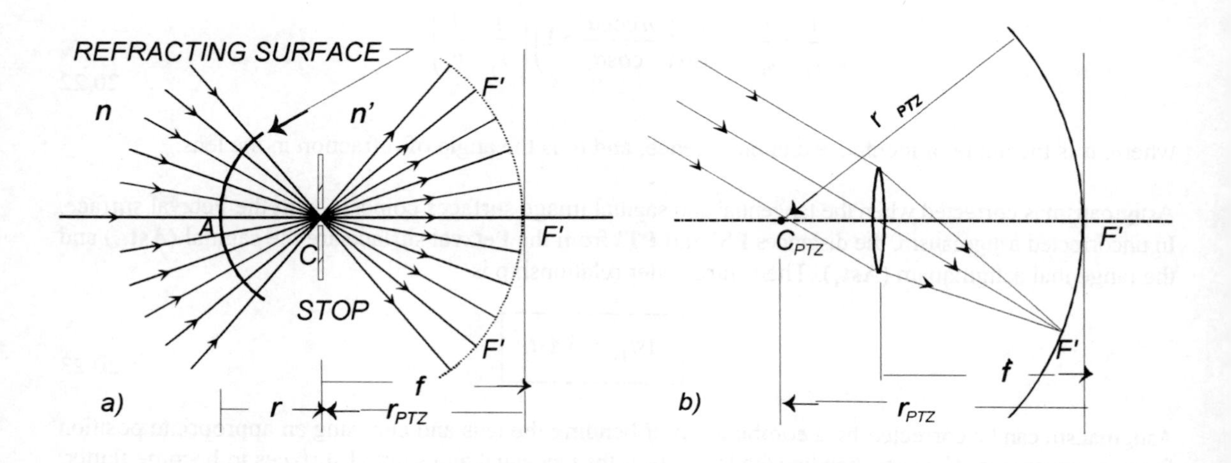

Figure 20.15 The Petzval radius of a) a spherical refracting surface; b) a thin lens.

20.6.2 Petzval Curvature of a Series of Surfaces

Petzval showed that the curvature of the image for a series of **k** surfaces is

$$R_{ptz} = n_k' \sum_{j=1}^{k} \frac{n_j - n_j'}{n_j n_j' r_j}$$ 20.26

or $$R_{ptz} = -n_k' \sum_{j=1}^{k} \frac{F_j}{n_j n_j'}$$ 20.27

where, j denotes a surface. The equations say that the total curvature of the image by a series of refracting surfaces is equal to the sum of the curvatures of the individual surfaces.

20.6.3 Petzval Curvature of a Thin Lens in Air

The thin lens has two surfaces, **k** =2. Thus, at the first surface, where j = 1, $n_j = 1$ and $n_j' = n$ (the lens index). At the second surface j = k, thus, $n_k = n$, and $n_k' = 1$. Substitution into Eq. 20.27 gives

$$R_{ptz} = -1[\frac{F_1}{n} + \frac{F_2}{n}] = -\frac{F}{n} \qquad 20.28$$

and the radius of the Petzval surface is

$$\boxed{r_{ptz} = -\frac{n}{F} = -nf} \qquad 20.29$$

EXAMPLE 8: Find the Petzval radius of a thin lens in air, where, f = 200 mm, n = 1.5.

$\mathbf{r_{ptz}}$ = -1.5(200) = -300 mm. See Figure 20.15b.

20.6.4 The Petzval Condition for Flattening the Field

Negative lenses produce field curvatures opposite that of positive lenses. Flattening the image by combining lenses that satisfy the Petzval condition is possible:

$$\boxed{\frac{1}{n_1 f_1} + \frac{1}{n_2 f_2} = 0} \qquad 20.30$$

The interesting point of this condition is that it does not depend on the separation of the lenses. In other words the field curvature of a positive lens can be corrected by placing a negative lens in or very near the image plane. Such lenses are called *field flatteners*. By being in the image plane they do not affect the power of the positive lens, and minimally affect the other aberrations. *The Petzval surface is an image surface only if astigmatism is corrected.*

20.7 ASTIGMATISM AND PETZVAL CURVATURE POLYNOMIAL TERMS

The ray intercept curves in the presence of third order astigmatism are given by the following terms of the Seidel polynomial, Eq. 20.2

$$\delta_y = (3k_3 + k_4)hy^2\cos\psi \; ; \qquad \delta_z = (k_3 + k_4)hy^2\sin\psi$$

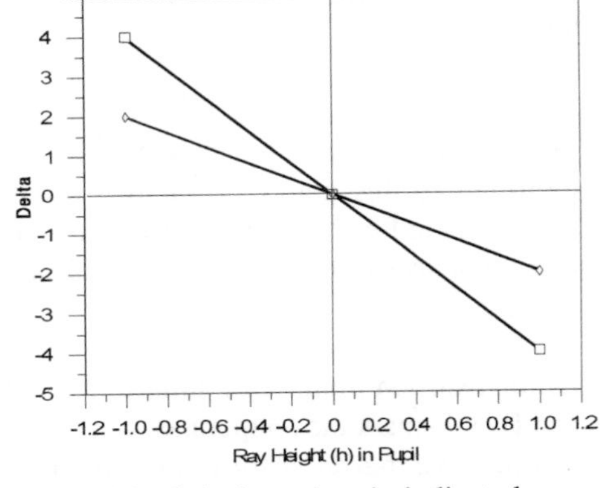

Figure 20.16 Astigmatism is indicated when the slopes of the tangential □ and sagittal ray ◊ intercept curves differ.

Typical ray intercept curves of the astigmatism of an off-axis object point are shown in Figure 20.16. This figure is based on hypothetical values for k_3, k_4, and y each equal to 1, and ray heights in the pupil from +1 to -1. Only rays in the meridional and sagittal sections are computed, that is, $\psi = 0°$ and 90°. The slope of the tangential ray intercept curve is twice that of the sagittal. Astigmatic line foci are indicated when the tangential and sagittal ray intercept curves are straight lines with different slopes. When the slopes are the same and the curves coincide, third order astigmatism is corrected and the image is a point on the Petzval surface. If the slopes are zero, the Petzval surface coincides with the paraxial image plane.

Note that astigmatism and Petzval curvature vary directly with a pupil diameter and with the square of the object height or field angle.

20.8 DISTORTION

If, somehow, all the other monochromatic aberrations of an optical system are corrected, and a sharp flat image is formed on a plane an image defect may remain. *Distortion is a variation in lateral magnification with object height.* It results in an image that is not the same shape as the object. A common example of a highly corrected optical system that nevertheless exhibits distortion is the wide angle camera lens. The photographs it takes are sharp but they are not a scaled down representation of the object.

Distortion results in image points not being imaged at their ideal positions. *Barrel distortion* exists if the image heights are less than the ideal. In *pincushion distortion* the image points lie further from the axis than ideal. See Figure 20.17. A positive lens with a front stop will exhibit barrel distortion. Move the stop to the rear and pincushion distortion is produced. The severity of distortion increases with the field angle.

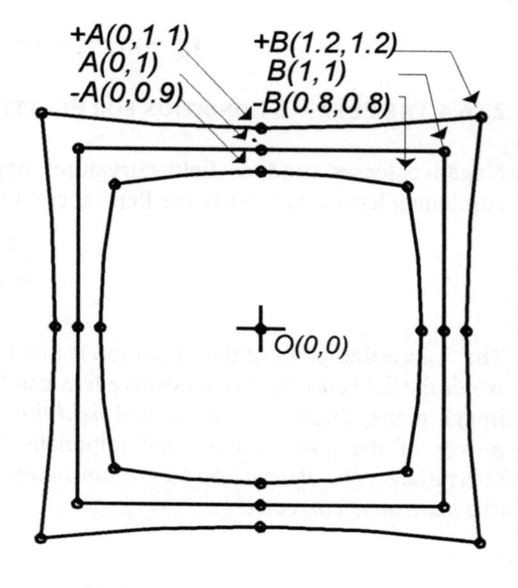

Figure 20.17 Pincushion and barrel distortion.

20.8.1 THE THIRD ORDER DISTORTION POLYNOMIAL TERM

Because we have assumed that the object lies in the meridional plane at a height **y**, the distortion will also be confined to this plane, i.e., it will produce a δ_y error. According to Eq. 20.2a, for pure distortion, the image will be a point displaced from the ideal image point as a cubic function of the height of the object or field angle, namely,

$$\delta_y = \mathbf{k_5 y^3}.$$

An undistorted image of a square with sides equal to two units is shown in Figure 20.17. Points A (0,1) and B (1,1) mark the midpoint of a side and a corner of the square. Pincushion distortion will result when k is positive; barrel distortion when k is negative. Assuming k = ±0.1, δ will equal ±0.1 for the point at the side. Pincushion distortion will cause the coordinates at +A to become (0,1.1) and at -A the coordinates are (0,0.9). The corner of the square, point B, is 1.414 units from the center. δ will equal ±0.2828. Pincushion distortion will place +B at 1.697 units from the center; -B will be at 1.1313 units from the center. Their Cartesian coordinates will be +B (1.2,1.2) and -B (0.8,0.8). Because the corners are 1.4 times as far from the centers as the sides of the squares, the distortion (δ) is 1.4^3 or 2.83 times as much as at the sides.

20.8.2 THE TANGENT CONDITION

Distortion will be corrected if the optical system satisfies the tangent condition. The tangent condition requires that the magnification by each zone of the lens be equal for all object heights, namely,

$$\frac{y_1'}{y_1} = \frac{y_2'}{y_2} \qquad 20.31$$

According to Figure 20.18.

$y_1' = E'M' \tan\theta_1'$

$y_1 = ME \tan\theta_1$

$y_2' = E'M' \tan\theta_2'$

$y_2 = ME \tan\theta_2$

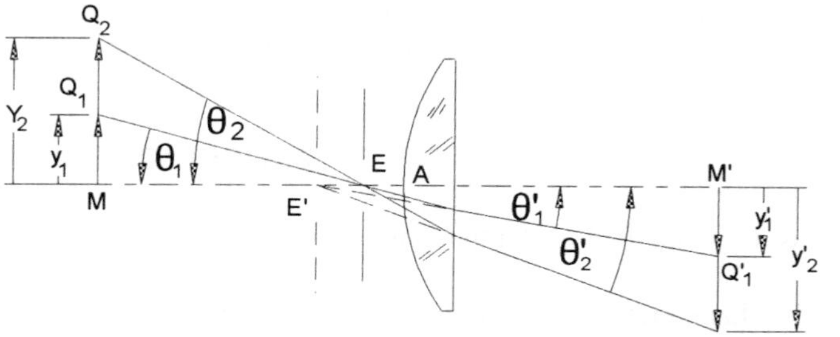

Figure 20.18 The tangent condition for the correction of distortion.

Substitution of these equations into Eq.20.31 results in $\dfrac{\tan\theta_1'}{\tan\theta_1} = \dfrac{\tan\theta_2'}{\tan\theta_2}$.

or the tangent condition says that the ratio of the tangents is a constant, i.e.,

$$\boxed{\frac{\tan\theta'}{\tan\theta} = constant} \qquad 20.32$$

A microscope objective lens that satisfies the tangent condition is *orthoscopic*. *Rectilinear* camera lenses are highly corrected for distortion.

A thin lens, with an aperture stop in contact with it, is free of distortion, but will have other aberrations. The pupil of the eye is the stop for a spectacle lens. It is a rear stop, consequently, the lenses for hyperopes produce pincushion distortion; for myopes it is barrel distortion.

Table V. Summary of Third Order Aberrations.

ABERRATION	APERTURE DEPENDENCE	FIELD DEPENDENCE
LSA	h^2	none
TSA	h^3	none
Coma	h^2	y
Astigmatism	h	y^2
Petzval Curvature	h	y^2
Distortion	none	y^3

QUESTIONS

1. What factors of lenses and objects affect spherical aberration?
2. What is zonal spherical aberration?
3. What is coma?
4. How are 3rd order sagittal and tangential coma related?
5. What does the sine condition require?
6. What aberrations are corrected in a good objective lens for telescopes?
7. What happens to the tangential and sagittal image surfaces when 3rd order astigmatism is corrected?
8. Why isn't a sharp image on the Petzval surface a good correction?
9. What kind of distortion is produced by a myope's spectacle lens?

PROBLEMS

1. What is the shape factor of a thin lens with r_1 = +75 mm and r_2 = +25 mm? Ans.: -2

2. Find the shape factor of a thin -20 D lens of n = 1.5, with r_1 = -29.1667mm. Ans.: s = +0.7143

3. What is the position factor for an object 200 mm in front of an equiconvex +10 D lens? Ans.: p = 0.

4. Find the constants A, B, C and D for an index of 2. Ans.: A = 0.25; B = 0.75; C = 0.5; D = 0.5.

5. Using the data and findings of Problems 3 and 4, find L', LSA and TSA for a ray incident at a height of 10 mm. Ans.: L' = 198.0198 mm ; LSA = -1.98019 mm; TSA = 0.1 mm.

6. What is the best form of the + 10 D lens of index 2 for an object at -200 mm? What are the radii of this best form lens? Ans.: s = 0 ; r_1 = +200 mm; r_2 = -200 mm.

7. A ray at a height of 20 mm on a lens undergoes -0.5 mm of 3rd order LSA. What is the LSA of a ray at a height of 10 mm? Ans.: -0.125 mm.

8. A cx-pl +10 D lens (n = 1.5) forms an image of a distant object at y'_{ideal} = 10 mm. Find tangential coma if the aperture diameter of the lens is 20 mm. Ans.: 0.0015 mm.

9. Recompute the coma for the above lens with an aperture diameter of 40 mm and an image height of 20 mm. Ans.: 0.012 mm.

10. Find the tangential and sagittal foci of a spherical refracting surface (r = +100 mm, n = 1, n'= 1.5) for an infinitely distant object at a field angle of 30°. Ans.: s' = 273.629 mm; t' = 243.225 mm.

11. What is the Petzval radius of the above spherical refracting surface? Ans.: -200 mm

12. Find the Petzval radius and curvature of +10 D lens made of 1.5 index glass.

Ans.: r_{ptz} = -150 mm; R_{ptz} = -6.67 m^{-1}

13. Find the focal length of a lens of index 2, that will flatten the field curvature of the Prob. 12 lens.

Ans.: -75 mm.

14. The image of a square object suffers pincushion distortion. By what factor is the distortion of the corners of the square greater than the sides?

Ans.: 2.83x

NOTES:

BIBLIOGRAPHY

Primary References

Conrady, A.E. *Applied Optics and Optical Design. vol I*, New York:Dover, 1957.

Conrady, A.E. *Applied Optics and Optical Design. vol II*, New York:Dover, 1960.

Goldwasser, E.L. *Optics, Waves, Atoms and Nuclei*, New York:W. A. Benjamin, 1965. P. 28-36

Jenkins, Francis Arthur and Harvey E. White. *Fundamentals of optics*. 4th ed. New York: McGraw-Hill, 1976.

Kingslake, R. *The Development of the Zoom Lens. J.Soc.Motion Pict.Telev.Eng.* 69:534-544, 1960.

Kingslake, R. *Lens Design Fundamentals*, New York:Academic, 1978.

Opticalman: Navy Training Course: NAVPERS 10205, Washington, D. C.: Bureau of Naval Personnel, 1966.

Sinclair, D.C. *Contemporary Optical Engineering. Vol. I: Image Forming Optics*, Rochester:Univ. of Rochester, 1974.

Smith, Warren J. *Modern optical engineering*: the design of optical systems. New York: McGraw-Hill, 1966.

Smith, W.J. Optics. In: *Handbook of Military Infrared Technology*, edited by Wolfe, W.L. Washington, D.C.: Office of Naval Research. Dept of Navy, 1965,

Southall, J.P.C. *Mirrors, Prisms and Lenses*, New York:Macmillan, 1933. Ed. 3rd

General References

Blaker, J. Warren. *Geometric optics;* the matrix theory. New York, M. Dekker, 1971.

Brewster, David. *The Kaleidoscope:* its history, theory and construction with its application to the fine and useful arts. 2d ed., London, J. Murray, 1858.

Brown, E.B. *Modern Optics*, Huntington:Robert E. Krieger, 1974.

Cox, Arthur. *A system of optical design*; the basics of image assessment and of design techniques with a survey of current lens types. London, New York, Focal Press [1964]

Curry, Colin. *Geometrical optics*. London : E. Arnold, 1953.

Ditchburn, R. W. *Light.* 3d ed. London ; New York : Academic Press, 1976.

Falk, David S. , Dieter R. Brill and David G. Stork. *Seeing the light* : optics in nature, photography, color, vision, and holography. New York : Harper & Row, c1986.

Freeman, M. H. *Optics.* 10th ed. London ; Boston : Butterworths, 1990.

Hardy A.and F.Perrin, *The Principles of Optics*, New York:McGraw Hill, 1932.

Heavens, O. S. and R.W. Ditchburn. *Insight into optics*. Chichester; New York: Wiley, 1991.

Hecht, Eugene. *Schaum"s outline of theory and problems of optics.* New York: McGraw-Hill, 1975

Hecht, Eugene and Alfred Zajac. *Optics*. 2nd ed. Reading, Mass. : Addison-Wesley Pub. Co., c1987.

Hopkins, H. *Wave Theory of Aberrations*, London:Oxford, 1950.

Hopkins, R.E. *Fundamentals of Geometric Optics*. In: *MIL-HDBK-141: Handbook of Optical Design*, Washington, D. C.: U. S Government Printing Office, 1962,

Iizuka, Keigo, *Engineering optics*. Berlin ; New York : Springer-Verlag, 1984.

Jalie, M. *The Principles of Ophhalmic Lenses*, London:The Assoc. of Dispensing Opticians, 1977. Ed. 3rd pp. 374-402.

Kingslake, R. *Optical System Design*, Orlando:Academic Press, 1983.

Klein, Miles V. and Thomas E. Furtak. *Optics*. 2nd ed. New York : Wiley, c1986.

Martin, Louis Claude. *An introduction to applied optics*. London, Sir I. Pitman & sons, ltd., 1930

Martin, Louis Claude. *Technical optics*. 2d ed.; edited by L. C. Martin and W. T. Welford. London, Pitman, 1966.

Meyer-Arendt, Jurgen R. *Introduction to classical and modern optics*. 3rd ed. Englewood Cliffs, N.J. Prentice-Hall, 1989.

Michelson, Albert Abraham. *Light waves and their uses*. Chicago, The University of Chicago Press ; Ann Arbor, Mich. 1903.

Miller, David. *Optics and refraction* : a user-friendly guide. New York: Gower Medical Pub. ; Philadelphia, PA, 1991.

Nussbaum, Allen. *Geometric optics*. Reading, Mass., Addison-Wesley Pub. Co. [1968]

Nussbaum, Allen and Richard A. Phillips. *Contemporary optics for scientists and engineers* Englewood Cliffs, N.J. : Prentice-Hall, c1976.

Ogle, Kenneth Neil. Optics, an introduction for ophthalmologists. 2d ed. Springfield, Ill., Thomas [1968]

Perrin, F. *Methods of Appraising Photographic Systems. SMPTE* 69:239-248, 1960.

Southall, James P. C. *The principles and methods of geometrical optics.* 3d ed. New York, Macmillan 1933.

Strong, John, *Procedures in applied optics.* New York : M. Dekker, c1989.

Welford, W. T. *Optics.* 3rd ed. Oxford ; New York : Oxford University Press, 1988.

Wolfe, W.L. *Handbook of Military Infrared Technology*, Washington, D. C.:Office of Naval Resarch. Dept of Navy, 1965.

Wood, Robert W. *Physical optics.* 3d ed. New York : Macmillan, 1934.

Young, Matt. *Optics and lasers* : including fibers and integrated optics. 3d. ed. Berlin; New York : Springer-Verlag, c1986.

Zimmer, Hans-Georg. *Geometrical optics.* [Translated from the German by R. N. Wilson] Berlin, New York, Springer-Verlag, 1970.

Figure References

Figures 1.11, 2.24, 3.6 and 3.9 were adapted from Southall, J.P.C. *Mirrors, Prisms and Lenses.*

Figures 3.3-5 were adapted from Goldwasser, E.L. *Optics, Waves, Atoms and Nuclei.*

Figures 10.2 and 20.12 were adapted from Katz, Milton, *Optics and Refraction* Ed. by A. Safir.

Figure 13.10 was adapted from Smith, Warren J. *Modern optical engineering.*

Figure 18.15-22 were adapted from MIL-HDBK-141 *Militatry standardization handbook: Optical Design.*

Figure 20.1 was adapted from Sinclair, D.C. *Contemporary Optical Engineering. Vol. I: Image Forming Optics.*

Figures 20.2, 20.3, 20.7 and 20.8 were based on Jenkins, Francis Arthur and Harvey E. White. *Fundamentals of optics.*

INDEX

Abbe illumination, 232
Abbe value, 253
Aberration
astigmatism, 239, 295–297, 299
chromatic, 255–263
coma, 291–293
curvature, 297–299
distortion, 300, 301
spherical, 184, 220, 277, 287–290
of thin lens, 287
Aberration polynomial, 286
Absorption, 252
Accommodation, 195, 200, 244
Achromatic
lens, 260, 263
prism, 256
Afocal system
Airy disk, 207
Ametropia, 200, 245, 246
Amplitude, 1, 2, 190
Angle
critical, 40, 41
deviation, 20, 21, 48
incidence, 1, 15–20, 29
rotation of mirror, 16
reflection, 1, 15
Aperture stop, 26, 82, 107, 213, 216
Aplanatic
lens, 184, 295
surface, 294
Astigmatic
lenses, 124
power, 123
surfaces, 119
Astigmatism, See Aberrations

Back focal length, 152, 159
Back vertex power, 157
Badal
lens, 244–249
optometer, 242
Baffles, 11
Beam, 17
Binoculars, 216–219
Biomicroscope, 247
Birefringence, 254
Blur, 8, 191, 212, 226, 278

Camera,
depth of field and focus, 225
hyperfocal distance, 227
obscura
panoramic, 224
pinhole
twin lens reflex, 224
single lens reflex, 224
Cardinal points, 139
Caustic, 44
Centrads, 59, 107
Chief ray, 95, 175, 176
Chromatic aberration, 255–263
Circle of least confusion, 121, 128
Coddington, 298
Coma, 291–293
Condensers, 233, 249
Conjugate points, 53, 74
Compound microscope, See Microscope
Constant deviation, 20
Contrast, 188
Convergence, 84, 93
Corpuscular theory, 1, 10, 28
Critical angle, 40, 50, 191
Curvature, 85, 119–125
Cylindrical lenses, 127
Cylindrical surfaces, 119

Depth of field and focus, 226, 227
of microscope, 205
Deviation, 38, 58–61
Diffraction, 9, 185
Diopters, 69, 100
Dispersion, 251–253
Displacement by a plate, 55–57
Distortion, 300
Divergence, 12, 84, 85, 93
Doublet lens, 261
Doubling principle, 238

Eclipses, 8
Effective focal length. See Equivalent focal length
Effective power, 158

Electromagnetic radiation, 1, 3
Electron microscope, 205
Emmetropia, 142
Empty magnification, 204
Entrance port, 174–178
Entrance pupil, 173, 174
Equivalent focal length, 160, 167
Equivalent power, 149–151
Exit port, 175
Exit pupil, 174
Eye point of microscope, 203
Eye relief, 202
Eye, schematic, 161–164
Eyepieces,
 Huygens, 264–266
 Ramsden, 266, 267
 Kellner, 267

f-number, 183, 184
Fermat's principle, 40
Fiber optics, 191–194
Field of view, 28, 82, 108, 175–180, 213
Field stop, 10, 175
Fluorescent lamp, 6
Focal length,
 of reflecting surface, 76
 of refracting surface, 75
 of thin lens, 96
 of thick lens, 141
Focal lines, 120, 130, 295
Focal planes, 76, 105
Focal points, 67, 93, 94, 139–142
Focimeter, 245
Frequency,
 electromagnetic, 1
 spatial, 188–191

Galilean telescope, 215
Galvanometer, 16
Glass, 6, 15, 253
Glass chart, 255
Graphical ray trace, 282
Grazing incidence and emergence, 48
Gullstrand equations, 149–152
Gullstrand schematic eye, 161

Huygens wave theory, 29
Huygens construction for refraction, 37
Huygens eyepiece, 264
Hyperfocal distance, 227
Hyperopia, 161, 242

Image space, 100
Images,
 conjugate, 72, 101
 construction of off-axis, 71, 79
 erect, 19, 22–27, 73, 81–84
 inversion and reversion of reflected, 18
 real, 14, 18
 virtual, 12, 18, 27, 44, 54, 72
Images, evaluation, 278
Incandescence, 5
Index of refraction
 absolute index, 36, 253
 relative index, 33–38
Infrared light, 2
Interval of Sturm, 120

Kaleidoscope, 25
Kellner eyepiece, 267
Keratometer, 235
Kohler illumination, 231
Knife edge distribution, 282

LaGrange Invariant, 76
Lamps, 5
Lasers, 6
Law of
 reflection, 1, 15
 refraction, 1, 33
Least distance of distinct vision, 195
Least time, principle of, See Fermat's principle
Lens bending, 287
Lens measure, 87
Lenses,
 achromatic, 260
 camera, 225
 condensers, 249
 cylindrical, 126
 field, 213
 panoramic, 184
 relay, 213
 telephoto and wide angle, 158
 thick, 149
 thin, 93
 toric, 126
 zoom, 229
Lensmaker's formula, 96
Lensometer, 245
Light Sources, 5

Magnification,
angular, 83, 146
empty, 204
lateral,
refracting surface, 73
reflecting surface, 81
thin lens, 99
thick lens, 146
longitudinal, 86, 145
Magnifier,
compound, 200
field of view, 199
maximum magnification, 197
nominal magnification, 196
Marginal ray, 271
Mean dispersion, 253
Meridional rays, 44
Microscope,
as a simple magnifier, 202
depth of focus of, 205
electron, 205
magnification, 201
resolution, 205
stops, 202
Mirages, 10
Mires, 235
Mirrors,
constant deviation, 19
ellipsoidal, 209
focal length of, 234
plane, 15
paraboloidal, 209
spherical, 76
two inclined, 21
Modulation transfer function, 189
Myopia, 242

Newton's corpuscular theory, 7, 28, 34
Newtonian Equations
spherical mirror, 81
spherical refracting surface, 74
thin lens, 103
Nodal points,
construction for lenses, 95, 141
See Cardinal points, 139
Normal, 15
Normal adjustment of telescope, 210
Numerical aperture, 183
microscope, 203

Objectives,
telescope, 209
microscope, 201
Ophthalmometer, 235–237
Optical
axis, 54, 65, 94
center of a lens, 95
cross, 126
Optical path length, 39
Optical systems, 11, 139
Optometer, 248

Parallel plates, 44, 54
Paraxial rays, 53
Pencils, 12
Penumbra, 8
Period, 3
Petzval curvature, 298
Phosphors, 6
Photographic lenses, 158, 223
Photons, 1
Pincushion distortion, 300–302
Pinhole camera, 8, 223
Plane surface
mirror, 15
refraction, 33
Point source, 12
Ports, 175
Position factor, 288
Power
back vertex, 157, 160
effective, 158
equivalent, 149
front vertex, 151
in oblique meridians, 123
thin lens, 100
Principal points and planes, 139, 149
Principal sections of a toric surface, 121
Prismatic effect of a lens, 104
Prisms,
achromatic prism, 258
deviation, 50
minimum deviation, 49
obliquely combined, 59
ophthalmic, 57
Pechan, 216
Porro, 21
Risley, 61
Schmidt, 216
thick, 47
thin, 58

Prism binocular, 216
Prism diopter, 57
Projectors, 231
Pupils,
plane mirror, 26
spherical mirror, 82
thin lens, 108
types of stops, 173
Purkinje images, 164

Quanta, 1

Radial energy diagram, 281
Radius of curvature, 65
Rays,
marginal ray, 271
meridional rays, 44
paraxial, 45
zonal ray, 271
Ray intercept curves, 280
Ray tracing,
paraxial constructions, 70–72
trigonometric, 271
Ramsden circle, 211
Ramsden eyepiece, 266
Rectilinear propagation, 6
Reduced thickness, 56, 150
Reduced vergence, 84
Reflection,
plane mirror, 15
spherical mirror, 66
two inclined mirrors, 23
total internal, 38
Refraction,
law, 33
plane surfaces, 53
spherical surface, 66
Resolution, 185
Resolution charts, 187
Reversibility of light, 33
Risley prisms, 61

Schematic eye, 151
Sagitta, 85
Sagittal coma, 291
Sagittal power of ophthalmic lens, 126
Seidel aberrations, 285
Sextant, 17
Shadows, 8
Shape factor, 288
Sign convention,
thick prisms, 47
paraxial image formation, 54
spherical surfaces, 65
trigonometric ray tracing, 271
Simple microscope, See Magnifier
Sine condition, 293
Slit lamp, 247
Smith-Helmholtz, 76
Snell's law, 36
Snellen chart, 188
Spectrum, 251
Spherical aberration, 44, 277, 287
Spherical equivalent lens, 128
Spherical surfaces, 65
Spherical wavefronts, 11
Spot diagrams, 280
Step along calculations, 115
Stereoscopic range, 217
Stops,
plane mirror, 26
procedure to find, 176
spherical mirror, 82
thin lens, 105

Tangent condition, 301
Telecentric systems, 244
Telephoto lenses, 155, 231
Telescopes, 209
astronomical, 209
Cassagrainian, 219
catadioptric, 219
Galilean, 215
Gregorian, 220
Keplerian, 209
Newtonian, 220
reflecting, 209, 217
terrestrial, 214
Thick lenses, 139, 149
Toric, 119, 239
Toric lenses, 124
Total internal reflection, 40
Transposition of flat prescriptions, 127
Transverse chromatic aberration, 264
Transverse spherical aberration, 277
Trigonometric ray trace, 271

Ultra-violet light, 216
Umbra, 8

V-value, 253
Velocity of light, 4
Vergence, 84, 100
Vertex, 43
Vertex power, 157
Vignetting, 175
Virtual
    object, 12
    image, 12
Visual angle, 12, 195

Wavefront, 11, 29, 68
Wavelength, 1
Wave theory, 1, 29
Wide-angle lenses, 155

Zonal ray, 271
Zoom lens, 229